Zoltán Ésik, Carlos Martín-Vide, Victor Mitrana (Eds.)

Recent Advances in Formal Languages and Applications

Studies in Computational Intelligence, Volume 25

Editor-in-chief
Prof. Janusz Kacprzyk
Systems Research Institute
Polish Academy of Sciences
ul. Newelska 6
01-447 Warsaw
Poland
E-mail: kacprzyk@ibspan.waw.pl

Further volumes of this series can be found on our homepage: springer.com

Vol. 9. Tsau Young Lin, Setsuo Ohsuga, Churn-Jung Liau, Xiaohua Hu (Eds.)
Foundations and Novel Approaches in Data Mining, 2005
ISBN 3-540-28315-3

Vol. 10. Andrzej P. Wierzbicki, Yoshiteru Nakamori
Creative Space, 2005
ISBN 3-540-28458-3

Vol. 11. Antoni Ligęza
Logical Foundations for Rule-Based Systems, 2006
ISBN 3-540-29117-2

Vol. 12. Jonathan Lawry
Modelling and Reasoning with Vague Concepts, 2006
ISBN 0-387-29056-7

Vol. 13. Nadia Nedjah, Ajith Abraham, Luiza de Macedo Mourelle (Eds.)
Genetic Systems Programming, 2006
ISBN 3-540-29849-5

Vol. 14. Spiros Sirmakessis (Ed.)
Adaptive and Personalized Semantic Web, 2006
ISBN 3-540-30605-6

Vol. 15. Lei Zhi Chen, Sing Kiong Nguang, Xiao Dong Chen
Modelling and Optimization of Biotechnological Processes, 2006
ISBN 3-540-30634-X

Vol. 16. Yaochu Jin (Ed.)
Multi-Objective Machine Learning, 2006
ISBN 3-540-30676-5

Vol. 17. Te-Ming Huang, Vojislav Kecman, Ivica Kopriva
Kernel Based Algorithms for Mining Huge Data Sets, 2006
ISBN 3-540-31681-7

Vol. 18. Chang Wook Ahn
Advances in Evolutionary Algorithms, 2006
ISBN 3-540-31758-9

Vol. 19. Ajita Ichalkaranje, Nikhil Ichalkaranje, Lakhmi C. Jain (Eds.)
Intelligent Paradigms for Assistive and Preventive Healthcare, 2006
ISBN 3-540-31762-7

Vol. 20. Wojciech Penczek, Agata Półrola
Advances in Verification of Time Petri Nets and Timed Automata, 2006
ISBN 3-540-32869-6

Vol. 21. Cândida Ferreira
Gene Expression on Programming: Mathematical Modeling by an Artificial Intelligence, 2006
ISBN 3-540-32796-7

Vol. 22. N. Nedjah, E. Alba, L. de Macedo Mourelle (Eds.)
Parallel Evolutionary Computations, 2006
ISBN 3-540-32837-8

Vol. 23. M. Last, Z. Volkovich, A. Kandel (Eds.)
Algorithmic Techniques for Data Mining, 2006
ISBN 3-540-33880-2

Vol. 24. Alakananda Bhattacharya, Amit Konar, Ajit K. Mandal
Parallel and Distributed Logic Programming, 2006
ISBN 3-540-33458-0

Vol. 25. Zoltán Ésik, Carlos Martín-Vide Victor Mitrana (Eds.)
Recent Advances in Formal Languages and Applications, 2006
ISBN 3-540-33460-2

Zoltán Ésik
Carlos Martín-Vide
Victor Mitrana

Recent Advances in Formal Languages and Applications

With 122 Figures and 12 Tables

Springer

Dr. Zoltán Ésik
Department of Informatics
University of Szeged
P.O. Box 652, H-6701 Szeged
Hungary
and
Rovira i Virgili University
Research Group on Mathematical
Linguistics
Pl. Imperial Tàrraco, 1
43005 Tarragona
Spain
E-mail: ze@inf.u-szeged.hu; zoltan.esik@urv.cat

Carlos Martín-Vide
Rovira i Virgili University
Research Group on Mathematical
Linguistics
Pl. Imperial Tàrraco, 1
43005 Tarragona
Spain
E-mail: carlos.martin@urv.cat

Dr. Victor Mitrana
Faculty of Mathematics and Computer Science
University of Bucharest
Str. Academiei 14
70109 Bucharest
Romania
and
Rovira i Virgili University
Research Group on Mathematical Linguistics
Pl. Imperial Tàrraco, 1
43005 Tarragona
Spain
E-mail: mitrana@fmi.unibuc.ro; vmi@urv.cat

ISSN print edition: 1860-949X
ISSN electronic edition: 1860-9503

ISBN 978-3-642-07009-9 e-ISBN 978-3-540-33461-3

Springer is a part of Springer Science+Business Media
springer.com

Printed in The Netherlands

Cover design: deblik, Berlin

Preface

The theory of formal languages is widely accepted as the backbone of theoretical computer science. It mainly originated from mathematics (combinatorics, algebra, mathematical logic) and generative linguistics. Later, new specializations originated from areas of either computer science (concurrent and distributed systems, computer graphics, artificial life), biology (plant development, molecular genetics), linguistics (parsing, text searching), or mathematics (cryptography). All human problem solving capabilities can be considered in a certain sense as a manipulation of symbols and structures composed by symbols, which is actually the stem of formal language theory. Language – in its two basic forms, natural and artificial – is a particular case of a symbol system.

This wide range of motivations and inspirations explains the diverse applicability of formal language theory – and all these together explain the very large number of monographs and collective volumes dealing with formal language theory.

In 2004 Springer-Verlag published the volume *Formal Languages and Applications*, edited by C. Martín-Vide, V. Mitrana and G. Păun, which was aimed to serve as an overall course-aid and self-study material especially for the PhD students in formal language theory and applications. Actually, the volume has emerged in such a context: it contains the core information from most of the lectures delivered to the students of the International PhD School in Formal Languages and Applications organized since 2002 by the Research Group on Mathematical Linguistics from Rovira i Virgili University, Tarragona, Spain. During the editing process of the aforementioned volume two situations have appeared:

1. Some important aspects, mostly extensions and applications of the classic formal language theory to different scientific areas, could not be covered, by different reasons.

2. New courses have been promoted in the next editions of the PhD school mentioned above.

The present volume is intended to fill up this gap. Having in mind that the book is addressed mainly to young researchers, the contributors present the main results and techniques of their areas of specialization in an easily accessible way accompanied with many references having a multiple role: historical, hints for complete proofs or solutions to exercises, directions for further research where the reader may identify attractive problems. This volume contains areas, mainly applications, which have not appeared in any collection of this type. We believe that the volume, besides accomplishing its main goal, to complement the aforementioned volume, both representing "a gate to formal language theory and its applications", will be also useful as a general source of information in computation theory, both at the undergraduate and research level.

For the sake of uniformity, the introductory chapter of the first volume that presents the mathematical prerequisites as well as most common concepts and notations used throughout all chapters appears in the present volume as well. However, it may happen that terms other than those in the introductory chapter have different meanings in different chapters or different terms have the same meaning. In each chapter, the subject is treated relatively independent to the other chapters, even if several chapters are related. In this way, the reader gets in touch with different points of view on a common aspect to two or more chapters. We are convinced on the usefulness of such an opportunity to a young researcher.

Acknowledgements

Our deep gratitude are due to all the contributors, for their professional and friendly cooperation, as well as to Springer-Verlag, for the efficient and pleasant cooperation.

October 2005

Zoltán Ésik
Carlos Martín-Vide
Victor Mitrana

Contents

1

Basic Notation and Terminology

This chapter presents the basic mathematical and formal language theory notations and terminology used throughout the book.

1.1 General Mathematical Notations

The notations are those provided by standard Latex and customary in mathematics.

Set theory: $\in$ denotes the membership (of an element to a set), $\subseteq$ denotes the inclusion (not necessarily proper) and $\subset$ denotes the strict inclusion; the union, intersection, and difference of two sets are denoted by $\cup, \cap, -$, respectively. (We do not use $\setminus$ for the difference, because $\setminus$ denotes the left quotient of languages.) The empty set is denoted by $\emptyset$, the power set of a set X is denoted by 2^X, while the cardinality of a set X is denoted by $\mathrm{card}(X)$. A singleton set is often identified with its single element, and hence we also write a for $\{a\}$. Two sets X and Y are said to be *incomparable* if both $X - Y$ and $Y - X$ are non-empty.

Sets of numbers: the set of natural numbers (zero included) is denoted by $\mathbf{N}$, while the sets of integer, rational, and real numbers are denoted by $\mathbf{Z}$, $\mathbf{Q}$, $\mathbf{R}$, respectively. The subsets of these sets consisting of strictly positive numbers are denoted by $\mathbf{N}_+, \mathbf{Z}_+, \mathbf{Q}_+, \mathbf{R}_+$, respectively.

1.2 Basic String Notions and Notation

An *alphabet* is a finite nonempty set of abstract symbols. For an alphabet V we denote by V^* the set of all *strings* (we also say *words*) of symbols from V. The empty string is denoted by λ. The set of nonempty strings over V, that is $V^* - \{\lambda\}$, is denoted by V^+. Each subset of V^* is called a *language* over V. A language which does not contain the empty string (hence being a subset of V^+) is said to be λ-*free*.

Basic Notation and Terminology, Studies in Computational Intelligence (SCI) **25**, 1–9 (2006)
www.springerlink.com

If $x = x_1x_2$, for some $x_1, x_2 \in V^*$, then x_1 is called a *prefix* of x and x_2 is called a *suffix* of x; if $x = x_1x_2x_3$ for some $x_1, x_2, x_3 \in V^*$, then x_2 is called a *substring* of x. The sets of all prefixes, suffixes, and substrings of a string x are denoted by Pref(x), Suf(x), and Sub(x), respectively. The sets of proper (that is, different from λ and from the string itself) prefixes, suffixes, and subwords of x are denoted by PPref(x), PSuf(x), and PSub(x), respectively.

The *length* of a string $x \in V^*$ (the number of occurrences of symbols from V in x) is denoted by $|x|$. The number of occurrences of a given symbol $a \in V$ in $x \in V^*$ is denoted by $|x|_a$. If $x \in V^*$ and $U \subseteq V$, then by $|x|_U$ we denote the length of the string obtained by erasing from x all symbols not in U, that is,

$$|x|_U = \sum_{a \in U} |x|_a.$$

For a language $L \subseteq V^*$, the set $length(L) = \{|x| \mid x \in L\}$ is called the *length set* of L.

The set of symbols occurring in a string x is denoted by $alph(x)$. For a language $L \subseteq V^*$, we denote $alph(L) = \bigcup_{x \in L} alph(x)$.

The *Parikh vector* associated with a string $x \in V^*$ with respect to the alphabet $V = \{a_1, \ldots, a_n\}$ is $\Psi_V(x) = (|x|_{a_1}, |x|_{a_2}, \ldots, |x|_{a_n})$ (note that the ordering of the symbols from V is relevant). For $L \subseteq V^*$ we define $\Psi_V(L) = \{\Psi_V(x) \mid x \in L\}$; the mapping $\Psi_V : V^* \longrightarrow \mathbf{N}^n$ is called the *Parikh mapping* associated with V.

A set M of vectors in $\mathbf{N}^n$, for some $n \geq 1$, is said to be *linear* if there are $m \geq 0$ and the vectors $v_i \in \mathbf{N}^n$, $0 \leq i \leq m$, such that

$$M = \{v_0 + \sum_{i=1}^{m} \alpha_i v_i \mid \alpha_1, \ldots, \alpha_m \in \mathbf{N}\}.$$

A finite union of linear sets is said to be *semilinear.*

A language $L \subseteq V^*$ is *semilinear* if $\Psi_V(L)$ is a semilinear set. The family of semilinear languages is denoted by $SLIN$.

1.3 Operations with Strings and Languages

The operations of union, intersection, difference, and complement are defined for languages in the standard set-theoretical way.

The *concatenation* of two languages L_1, L_2 is $L_1L_2 = \{xy \mid x \in L_1, y \in L_2\}$.

We define further:

$$L^0 = \{\lambda\},$$
$$L^{i+1} = LL^i, \ i \geq 0,$$
$$L^* = \bigcup_{i=0}^{\infty} L^i \ \text{(the Kleene } *\text{-closure)},$$
$$L^+ = \bigcup_{i=1}^{\infty} L^i \ \text{(the Kleene } +\text{-closure)}.$$

A mapping $s : V \longrightarrow 2^{U^*}$, extended to $s : V^* \longrightarrow 2^{U^*}$ by $s(\lambda) = \{\lambda\}$ and $s(x_1x_2) = s(x_1)s(x_2)$, for $x_1, x_2 \in V^*$, is called a *substitution*. If for all $a \in V$ we have $\lambda \notin s(a)$, then h is a *λ-free* substitution. If $\text{card}(s(a)) = 1$ for all $a \in V$, then s is called a *morphism* (we also say homomorphism).

A morphism $h : V^* \longrightarrow U^*$ is called a *coding* if $h(a) \in U$ for each $a \in V$ and a *weak coding* if $h(a) \in U \cup \{\lambda\}$ for each $a \in V$. If $h : (V_1 \cup V_2)^* \longrightarrow V_1^*$ is the morphism defined by $h(a) = a$ for $a \in V_1$, and $h(a) = \lambda$ otherwise, then we say that h is a *projection* (associated with V_1) and we denote it by pr_{V_1}. For a morphism $h : V^* \longrightarrow U^*$, we define a mapping $h^{-1} : U^* \longrightarrow 2^{V^*}$ (and we call it an *inverse morphism*) by $h^{-1}(w) = \{x \in V^* \mid h(x) = w\}$, $w \in U^*$.

The substitutions (hence also the morphisms and inverse morphisms) are extended to languages in the natural way.

For $x, y \in V^*$ we define their *shuffle* by

$$x \sqcup\!\sqcup\, y = \{x_1y_1 \dots x_ny_n \mid x = x_1 \dots x_n, y = y_1 \dots y_n, \\ x_i, y_i \in V^*, 1 \leq i \leq n, n \geq 1\}.$$

The *left quotient* of a language $L_1 \subseteq V^*$ with respect to $L_2 \subseteq V^*$ is

$$L_2 \backslash L_1 = \{w \in V^* \mid \text{there is } x \in L_2 \text{ such that } xw \in L_1\}.$$

The *left derivative* of a language $L \subseteq V^*$ with respect to a string $x \in V^*$ is

$$\partial_x^l(L) = \{w \in V^* \mid xw \in L\}.$$

The *right quotient* and the *right derivative* are defined in a symmetric manner:

$$L_1/L_2 = \{w \in V^* \mid \text{there is } x \in L_2 \text{ such that } wx \in L_1\},$$
$$\partial_x^r(L) = \{w \in V^* \mid wx \in L\}.$$

Let $\mathcal{F}$ be a family of languages and $\circ$ be an n-ary operation with languages from $\mathcal{F}$. The family $\mathcal{F}$ is *closed under* $\circ$ if $\circ(L_1, L_2, \dots, L_n) \in \mathcal{F}$ for any choice of the languages $L_i \in \mathcal{F}$, $1 \leq i \leq n$. The family $\mathcal{F}$ is *closed under substitution with languages from* the family $\mathcal{C}$ if for any language $L \subseteq V^*$, $L \in \mathcal{F}$, and any substitution $s : V^* \longrightarrow 2^{U^*}$ such that $s(a) \in \mathcal{C}$ for all $a \in V$, the language $s(L) = \bigcup_{x \in L} s(x)$ still lies in $\mathcal{F}$. If $\mathcal{C} = \mathcal{F}$, we simply say that $\mathcal{F}$ is closed under substitution.

A family of languages closed under (arbitrary) λ-free morphisms, inverse morphisms and intersection with regular languages is called *(full) trio* - known also as *(cone) faithful cone*. If a (full) trio is further closed under union, then it is called *(full) semi-AFL*. The abbreviation AFL comes from *Abstract Family of Languages*. A (full) semi-AFL closed under concatenation and Kleene (*-) +-closure is called a *(full) AFL*.

1.4 Chomsky Grammars

A *Chomsky grammar* is a quadruple $G = (N, T, S, P)$, where N, T are disjoint alphabets, $S \in N$, and P is a finite subset of $(N \cup T)^*N(N \cup T)^* \times (N \cup T)^*$.

The alphabet N is called the *nonterminal alphabet*, T is the *terminal alphabet*, S is the *axiom* (start symbol), and P is the set of *production rules* of G. The rules (we also say *productions*) (u, v) of P are written in the form $u \to v$. Note that $|u|_N \geq 1$. Sometimes, one uses to denote by V_G the total alphabet of G, that is, $V_G = N \cup T$.

For $x, y \in (N \cup T)^*$ we write

$$x \Longrightarrow_G y \text{ iff } x = x_1 u x_2, y = x_1 v x_2, \quad \text{for some } x_1, x_2 \in (N \cup T)^* \text{ and } u \to v \in P.$$

One says that x *directly derives* y (with respect to G). When G is understood we write $\Longrightarrow$ instead of $\Longrightarrow_G$. The reflexive closure of the relation $\Longrightarrow$ is denoted by $\Longrightarrow^+$, and the reflexive and transitive closure by $\Longrightarrow^*$. Each string $w \in (N \cup T)^*$ such that $S \Longrightarrow_G^* w$ is called a *sentential form*.

The language generated by G, denoted by $L(G)$, is defined by

$$L(G) = \{x \in T^* \mid S \Longrightarrow^* x\}.$$

Two grammars G_1, G_2 are called *equivalent* if $L(G_1) - \{\lambda\} = L(G_2) - \{\lambda\}$ (the two languages coincide modulo the empty string).

According to the form of their rules, the Chomsky grammars are classified as follows. A grammar $G = (N, T, S, P)$ is called:

- *length-increasing* (one also says *monotonous*), if for all $u \to v \in P$ we have $|u| \leq |v|$.
- *context-sensitive*, if each $u \to v \in P$ has $u = u_1 A u_2, v = u_1 x u_2$, for $u_1, u_2 \in (N \cup T)^*, A \in N$, and $x \in (N \cup T)^+$. (In length-increasing and context-sensitive grammars the production $S \to \lambda$ is allowed, provided that S does not appear in the right-hand members of rules in P.)
- *context-free*, if each production $u \to v \in P$ has $u \in N$.
- *linear*, if each rule $u \to v \in P$ has $u \in N$ and $v \in T^* \cup T^*NT^*$.
- *right-linear*, if each rule $u \to v \in P$ has $u \in N$ and $v \in T^* \cup T^*N$.
- *left-linear*, if each rule $u \to v \in P$ has $u \in N$ and $v \in T^* \cup NT^*$.
- *regular*, if each rule $u \to v \in P$ has $u \in N$ and $v \in T \cup TN \cup \{\lambda\}$.

The arbitrary, length-increasing, context-free, and regular grammars are also said to be of *type* 0, *type* 1, *type* 2, and *type* 3, respectively.

We denote by *RE, LI, CS, CF, LIN, RLIN, LLIN*, and *REG* the families of languages generated by arbitrary, length-increasing, context-sensitive, context-free, linear, right-linear, left-linear, and regular grammars, respectively (RE stands for *recursively enumerable*). By FIN we denote the family of finite languages, and by ARB the family of arbitrary languages.

The following equalities and strict inclusions hold:

$$FIN \subset REG = RLIN = LLIN \subset LIN \subset CF \subset CS = LI \subset RE \subset ARB.$$

We call this the *Chomsky hierarchy.*

1.5 Decision Problems

The goal of this section is to give an informal description of a decision problem and to mention the most common decision problems in formal language theory.

Roughly speaking, a *decision problem* requires an output YES/NO to any of its instances. For example, "Is the natural number n prime?" is a decision problem; further, "Is 3 prime?" is an instance of the problem which is true while "Is 4 prime?" is a false instance of the same problem. A decision problem is *(algorithmically/recursively) decidable* if there exists an algorithm, which for any instance of the problem given as input, outputs YES or NO, provided that the input is true or not, respectively.

The most common decision problems in formal language theory are:

- *Emptiness*: Is a given language empty?
- *Finiteness*: Is a given language a finite set?
- *Membership*: Does $w \in L$ hold for a given word w and a language L?
- *Inclusion*: Does $L_1 \subseteq L_2$ hold for two given languages L_1 and L_2?
- *Equivalence*: Does $L_1 = L_2$ hold for two given languages L_1 and L_2?

Clearly, a decision problem is proved to be decidable if one provides an algorithm as above. Generally, a decision problem is proved to be undecidable by reducing it to a problem known to be undecidable. The following combinatorial problem, known as the *Post Correspondence Problem* (PCP), is undecidable. An instance of the PCP consists of an alphabet V with at least two letters and two lists of words over V

$$u = (u_1, u_2, \ldots, u_n) \quad \text{and} \quad v = (v_1, v_2, \ldots, v_n).$$

The problem asks whether or not a sequence $i_1, i_2, \ldots, i_k$ of positive integers exists, each between 1 and n, such that $u_{i_1}u_{i_2}\ldots u_{i_k} = v_{i_1}v_{i_2}\ldots v_{i_k}$.

We do not give here further elements of formal language theory. They will be elaborated in the subsequent chapters.

For the reader's convenience, we end this section with a list of monographs and collective volumes directly or partially related to formal language theory.

1.6 Books on Formal Language Theory

1. A.V. Aho, J.D. Ullman, *The Theory of Parsing, Translation, and Compiling*, Prentice Hall, Englewood Cliffs, N.J., vol. I: 1971, vol. II: 1973.
2. A.V. Aho, J.D. Ullman, *Principles of Compiler Design*, Addison-Wesley, Reading, Mass., 1977.
3. I. Alexander, F.K. Hanna, *Automata Theory: An Engineering Approach*, Crane Russak, 1975.
4. J. Berstel, *Transductions and Context-Free Languages*, Teubner, Stuttgart, 1979.
5. R.V. Book, ed., *Formal Language Theory. Perspectives and Open Problems*, Academic Press, New York, 1980.
6. W. Brauer, *Automatentheorie*, B.G. Teubner, Stuttgart, 1984.
7. C. Choffrut, ed., *Automata Networks, Lecture Notes in Computer Science*, Springer-Verlag, Berlin, 1988.
8. D.I.A. Cohen, *Computer Theory*, 2nd edition, John Wiley, 1997.
9. E. Csuhaj-Varju, J. Dassow, J. Kelemen, Gh. Păun, *Grammar Systems. A Grammatical Approach to Distribution and Cooperation*, Gordon and Breach, London, 1994.
10. J. Dassow, Gh. Păun, *Regulated Rewriting in Formal Language Theory*, Springer-Verlag, Berlin, Heidelberg, 1989.
11. J. Dassow, G. Rozenberg, A. Salomaa, eds., *Developments in Language Theory*, World Scientific, Singapore, 1995.
12. M.D. Davis, E.J. Weyuker, *Computability, Complexity, and Languages*, Academic Press, New York, 1983.
13. P.J. Denning, J.B. Dennis, J.E. Qualitz, *Machines, Languages, and Computation*, Prentice-Hall, Englewood Cliffs, N.J., 1978.
14. D.-Z. Du, K.-I Ko, *Problem Solving in Automata, Languages and Complexity*, John Wiley, 2001.
15. H. Ehrig, G. Engels, H-J. Kreowski, G. Rozenberg, eds., *Handbook of Graph Grammars and Computing by Graph Transformation*, World Scientific, Singapore, 1999.
16. S. Eilenberg, *Automata, Languages, and Machines*, Academic Press, New York, vol. A: 1974, vol. B: 1976.
17. E. Engeler, *Formal Languages*, Markham, Chicago, 1968.
18. K.S. Fu, *Syntactic Pettern Recognition. Applications*, Springer-Verlag, Heidelberg, 1977.
19. M.R. Garey, D.S. Johnson, *Computers and Intractability. A Guide to the Theory of NP-completeness*, W.H. Freeman, San Francisco, 1979.
20. F. Gécseg, *Products of Automata*, Springer-Verlag, Berlin, 1986.
21. F. Gécseg, I. Peak, *Algebraic Theory of Automata*, Akademiai Kiado, Budapest, 1972.
22. F. Gécseg, M. Steinby, *Tree Automata*, Akademiai Kiado, Budapest, 1984.
23. S. Ginsburg, *The Mathematical Theory of Context-Free Languages*, McGraw-Hill Book Comp., New York, 1966.

24. S. Ginsburg, *Algebraic and Automata-Theoretic Properties of Formal Languages*, North-Holland, Amsterdam, 1975.
25. A. Ginzburg, *Algebraic Theory of Automata*, Academic Press, New York, 1968.
26. M. Gross, A. Lentin, *Notions sur les grammaires formelles*, Gauthier-Villars, Paris, 1967.
27. M. Harrison, *Introduction to Formal Language Theory*, Addison-Wesley, Reading, Mass., 1978.
28. G.T. Herman, G. Rozenberg, *Developmental Systems and Languages*, North-Holland, Amsterdam, 1975.
29. J.E. Hopcroft, J.D. Ullman, *Formal Languages and Their Relations to Automata*, Addison-Wesley, Reading, Mass., 1969.
30. J.E. Hopcroft, J.D. Ullman, *Introduction to Automata Theory, Languages, and Computing*, Addison-Wesley, Reading, Mass., 1979.
31. J.E. Hopcroft, R. Motwani, J.D. Ullman, *Introduction to Automata Theory, Languages, and Computation*, Addison-Wesley, Boston, 2001.
32. M. Ito, ed., *Words, Languages, and Combinatorics*, World Scientific, Singapore, 1992.
33. M. Ito, Gh. Păun, S. Yu, eds., *Words, Semigroups, and Transductions*, World Scientific, Singapore, 2001.
34. J. Karhumäki, H.A. Maurer, Gh. Păun, G. Rozenberg, eds., *Jewels are Forever*, Springer-Verlag, Berlin, 1999.
35. D. Kelley, *Automata and Formal Languages. An Introduction.* Prentice-Hall, New Jersey, 1995.
36. Z. Kohavi, *Switching and Finite Automata Theory*, McGraw-Hill Book Comp., New York, 1978.
37. D.C. Kozen, *Automata and Computability*, Springer-Verlag, New York, 1997.
38. W. Kuich, A. Salomaa, *Semirings, Automata, Languages*, Springer-Verlag, Berlin, Heidelberg, New York, 1986.
39. P. Linz, *An Introduction to Formal Languages and Automata*, D.C. Heath and Co., Lexington, Mass., 1990.
40. M. Lothaire, *Combinatorics on Words*, Addison-Wesley, Reading, Mass., 1983.
41. M. Lothaire, *Algebraic Combinatorics on Words*, Cambridge University Press, 1997.
42. C. Martín-Vide, V. Mitrana, eds., *Where Mathematics, Computer Science, Linguistics, and Biology Meet*, Kluwer, Dordrecht, 2000.
43. C. Martín-Vide, V. Mitrana, eds., *Grammars and Automata for String Processing: From Mathematics and Computer Science to Biology, and Back*, Taylor and Francis, London, 2002.
44. C. Martín-Vide, Gh. Păun, eds., *Recent Topics in Mathematical and Computational Linguistics*, Ed. Academiei, Bucureşti, 2000.
45. C. Martín-Vide, V. Mitrana, Gh. Păun, eds., *Formal Languages and Applications*, Springer-Verlag, Berlin, 2004.

46. H. Maurer, *Theoretische Grundlagen der Programmiersprachen*, Hochschultaschenbücher 404, Bibliographisches Inst., 1969.
47. A. Meduna, *Automata and Languages*, Springer-Verlag, London, 2000.
48. M. Minsky, *Computation: Finite and Infinite Machines*, Prentice-Hall, Englewood Cliffs, NJ, 1967.
49. J.N. Mordenson, D.S. Malik, *Fuzzy Automata and Languages*, Chapman & Hall/CRC, London, 2002.
50. A. Paz, *Introduction to Probabilistic Automata*, Academic Press, New York, 1971.
51. Gh. Păun, *Recent Results and Problems in Formal Language Theory*, The Scientific and Encyclopaedic Publ. House, Bucharest, 1984 (in Romanian).
52. Gh. Păun, ed., *Mathematical Aspects of Natural and Formal Languages*, World Scientific, Singapore, 1994.
53. Gh. Păun, ed., *Mathematical Linguistics and Related Topics*, The Publ. House of the Romanian Academy, Bucharest, 1995.
54. Gh. Păun, *Marcus Contextual Grammars*, Kluwer, Dordrecht, 1997.
55. Gh. Păun, *Membrane Computing. An Introduction*, Springer-Verlag, Berlin, 2002.
56. Gh. Păun, G. Rozenberg, A. Salomaa, *DNA Computing. New Computing Paradigms*, Springer-Verlag, Berlin, 1998.
57. Gh. Păun, G. Rozenberg, A. Salomaa, eds., *Current Trends in Theoretical Computer Science. Entering the 21st Century*, World Scientific, Singapore, 2001.
58. Gh. Păun, A. Salomaa, eds., *New Trends in Formal Languages: Control, Cooperation, Combinatorics*, *Lecture Notes in Computer Science* 1218, Springer-Verlag, Berlin, 1997.
59. Gh. Păun, A. Salomaa, eds., *Grammatical Models of Multi-Agent Systems*, Gordon and Breach, London, 1999.
60. J.E. Pin, *Varieties of Formal Languages*, Plenum Press, Oxford, 1986.
61. G.E. Revesz, *Introduction to Formal Languages*, McGraw-Hill Book Comp., New York, 1983.
62. G. Rozenberg, A. Salomaa, *The Mathematical Theory of L Systems*, Academic Press, New York, 1980.
63. G. Rozenberg, A. Salomaa, *Cornerstones of Undecidability*, Prentice Hall, New York, 1994.
64. G. Rozenberg, A. Salomaa, eds., *Developments in Language Theory*, World Scientific, Singapore, 1994.
65. G. Rozenberg, A. Salomaa, eds., *Handbook of Formal Languages*, Springer-Verlag, Berlin, 3 volumes, 1997.
66. A. Salomaa, *Theory of Automata*, Pergamon, Oxford, 1969.
67. A. Salomaa, *Formal Languages*, Academic Press, New York, London, 1973.
68. A. Salomaa, *Jewels of Formal Language Theory*, Computer Science Press, Rockville, 1981.
69. A. Salomaa, *Computation and Automata*, Cambridge Univ. Press, Cambridge, 1985.

70. A. Salomaa, M. Soittola, *Automata-Theoretic Aspects of Formal Power Series*, Springer-Verlag, Berlin, New York, 1978.
71. A. Salomaa, D. Wood, S. Yu, eds., *A Half-Century of Automata Theory*, World Scientific, Singapore, 2001.
72. D. Simovici, R.L. Tenney, *Theory of Formal Languages With Applications*, World Scientific, Singapore, 1999.
73. S. Sippu, E. Soisalon-Soininen, *Parsing Theory. Vol. I: Languages and Parsing*, Springer-Verlag, Berlin, Heidelberg, 1988.
74. M. Sipser, *Introduction to the Theory of Computation*, PWS Publishing Company, Boston, 1997.
75. H.J. Shyr, *Free Monoids and Languages*, Hon Min Book Comp., Taichung, 1991.
76. R.G. Taylor, *Models of Computation and Formal Languages*, Oxford University Press, 1998.
77. D. Wood, *Grammar and L Forms. An Introduction*, Springer-Verlag, Berlin, 1980 (*Lecture Notes in Computer Science*, 91).
78. D. Wood, *Theory of Computation*, Harper and Row, New York, 1987.

2

Topics in Asynchronous Circuit Theory*

Janusz Brzozowski

School of Computer Science, University of Waterloo,
Waterloo, ON, Canada N2L 3G1
E-mail: brzozo@uwaterloo.ca
http://maveric.uwaterloo.ca

Summary. This is an introduction to the theory of asynchronous circuits — a survey of some old and new results in this area. The following topics are included: benefits of asynchronous design, gate circuits, Boolean analysis methods, the theory of hazards, ternary and other multi-valued simulations, behaviors of asynchronous circuits, delay-insensitivity, and asynchronous modules used in modern design.

2.1 Motivation

This paper is a survey of some aspects of asynchronous circuit theory. No background in digital circuits is assumed. The mathematics used includes sets, relations, partial orders, Boolean algebra, and basic graph theory; other concepts are introduced as needed. Only a general knowledge of theoretical computer science is required, in particular, automata and formal languages.

Most computers have a fixed-frequency signal, called a *clock*, that controls their operations; such circuits are called *synchronous*. Because of this central control, the design of synchronous circuits is relatively easy. Circuits operating without a clock are *asynchronous*. Here, the components communicate with each other using some sort of "hand shaking" protocol. This makes the design of asynchronous circuits more challenging.

Several reasons for studying asynchronous circuits are given below; for a more complete discussion of these reasons, as well as an informal general introduction to computing without a clock, we refer the reader to a recent article by Sutherland and Ebergen [41].

- **Higher speed**
  In a synchronous circuit, the clock frequency is determined by the slowest components. In an asynchronous one, each component works at its own pace; this may lead to a higher average speed, especially in applications in which slow actions are rare.

* This research was supported by the Natural Sciences and Engineering Research Council of Canada under Grant No. OGP0000871.

J. Brzozowski: *Topics in Asynchronous Circuit Theory*, Studies in Computational Intelligence (SCI) **25**, 11–42 (2006)
www.springerlink.com

- **Lower energy consumption**
  Up to 30% of the total energy may be used by the clock and its distribution system, and energy is wasted if nothing useful is being done. In asynchronous circuits, idle components consume negligible energy — an important feature in battery-operated systems.
- **Less radio interference**
  The fixed clock frequency in a synchronous circuit results in strong radio signals at that frequency and its harmonics; this interferes with cellular phones, aircraft navigation systems, *etc.* In asynchronous circuits, the radiated energy tends to be distributed over the radio spectrum.
- **Asynchronous interfaces**
  Computers with different clock frequencies are often linked in networks. Communication between such computers is necessarily asynchronous.
- **Better handling of metastability**
  An element that arbitrates between two requests goes to one state, if the first request is granted, and to another state, if the second request wins. If the two requests arrive simultaneously, the arbiter enters an intermediate *metastable* state of unstable equilibrium, and can remain there for an unbounded amount of time [14]. In a synchronous circuit, a long-lasting metastable state is likely to cause an error. An asynchronous circuit waits until the metastable state is resolved, thus avoiding errors.
- **Scaling**
  As integrated-circuit technology improves, components become smaller. However, delays in long wires may not scale down proportionally [25]. Under such circumstances, a synchronous design may no longer be correct, while an asynchronous circuit *can* tolerate such changes in relative delays.
- **Separation of Concerns**
  In asynchronous designs, it is possible to separate physical correctness concerns (*e.g.*, restrictions on delay sizes) in the design of basic elements from the mathematical concerns about the correctness of the behavior of networks of basic elements. This cannot be done in synchronous designs.

Another reason for studying asynchronous circuits is their generality. Synchronous circuits can be viewed as asynchronous circuits satisfying certain timing restrictions. Finally, asynchronous circuits present challenging theoretical problems. Building blocks more complex than gates must be used, *e.g.*, *rendezvous elements,* which detect when two actions have both been completed, and *arbiters,* which decide which of two competing requests should be served. Efficient methods are needed for detecting *hazards,* unwanted pulses that can cause errors. The presence of many concurrent actions leads to an exponential number of possible event sequences; this “state-space explosion” must be overcome. Methods are needed for the design of *delay-insensitive* networks whose behavior should be correct even in the presence of arbitrary component and wire delays.

2.2 Gates

Logic circuits have been analyzed and designed using Boolean algebra since Shannon's pioneering paper in 1938 [38]. A *Boolean algebra* [10] is a structure $\mathbf{B} = (\mathcal{A}, +, *, ^-, 0, 1)$, where $\mathcal{A}$ is a set, $+$ and $*$ are binary operations on $\mathcal{A}$, $^-$ is a unary operation on $\mathcal{A}$, $0, 1 \in \mathcal{A}$ are distinct, and the laws of Table 2.1 hold. The laws are listed in *dual* pairs: to find the dual of a law interchange $+$ and $*$, and 0 and 1. Multiplication has precedence over addition.

Table 2.1. Laws of Boolean algebra

B1	$a + a = a$	B1′	$a * a = a$
B2	$a + b = b + a$	B2′	$a * b = b * a$
B3	$a + (b + c) = (a + b) + c$	B3′	$a * (b * c) = (a * b) * c$
B4	$a + (a * b) = a$	B4′	$a * (a + b) = a$
B5	$a + 0 = a$	B5′	$a * 1 = a$
B6	$a + 1 = 1$	B6′	$a * 0 = 0$
B7	$a + \overline{a} = 1$	B7′	$a * \overline{a} = 0$
B8	$\overline{(\overline{a})} = a$		
B9	$a + (b * c) = (a + b) * (a + c)$	B9′	$a * (b + c) = (a * b) + (a * c)$
B10	$\overline{(a + b)} = \overline{a} * \overline{b}$	B10′	$\overline{(a * b)} = \overline{a} + \overline{b}$

The smallest Boolean algebra is $\mathbf{B}_0 = (\{0, 1\}, +, *, ^-, 0, 1)$, where $+$, $*$, and $^-$ are the OR, AND, and NOT operations of Table 2.2.

Table 2.2. The operations $+$, $*$, and $^-$ in $\mathbf{B}_0$

a_1	a_2	$a_1 + a_2$
0	0	0
0	1	1
1	0	1
1	1	1

a_1	a_2	$a_1 * a_2$
0	0	0
0	1	0
1	0	0
1	1	1

a	$\overline{a}$
0	1
1	0

A Boolean function $f(x_1, \ldots, x_n)$ is a mapping $f : \{0, 1\}^n \to \{0, 1\}$, for $n \geq 0$. Given $f : \{0, 1\}^n \to \{0, 1\}$ and $g : \{0, 1\}^n \to \{0, 1\}$, define, for each $a = (a_1, \ldots, a_n) \in \{0, 1\}^n$, $(f + g)(a) = f(a) + g(a)$, $(f * g)(a) = f(a) * g(a)$, and $\overline{f}(a) = \overline{f(a)}$, where $+$, $*$, and $^-$ on the right are the operations of $\mathbf{B}_0$.

Let $\mathbf{B}_n$ be the set of all 2^{2^n} Boolean functions of n variables. Then $(\mathbf{B}_n, +, *, ^-, \mathbf{0}, \mathbf{1})$, is a Boolean algebra where the operations $+$, $*$, and $^-$ are the operations on Boolean functions defined above, and $\mathbf{0}$ and $\mathbf{1}$ are the functions that are identically 0 and 1, respectively.

Let 0, 1, $x_1, \ldots, x_n$ be distinct symbols. A *Boolean expression* over $x_1, \ldots, x_n$ is defined inductively:

1. 0, 1, $x_1, \ldots, x_n$ are Boolean expressions.
2. If E and F are Boolean expressions, then so are $(E + F)$, $(E * F)$, and $\overline{E}$.
3. Any Boolean expression is obtained by a finite number of applications of Rules 1 and 2.

Boolean expressions are used to denote Boolean functions. Expressions 0, 1, and x_i denote **0**, **1**, and the function that is identically equal to x_i, respectively. The meaning of other expressions is obtained by structural induction.

Signals in digital circuits take two values, "low voltage" and "high voltage," represented by 0 and 1, respectively. A *gate* is a circuit component performing a Boolean function. It has one or more *inputs*, one *output*, and a *direction* from the inputs to the output. Symbols of some commonly used gates are shown in Fig. 2.1. In the first row are: a *buffer* or *delay* with the identity function, $y = x$; the AND gate, with multiplication in $\mathbf{B}_0$, $y = x_1 * x_2$; the OR gate, with addition in $\mathbf{B}_0$, $y = x_1 + x_2$; the XOR gate with the exclusive-or function $y = x_1 \not\equiv x_2$, which has value 1 if the inputs differ, and 0 if they agree. The gates in the second row perform the complementary functions: the *inverter* or NOT gate; the NAND gate; the NOR gate; and the equivalence gate, EQUIV.

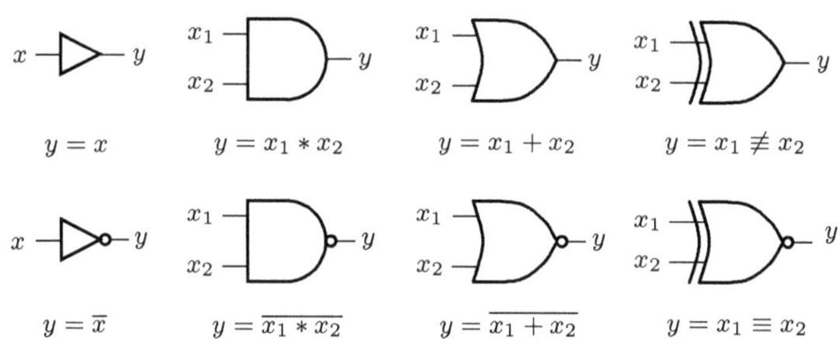

Fig. 2.1. Gate symbols

For some applications, the Boolean-function model of a gate is sufficient. To study general circuit behaviors, however, it is necessary to take gate delays into account. Thus, if the gate inputs change to produce a new output value, the output does not change at the same time as the inputs, but only after some delay. Therefore, a more accurate model of a gate consists of an ideal gate performing a Boolean function followed by a delay.

Figure 2.2(a) shows our conceptual view of the new model of an inverter. Signal Y is a fictitious signal that would be produced by an inverter with zero delay, the rectangle represents a delay, and y is the inverter output. Figure 2.2(b) shows some inverter waveforms, assuming a fixed delay. In our binary model, signals change directly from 0 to 1 and from 1 to 0, since we cannot represent intermediate values. The first waveform shows the input x as a function of time. The ideal inverter with output Y inverts this waveform with zero delay. If the delay were ideal, its output would have the shape $y'(t)$.

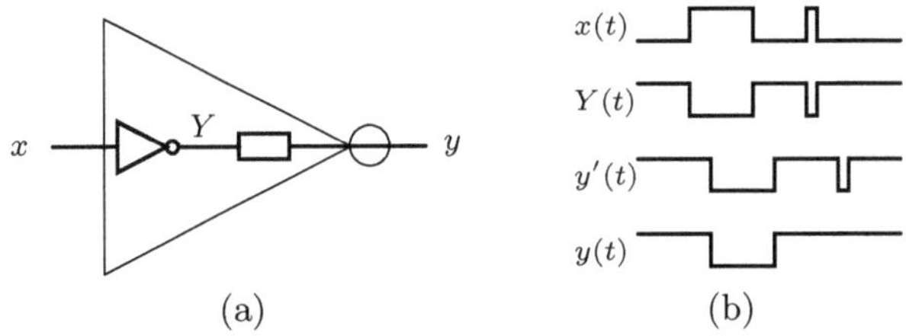

Fig. 2.2. Inverter: (a) delay model, (b) waveforms

However, if an input pulse is shorter than the delay, the pulse does not appear at the output, as in $y(t)$. In other words, the delays in our model are *inertial*. Also, we assume that the delay can change with time in an arbitrary fashion. This is a pessimistic model, but it has been used often, and it is one of the simplest. For a detailed discussion of delay models see [10].

Unless stated otherwise, we assume that each gate symbol represents a Boolean function followed by a delay. Thus, the buffer gate (with the identity function) is equivalent to a delay; hence we use a triangle to represent a delay.

2.3 Gate Circuits

A gate circuit consists of *external input terminals* (or simply *inputs*), gates, *external output terminals* (or simply *outputs*), *forks*, and *wires*. The inputs, gates, and outputs are labeled by variables $x_1, \dots, x_m$, $y_1, \dots, y_n$ and $z_1, \dots, z_p$, respectively. Let $\mathcal{X} = \{x_1, \dots, x_m\}$, $\mathcal{Y} = \{y_1, \dots, y_n\}$, and $\mathcal{Z} = \{z_1, \dots, z_p\}$. Forks are junctions with one input and two or more outputs. The environment supplies binary values to the input terminals. An external input terminal, a gate output and a fork output must be connected by a wire to a gate input, a fork input, or an external output terminal, and *vice versa*, except that we do not permit an input x to be connected to an output z, because x and z would be disjoint from the rest of the circuit. Note that a wire connects two points only; multiple connections are done through forks.

Figure 2.3(a) shows a circuit called the NOR *latch*. Forks are shown by small black dots (circled in this figure). Input x_1 is connected to an input of NOR gate y_1 by wire w_1, *etc.*

In analyzing a circuit, it is usually not necessary to assume that *all* components have delays. Several models with different delay assumptions have been used [10]. We use the *gate-delay* model. In this model, in the latch of Fig. 2.3(a), gate variables y_1 and y_2 are assumed to have nonzero delays, and wires have zero delays. The choice of delays determines the *(internal) state variables* of the circuit. In our example, we have decided that the state of the two gates suffices to give us a proper description of the behavior of the latch.

Having selected gates as state variables, we model the circuit by a *network graph*, which is a directed graph [1] (digraph) (V, E), where $V = \mathcal{X} \cup \mathcal{Y} \cup \mathcal{Z}$

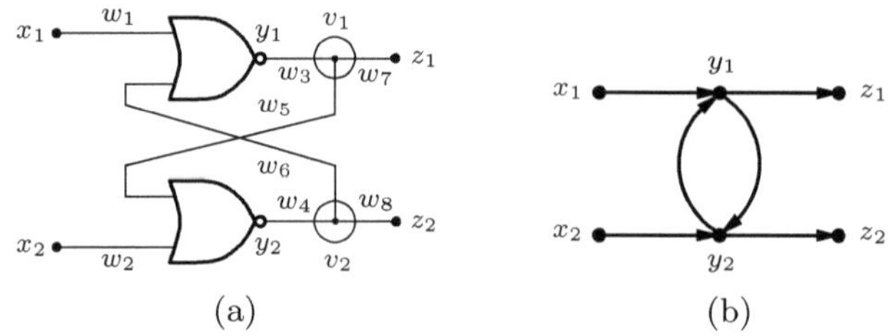

Fig. 2.3. The NOR latch: (a) circuit, (b) network

is the set of vertices, and $E \subseteq V \times V$ is the set of directed edges defined as follows. There is an edge (v, v') if vertex v is connected by a wire, or by several wires via forks, to vertex v' in the circuit, and vertex v' depends functionally on vertex v. The network graph of the latch is shown in Fig. 2.3(b).

Suppose there are incoming edges from external inputs $x_{j_1}, \ldots, x_{j_{m_i}}$ and gates $y_{k_1}, \ldots, y_{k_{n_i}}$ to a gate y_i, and the Boolean function of the gate is

$$f_i : \{0,1\}^{m_i+n_i} \to \{0,1\}. \tag{2.1}$$

The *excitation function* (or *excitation*) of gate y_i is

$$Y_i = f_i(x_{j_1}, \ldots, x_{j_{m_i}}, y_{k_1}, \ldots, y_{k_{n_i}}). \tag{2.2}$$

If the incoming edge of output variable z_i comes from vertex v, then the *output equation* is $z_i = v$.

In the example of the latch, we have the excitations and output equations:

$$Y_1 = \overline{x_1 + y_2}, \quad Y_2 = \overline{x_2 + y_1}; \qquad z_1 = y_1, \quad z_2 = y_2.$$

A *network* $N = (x, y, z, Y, E_z)$ of a circuit consists of an input m-tuple $x = (x_1, \ldots, x_m)$, a state n-tuple $y = (y_1, \ldots, y_n)$, an output p-tuple $z = (z_1, \ldots, z_p)$, an n-tuple $Y = (Y_1, \ldots, Y_n)$ of excitation functions, and a p-tuple $E_z = (z_1 = y_{h_1}, \ldots, z_p = y_{h_p})$ of output equations. In examples, we use circuit diagrams, and leave the construction of the networks to the reader.

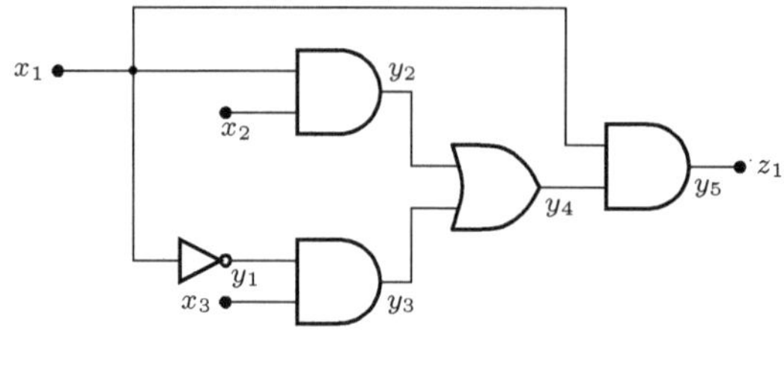

Fig. 2.4. A feedback-free circuit

A network is *feedback-free* if its network graph is acyclic; an example is given in Fig. 2.4. In a feedback-free network, one can define the concept of *level*. External input vertices are of level 0. A gate vertex is of level 1, if its inputs are of level 0. Inductively, a gate is of level k, if its inputs come from vertices of level $\leq k-1$ and at least one of them comes from a vertex of level $k-1$. An output $z_i = y_j$ has the level of y_j.

In the circuit of Fig. 2.4, y_1 and y_2 are of level 1, y_3 is of level 2, y_4 is of level 3, and y_5 and z_1 are of level 4. The excitations and output equations are:

$$Y_1 = \overline{x_1}, \quad Y_2 = x_1 * x_2, \quad Y_3 = y_1 * x_3, \quad Y_4 = y_2 + y_3, \quad Y_5 = x_1 * y_4; \qquad z_1 = y_5.$$

2.4 Binary Analysis

Our analysis of asynchronous circuits by Boolean methods follows the early work of Muller [28, 31, 32], but we use the terminology of [7, 10, 12, 19].

A *total state* $c = a \cdot b$ of a network $N = (x, y, z, Y, E_z)$ is an $(m+n)$-tuple of values from $\{0,1\}$, the first m values (m-tuple $a = (a_1, \ldots, a_m)$) being the inputs, and the remaining n (n-tuple $b = (b_1, \ldots, b_n)$), the state variables $y_1, \ldots, y_n$. The $\cdot$ is used for readability. The excitation function f_i of (2.1) is extended to a function $f'_i : \{0,1\}^{m+n} \to \{0,1\}$ as follows:

$$f'_i(a \cdot b) = f_i(a_{j_1}, \ldots, a_{j_{m_i}}, b_{k_1}, \ldots, b_{k_{n_i}}). \tag{2.3}$$

Usually we write f'_i simply as f_i; the meaning is clear from the context. A state variable y_i is *stable* in state $a \cdot b$ if $b_i = f_i(a \cdot b)$; otherwise it is *unstable*. The set of unstable state variables is $U(a \cdot b)$. A state $a \cdot b$ is *stable* if $U(a \cdot b) = \emptyset$.

We are interested in studying the behavior of a network started in a given state $a \cdot b$ with the input kept constant at $a \in \{0,1\}^m$. A state in which two or more variables are unstable is said to contain a *race*. We use the "general multiple-winner" (GMW) model, where any number of variables can "win" the race by changing simultaneously. With this in mind, we define a binary relation R_a on the set $\{0,1\}^n$ of states of N: For any $b \in \{0,1\}^n$,

bR_ab, if $U(a \cdot b) = \emptyset$, *i.e.*, if total state $a \cdot b$ is stable, and

bR_ab^K, if $U(a \cdot b) \neq \emptyset$, and K is any nonempty subset of $U(a \cdot b)$,

where b^K is b with all the variables in K complemented. Nothing else is in R_a. As usual, we associate a digraph with the relation R_a, and denote it G_a. The set of all states reachable from b in relation R_a is

$$\mathit{reach}(R_a(b)) = \{c \mid bR_a^*c\}, \tag{2.4}$$

where R_a^* is the reflexive-and-transitive closure of R_a. We denote by $G_a(b)$ the subgraph of G_a corresponding to $\mathit{reach}(R_a(b))$.

In examples, we write binary tuples as binary words, for simplicity. The circuit of Fig. 2.5 has binary input x, and state $y = (y_1, y_2, y_3)$. The excitations are $Y_1 = x + y_1$, $Y_2 = \overline{y_1}$, and $Y_3 = \overline{x * y_2 * y_3}$. Graph $G_1(011)$ for this circuit

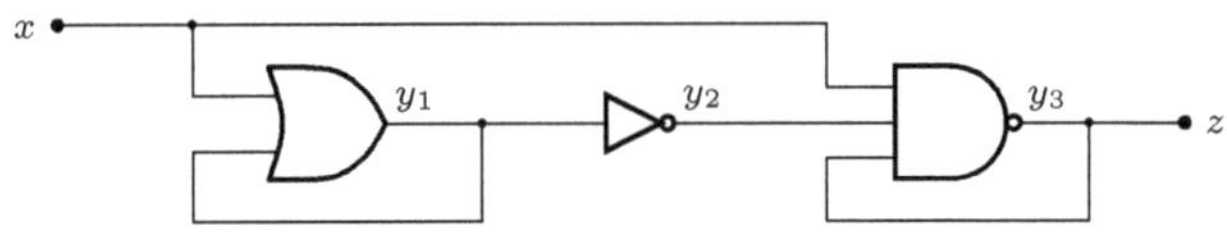

Fig. 2.5. A circuit with a transient oscillation

is shown in Fig. 2.6(a), where unstable entries are underlined. There are three cycles in $G_1(011)$: $(011, 010)$, $(111, 110)$, and (101). In general, a cycle of length 1 is a stable state, while a cycle of length > 1 is an *oscillation*. A cycle in which a variable y_i has the same value in all of the states of the cycle, and is unstable in the entire cycle is called *transient*. Under the assumption that the delay of each gate is finite, a circuit cannot remain indefinitely in a transient cycle. In the example above, both cycles $(011, 010)$ and $(111, 110)$ are transient.

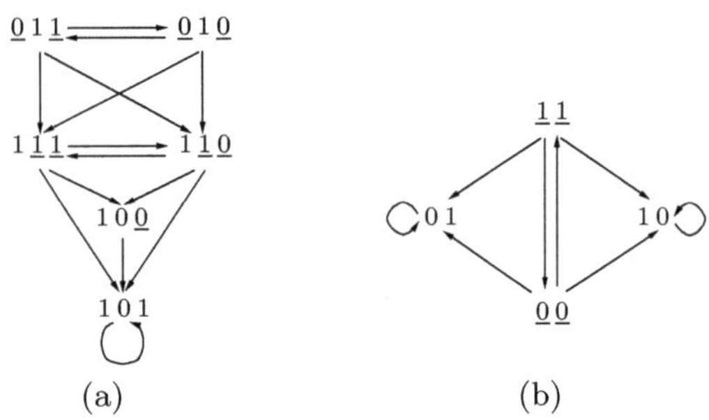

Fig. 2.6. Graphs: (a) $G_1(011)$ for Fig. 2.5, (b) $G_{00}(11)$ for Fig. 2.3

One of the important questions in the analysis of asynchronous circuits is: What happens after the input has been held constant for a long time? (This concept will be made precise.) In the example of Fig. 2.6(a), we conclude that the circuit can only be in stable state 101 after a long time, because unstable gates must eventually change, since they have finite delays.

A different kind of cycle is present in graph $G_{00}(11)$ of Fig. 2.6(b) for the NOR latch. In state 11 there is a two-way race. If y_1 wins the race, the state becomes 01, and the instability that was present in y_2 in state 11 has been removed by the change in y_1. This shows that our mathematical model captures the inertial nature of the delays: the instability in y_2 did not last long enough, and was ignored. Similarly, if y_2 wins the race, stable state 10 is reached. Here the race is *critical,* since its outcome depends on the relative sizes of the delays. A third possibility exists, namely, the cycle $(11, 00)$. The circuit can remain in this cycle only if both delays are perfectly matched; any imbalance will cause the circuit to leave the cycle and enter one of the two stable states. For this reason, such cycles are called *match-dependent*. In a crude way they represent the phenomenon of metastability [10, 14]. Let the set of *cyclic states* reachable from b in the graph of R_a be

$$cycl(R_a(b)) = \{s \in \{0,1\}^n \mid bR_a^* s \text{ and } sR_a^+ s\}, \tag{2.5}$$

R_a^+ being the transitive closure of R_a. Let the set of *nontransient cyclic states* be $cycl\_nontrans(R_a(b))$. The *outcome* of state b under input a is the set of all the states reachable from some state that appears in a nontransient cycle of $G_a(b)$. Mathematically, we have

$$out(R_a(b)) = \{s \mid bR_a^* c \text{ and } cR_a^* s, \text{ where } c \in cycl\_nontrans(R_a(b))\}. \tag{2.6}$$

This definition captures the intuitive notion that a circuit can be in a state d after a sufficiently long time only if d is in the outcome. For a more detailed discussion of these issues, see [10], where the following result is proved:

Theorem 1. *Let N be a network with n state variables with delay sizes bounded by D. If N is in state b at time 0 and the input is held constant in the interval 0 to t, $t \geq (2^n - 2)D$, then the state at time t is in $out(R_a(b))$.*

In the examples of Fig. 2.6 (a) and (b), the outcomes are $\{101\}$ and $\{00, 01, 10, 11\}$ respectively. We leave it to the reader to verify that the outcome of graph $G_{111}(10110)$ of the circuit of Fig. 2.4 consists of the stable state 01011.

The size of graph $G_a(b)$ is exponential in the number of state variables, and this makes the computation of outcome inefficient. Better methods for determining properties of the outcome will be given later.

2.5 Hazards

In calculating the outcome, we find the final states in which the network may end up after an input change; this is a "steady-state" property. We now turn our attention to "transient properties", *i.e.*, the sequences of values that variables go through before reaching their final values in the outcome.

Hazards are unwanted pulses occurring in digital circuits because of the presence of "stray" delays. It is important to detect and eliminate them, because they can result in errors. Early work on hazards was done by Huffman [24], McCluskey [26], and Unger [45], among others. For more details and references, see [6], where the use of multi-valued algebras for hazard detection is discussed. Recent results of Brzozowski and Esik [5] unify the theory of hazard algebras. We begin by a brief outline of this theory, which has its roots in Eichelberger's work on ternary simulation [10, 18]; see also [7].

Consider the feedback-free circuit of Fig. 2.4 redrawn in Fig. 2.7. Suppose the input $x = (x_1, x_2, x_3)$ has the constant value 011; one verifies that the state $y = (y_1, \ldots, y_5)$ becomes 10110, as shown by the first value of each variable in the figure. If the input becomes 111, the following changes take

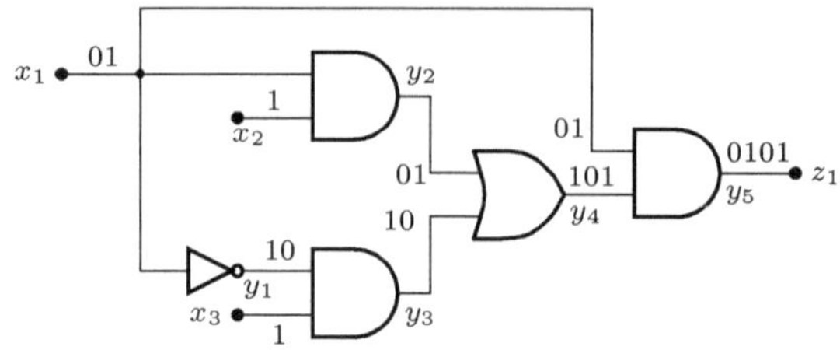

Fig. 2.7. Illustrating hazards

place: $y_1 : 1 \to 0$, $y_2 : 0 \to 1$, and $y_3 : 1 \to 0$, as shown by the second values. Two possibilities exist for y_4. If y_2 changes first, or at the same time as y_3, then y_4 remains 1. If y_3 changes first, however, there is a period during which both y_2 and y_3 are 0, and y_4 may have a 0-pulse. Such a pulse is called a *static hazard*—static, because y_4 has the same value before and after the input change, and hence is not supposed to change. For y_5, if the change in x_1 is followed by a double change in y_4 as shown, then the output has a *dynamic hazard:* y_5 should change only once from 0 to 1, but changes three times. The 1-pulse can lead to an error. In summary, hazards are unexpected signal changes: In the case of a static hazard, there is a nonzero even number of changes where there should be none; in case of a dynamic hazard, there is a nonzero even number of changes in addition to the one expected change.

Hazards can be detected using binary analysis by examining all paths in graph $G_a(b)$. For example, one verifies that the following path exists in graph $G_{111}(10110)$ of the circuit in Fig. 2.7: $\pi = \underline{1}\underline{0}11\underline{0} \to 0\underline{0}\underline{1}11 \to 0\underline{0}0\underline{1}1 \to 0\underline{0}00\underline{1} \to 0\underline{0}000 \to 010\underline{0}0 \to 0101\underline{0} \to 01011$. Along this path, we have the changes $y_4 : 1 \to 0 \to 1$, and $y_5 : 0 \to 1 \to 0 \to 1$.

In general, graph $G_a(b)$ may have as many as 2^n nodes, and the binary method is not practical for hazard detection; hence alternate methods have been sought [6]. We describe one such method, which uses "transients."

A *transient* is a nonempty binary word in which no two consecutive letters are the same. The set of all transients is $\mathbf{T} = \{0, 1, 01, 10, 010, 101, \ldots\}$. Transients represent signal waveforms in the obvious way. For example, waveform $x(t)$ in Fig. 2.2(b) corresponds to the transient 01010.

By *contraction* of a binary word s, we mean the operation of removing all letters, say a, repeated directly after the first occurrence of a, thus obtaining a word $\hat{s}$ of alternating 0s and 1s. For example, the contraction of the sequence 01110001 of values of y_5 along path π above is 0101. For variable y_i, such a contraction is the *history* of that variable along π and is denoted $\boldsymbol{\sigma}_i^\pi$.

We use boldface symbols to denote transients. If $\mathbf{t}$ is a transient, $\alpha(\mathbf{t})$ and $\omega(\mathbf{t})$ are its first and last letters, respectively. Also, $z(\mathbf{t})$ and $u(\mathbf{t})$ denote the number of 0s and 1s in $\mathbf{t}$, respectively. We write $\mathbf{t} \le \mathbf{t}'$ if $\mathbf{t}$ is a prefix of $\mathbf{t}'$; $\le$ is a partial order on $\mathbf{T}$. Let $[n]$ denote the set $\{1, \ldots, n\}$. We extend the partial order $\le$ to n-tuples of transients: for $\mathbf{t} = (\mathbf{t}_1, \ldots, \mathbf{t}_n)$ and $\mathbf{t}' = (\mathbf{t}'_1, \ldots, \mathbf{t}'_n)$,

$\mathbf{t} \le \mathbf{t}'$, if $\mathbf{t}_i \le \mathbf{t}'_i$, for all $i \in [n]$. We denote by $\mathbf{t} \circ \mathbf{t}'$ concatenation followed by contraction [7], *i.e.*, $\mathbf{t} \circ \mathbf{t}' = \widehat{\mathbf{t}\mathbf{t}'}$. The suffix relation $\preceq$ is also a partial order.

In the algebra we are about to define, gates process transients instead of 0s and 1s. Thus it is necessary to extend Boolean functions to the domain of transients. Let $f : \{0,1\}^n \to \{0,1\}$ be any Boolean function. Its *extension* to $\mathbf{T}$ is the function $\mathbf{f} : \mathbf{T}^n \to \mathbf{T}$ defined as follows: For any tuple $(\mathbf{t}_1, \ldots, \mathbf{t}_n)$ of transients, $\mathbf{f}(\mathbf{t}_1, \ldots, \mathbf{t}_n)$ is the longest transient produced when $\mathbf{t}_1, \ldots, \mathbf{t}_n$ are applied to the inputs of a gate performing function f. A method for computing the extension of any function is given in [5], as well as formulas for the extensions of some common functions. Here we give only the extensions of complementation, and the 2-input AND and OR functions. If $\mathbf{t} = a_1 \ldots a_n$ is any transient, where the a_i are in $\{0,1\}$, and f is complementation, then $\mathbf{f}(\mathbf{t}) = \overline{\mathbf{t}} = \overline{a_1} \ldots \overline{a_n}$. If f is the 2-input OR function, then $\mathbf{s} = \mathbf{f}(\mathbf{t}_1, \mathbf{t}_2)$ is the word in $\mathbf{T}$ determined by the conditions[2]

$$\begin{aligned}
\alpha(\mathbf{s}) &= \alpha(\mathbf{t}_1) + \alpha(\mathbf{t}_2) \\
\omega(\mathbf{s}) &= \omega(\mathbf{t}_1) + \omega(\mathbf{t}_2) \\
z(\mathbf{s}) &= \begin{cases} 0 & \text{if } \mathbf{t}_1 = 1 \text{ or } \mathbf{t}_2 = 1 \\ z(\mathbf{t}_1) + z(\mathbf{t}_2) - 1 & \text{otherwise.} \end{cases}
\end{aligned}$$

Dually, if f is the 2-input AND function, then $\mathbf{s} = \mathbf{f}(\mathbf{t}_1, \mathbf{t}_2)$ is given by

$$\begin{aligned}
\alpha(\mathbf{s}) &= \alpha(\mathbf{t}_1) * \alpha(\mathbf{t}_2) \\
\omega(\mathbf{s}) &= \omega(\mathbf{t}_1) * \omega(\mathbf{t}_2) \\
u(\mathbf{s}) &= \begin{cases} 0 & \text{if } \mathbf{t}_1 = 0 \text{ or } \mathbf{t}_2 = 0 \\ u(\mathbf{t}_1) + u(\mathbf{t}_2) - 1 & \text{otherwise.} \end{cases}
\end{aligned}$$

We denote by $\otimes$ and $\oplus$ the extensions of the 2-argument Boolean AND $(*)$ and OR $(+)$ functions, respectively. Algebra $C = (\mathbf{T}, \oplus, \otimes, ^{-}, 0, 1)$, is called the *change-counting algebra*, and is a commutative de Morgan bisemigroup [5], *i.e.*, it satisfies laws B2, B3, B5, B6, B8, B10 of Boolean algebra from Table 2.1 and their duals.

We refer to C as the *algebra of transients*, and use it to define efficient simulation algorithms for detecting hazards. There are two algorithms, A and B, and two versions, A and Ã, of the first algorithm. Algorithm A is general, whereas Ã applies only if the initial state is stable. It is proved in [7] that A and Ã produce the same result if the initial state is stable, under the condition that the network contains *input delays*, *i.e.*, delays in the wires leaving the input terminals; these delays can be viewed as gates performing identity functions.

For brevity we describe only Algorithm Ã. In this section, we ignore the output terminals and consider only inputs and state variables.

Let $N = (x, y, z, Y, E_z)$ be a network. The *extension* of N to the domain $\mathbf{T}$ of transients is $\mathbf{N} = (\mathbf{x}, \mathbf{y}, \mathbf{z}, \mathbf{Y}, \mathbf{E_z})$, where $\mathbf{x}$, $\mathbf{y}$, and $\mathbf{z}$, are variables taking

[2] In $\alpha(\mathbf{s})$ and $\omega(\mathbf{s})$, $+$ represents Boolean OR, whereas in $z(\mathbf{s})$ and $u(\mathbf{s})$ it is addition.

values from $\mathbf{T}$, and the Boolean excitation functions are replaced by their extensions to $\mathbf{T}$.

Let $a \cdot b$ be a (binary) total state of $\mathbf{N}$. Algorithm $\tilde{\mathrm{A}}$ is defined as follows:

Algorithm $\tilde{\mathbf{A}}$
$\mathbf{a} = \tilde{a} \circ a$;
$\mathbf{s}^0 := b$;
$h := 1$;
$\mathbf{s}^h := \mathbf{S}(\mathbf{a} \cdot \mathbf{s}^0)$;
<u>while</u> $(\mathbf{s}^h <> \mathbf{s}^{h-1})$ <u>do</u>
    $h := h + 1$;
    $\mathbf{s}^h := \mathbf{S}(\mathbf{a} \cdot \mathbf{s}^{h-1})$;

We illustrate the algorithm in Table 2.3 using the circuit of Fig. 2.7. The initial state $011 \cdot 10110$ is stable. We change the input to $(1, 1, 1)$ and concatenate $(0, 1, 1)$ and $(1, 1, 1)$ component-wise with contraction, obtaining $(01, 1, 1)$ as the new input. In the first row with this input, y_1, y_2, and y_5 are unstable in the algebra of transients; they are all changed *simultaneously* to get the second row. Now y_3 is unstable, and changes to 10 in the third row. In the third row, y_4 is unstable; according to the definition of the extended OR, the output becomes 101. In the fifth row, y_5 becomes 0101. At this point, the state $(10, 01, 10, 101, 0101)$ is stable, and the algorithm stops. Note that the last row shows that y_1, y_2, and y_3 have single, hazard-free changes, that y_4 has the static hazard 101, and y_5, the dynamic hazard 0101. The final state, 01011, is shown by the last bits of the transients.

Table 2.3. Simulation in Algorithm $\tilde{\mathrm{A}}$

	$\mathbf{x}_1$	$\mathbf{x}_2$	$\mathbf{x}_3$	$\mathbf{y}_1$	$\mathbf{y}_2$	$\mathbf{y}_3$	$\mathbf{y}_4$	$\mathbf{y}_5$
initial state	0	1	1	1	0	1	1	0
	01	1	1	1	0	1	1	0
	01	1	1	10	01	1	1	01
	01	1	1	10	01	10	1	01
	01	1	1	10	01	10	101	01
result $\tilde{\mathrm{A}}$	01	1	1	10	01	10	101	0101

Algorithm $\tilde{\mathrm{A}}$ results in a state sequence that is nondecreasing with respect to the prefix order [5]:

$$\mathbf{s}^0 \le \mathbf{s}^1 \le \ldots \le \mathbf{s}^h \le \ldots$$

Algorithm $\tilde{\mathrm{A}}$ may not terminate, for example, for the NOR latch with initial state $11 \cdot 00$ and inputs changing to 00. If Algorithm $\tilde{\mathrm{A}}$ does terminate, let its result be $\mathbf{s}^{\tilde{A}}$. Note that $\mathbf{s}^{\tilde{A}} = \mathbf{S}(\mathbf{a}, \mathbf{s}^{\tilde{A}})$, *i.e.*, the last state is stable.

The next theorem shows that it is possible to reduce the set of state variables and still obtain the same result using simulation with the remaining variables. A set $\mathcal{F}$ of vertices of a digraph G is a *feedback-vertex set* if every cycle in G contains at least one vertex from $\mathcal{F}$.

To illustrate variable removal, consider the NOR latch of Fig. 2.3. If we choose y_1 as the only feedback variable, we eliminate y_2 by substituting $\overline{y_1 + x_2}$ for it, obtaining $Y_1 = \overline{x_1 + y_2} = \overline{x_1 + \overline{y_1 + x_2}} = \overline{x_1} * (y_1 + x_2)$. In this way we obtain a *reduced* network[3] $\dot{N}$ of N. We perform similar reductions in the extended network $\mathbf{N}$ to obtain $\dot{\mathbf{N}}$. The proof of the following claim is similar to an analogous theorem proved in [7] for Algorithm A:

Theorem 2. *Let $\mathcal{F}$ be a feedback-vertex set of a network $\mathbf{N}$, and let $\dot{\mathbf{N}}$ be the reduced version of $\mathbf{N}$ with vertex set $\mathcal{X} \cup \mathcal{F}$. If Algorithm $\tilde{A}$ terminates on $\mathbf{N}$, the final state of $\mathbf{N}$ agrees with that of $\dot{\mathbf{N}}$ for the state variables in $\mathcal{F}$.*

We now compare the simulation algorithm to binary analysis. The next theorem [7] shows that the result of simulation "covers" the result of binary analysis in the following sense. Suppose π is a path of length h starting at b in $G_a(b)$, and the n–tuple of histories of the state variables along π is $\boldsymbol{\sigma}^{\pi}$. Then $\boldsymbol{\sigma}^{\pi}$ is a prefix of the simulation result after h steps. More formally, we have

Theorem 3. *For all paths $\pi = s^0, \dots, s^h$ in $G_a(b)$, with $s^0 = b$, we have $\boldsymbol{\sigma}^{\pi} \leq \mathbf{s}^h$, where $\mathbf{s}^h$ is the $(h+1)$st state in the sequence resulting from Algorithm $\tilde{A}$.*

Corollary 1. *If Algorithm $\tilde{A}$ terminates with state $\mathbf{s}^H$, then for any path π in $G_a(b)$, $\boldsymbol{\sigma}^{\pi} \leq \mathbf{s}^H$.*

Only partial results are known about the converse of Corollary 1. It has been shown by Gheorghiu [19, 20] that, for feedback-free circuits of 1- and 2-input gates, if appropriate wire delays are taken into account, there exists a path π in $G_a(b)$ such that $\boldsymbol{\sigma}^{\pi} = \mathbf{s}^H$.

Alternatives to simulation with an infinite algebra are described next.

2.6 Ternary Simulation

Algebra C permits us to count an arbitrary number of signal changes; an alternative is to count precisely only up to some threshold $k-1$, $k > 1$, and consider all transients of length $\geq k$ as equivalent [5]. For $k > 1$, define relation $\sim_k$ as follows; For $\mathbf{s}, \mathbf{t} \in \mathbf{T}$, $\mathbf{s} \sim_k \mathbf{t}$ if either $\mathbf{s} = \mathbf{t}$ or $\mathbf{s}$ and $\mathbf{t}$ are both of length $\geq k$. Thus the equivalence class $[\mathbf{t}]$ of $\mathbf{t}$ is the singleton $\{\mathbf{t}\}$ if the length of $\mathbf{t}$ is $< k$; all the remaining elements of $\mathbf{T}$ are in one class that we denote Φ_k, or simply Φ, if k is understood.

Relation $\sim_k$ is a congruence relation on C, meaning that for all $\mathbf{s}, \mathbf{t}, \mathbf{w} \in \mathbf{T}$, $\mathbf{s} \sim_k \mathbf{t}$ implies $(\mathbf{w} \oplus \mathbf{s}) \sim_k (\mathbf{w} \oplus \mathbf{t})$, and $\overline{\mathbf{s}} \sim_k \overline{\mathbf{t}}$. Thus, there is a unique algebra $C_k = C/\!\sim_k = (\mathbf{T}_k, \oplus, \otimes, ^-, 0, 1)$, where $\mathbf{T}_k = \mathbf{T}/\!\sim_k$ is the quotient set of

[3] After the substitution we must compute the outputs differently. We still have $z_1 = y_1$, but $z_2 = y_2 = \overline{x_2 + y_1}$ here.

equivalence classes of $\mathbf{T}$ with respect to $\sim_k$, such that the function from $\mathbf{T}$ to $\mathbf{T}_k$ taking a transient to its equivalence class is a homomorphism, *i.e.*, preserves the operations and constants. The operations in $\mathbf{T}_k$ are as follows: $\overline{[\mathbf{t}]} = [\overline{\mathbf{t}}]$, $[\mathbf{t}] \oplus [\mathbf{t}'] = [\mathbf{t} \oplus \mathbf{t}']$, and $[\mathbf{t}] \otimes [\mathbf{t}'] = [\mathbf{t} \otimes \mathbf{t}']$. The quotient algebra C_k is a commutative de Morgan bisemigroup with $2k-1$ elements [5].

For Algebras $\mathbf{T}_k$, Algorithm $\tilde{\mathrm{A}}$ always terminates; we denote its result by $\mathbf{s}^{\tilde{A}}$. Following Eichelberger [10, 18], we define a second simulation algorithm, Algorithm B:

Algorithm B
$\mathbf{t}^0 := \mathbf{s}^{\tilde{A}}$;
$h := 1$;
$\mathbf{t}^h := \mathbf{S}(a, \mathbf{t}^{h-1})$;
<u>while</u> $\mathbf{t}^h <> \mathbf{t}^{h-1}$ <u>do</u>
    $h := h + 1$;
    $\mathbf{t}^h := \mathbf{S}(a, \mathbf{t}^{h-1})$;

Now the network is started in the state which results from Algorithm $\tilde{\mathrm{A}}$. The input is set to its final binary value a, and Algebra $\mathbf{T}_k$ is used for the computation of the next state. It is easy to verify that Algorithm B results in a sequence of states that is nonincreasing in the suffix order $\succeq$:

$$\mathbf{s}^{\tilde{A}} = \mathbf{t}^0 \succeq \mathbf{t}^1 \succeq \ldots \succeq \mathbf{t}^B.$$

The network reaches a stable state, *i.e.*, $\mathbf{t}^B = \mathbf{S}(a, \mathbf{t}^B)$.

The smallest quotient algebra $\mathbf{T}_2$ is isomorphic to the well-known ternary algebra [10]. Its operations are shown in Table 2.4, and satisfy all the laws of Boolean algebra from Table 2.1, except B7 and its dual, and also the laws of Table 2.5. The third element, Φ_2 (denoted simply as Φ) can be interpreted as the *uncertain* value, whereas 0 and 1 are *certain* values. The following partial order, the *uncertainty partial order* [10, 30], reflects this interpretation: $\mathbf{t} \sqsubseteq \mathbf{t}$ for all $\mathbf{t} \in \{0, \Phi, 1)$, $0 \sqsubseteq \Phi$, and $1 \sqsubseteq \Phi$.

Table 2.4. The operations $\oplus$, $\otimes$, and $^-$ in $\mathbf{T}_2$

	$\mathbf{t}_1 \oplus \mathbf{t}_2$	$\mathbf{t}_2$: 0	Φ	1
	0	0	Φ	1
$\mathbf{t}_1$	Φ	Φ	Φ	1
	1	1	1	1

	$\mathbf{t}_1 \otimes \mathbf{t}_2$	$\mathbf{t}_2$: 0	Φ	1
	0	0	0	0
$\mathbf{t}_1$	Φ	0	Φ	Φ
	1	0	Φ	1

$\mathbf{t}$	$\overline{\mathbf{t}}$
0	1
Φ	Φ
1	0

Algorithm $\tilde{\mathrm{A}}$ in $\mathbf{T}_2$ for the circuit of Fig. 2.7 is shown in the first part of Table 2.6. Here, the inputs that change become uncertain, and this uncertainty "spreads" to some of the gates. Then, in Algorithm B, the inputs are set to their final values, and some (or all) of the uncertainty is removed.

Table 2.5. Ternary laws

T1 $\overline{\Phi} = \Phi$
T2 $(\mathbf{t} \oplus \overline{\mathbf{t}}) \oplus \Phi = \mathbf{t} \oplus \overline{\mathbf{t}}$ T2′ $(\mathbf{t} \otimes \overline{\mathbf{t}}) \otimes \Phi = \mathbf{t} \otimes \overline{\mathbf{t}}$

Table 2.6. Ternary simulation

	$\mathbf{x}_1$	$\mathbf{x}_2$	$\mathbf{x}_3$	$\mathbf{y}_1$	$\mathbf{y}_2$	$\mathbf{y}_3$	$\mathbf{y}_4$	$\mathbf{y}_5$
initial state	0	1	1	1	0	1	1	0
	Φ	1	1	1	0	1	1	0
	Φ	1	1	Φ	Φ	1	1	Φ
	Φ	1	1	Φ	Φ	Φ	1	Φ
result Ã	Φ	1	1	Φ	Φ	Φ	Φ	Φ
	1	1	1	Φ	Φ	Φ	Φ	Φ
	1	1	1	0	1	Φ	Φ	Φ
	1	1	1	0	1	0	1	Φ
result B	1	1	1	0	1	0	1	1

Eichelberger [18] did not formally relate the result of ternary simulation to binary analysis. Partial results in this direction were obtained by Brzozowski and Yoeli [12], where it was shown that the result of Algorithm B "covers" the outcome of binary analysis, and it was conjectured that the converse also holds if wire delays are taken into account. The conjecture was proved by Brzozowski and Seger [9]. We briefly outline these results.

By a *complete* network we mean one in which each wire has a delay. The following characterizes the result of Algorithm Ã, where *lub* denotes the least upper bound with respect to the uncertainty partial order $\sqsubseteq$:

Theorem 4. *Let $N = (x, y, z, Y, E_z)$ be a complete binary network, and let $\mathbf{N} = (\mathbf{x}, \mathbf{y}, \mathbf{z}, \mathbf{Y}, \mathbf{E_z})$ be its ternary counterpart. If N is started in total state $\tilde{a} \cdot b$ and the input changes to a, then the result $\mathbf{s}^{\tilde{A}}$ of Algorithm $\tilde{A}$ for $\mathbf{N}$ is equal to the least upper bound of the set of states reachable from the initial state in the GMW analysis of N, i.e.,*

$$\mathbf{s}^{\tilde{A}} = lub\, reach(R_a(b)).$$

The characterization of the result of Algorithm B is given by

Theorem 5. *Let $N = (x, y, z, Y, E_z)$ be a complete binary network, and let $\mathbf{N} = (\mathbf{x}, \mathbf{y}, \mathbf{z}, \mathbf{Y}, \mathbf{E_z})$ be its ternary counterpart. If N is started in total state $\tilde{a} \cdot b$ and the input changes to a, then the result $\mathbf{t}^B$ of Algorithm B is equal to the least upper bound of the outcome of the GMW analysis, i.e.,*

$$\mathbf{t}^B = lub\, out(R_a(b)).$$

Ternary simulation can also detect static hazards. For a detailed discussion of these issues see [10].

Theorem 6. *A complete network N started in total state $\tilde{a} \cdot b$ with the input changing to a has a static hazard on variable s_i if and only if its ternary extension* $\mathbf{N}$ *has the following property: The result $\mathbf{s}_i^{\tilde{A}}$ of Algorithm $\tilde{A}$ is Φ, while the result $\mathbf{t}_i^B$ of Algorithm B is equal to the initial value b_i.*

2.7 Simulation in Other Finite Algebras

Ternary simulation detects static hazards, but does not explicitly identify them. As k is increased, simulation in Algebra $\mathbf{T}_k$ provides more and more information. Tables 2.7 and 2.8 show simulations of the circuit of Fig. 2.7 in $\mathbf{T}_3$ and $\mathbf{T}_4$, which have five and seven elements, respectively. Quinary simulation (in $\mathbf{T}_3$) reveals in Algorithm Ã the signals which have no hazards, and detects both the static and dynamic hazards by the presence of a Φ; whereas septenary simulation (in $\mathbf{T}_4$) explicitly identifies the static hazard in y_4. For $k = 5$, the nonary Algorithm Ã (in $\mathbf{T}_5$) is identical in this case to simulation in Algebra C in Table 2.3, where the dynamic hazard is explicitly identified as 0101; consequently, Algorithm B is no longer needed.

Table 2.7. Quinary simulation

	x_1	x_2	x_3	y_1	y_2	y_3	y_4	y_5
initial state	0	1	1	1	0	1	1	0
	01	1	1	1	0	1	1	0
	01	1	1	10	01	1	1	01
	01	1	1	10	01	10	1	01
	01	1	1	10	01	10	Φ	01
result Ã	01	1	1	10	01	10	Φ	Φ
	1	1	1	10	01	10	Φ	Φ
	1	1	1	0	1	10	Φ	Φ
	1	1	1	0	1	0	1	Φ
result B	1	1	1	0	1	0	1	1

Several other multi-values algebras have been proposed for hazard detection over the years [6]. It was shown in [5] that all the successful algebras can be derived from Algebra C.

2.8 Fundamental-Mode Behaviors

Until now, we have considered only what happens in a network started in a particular total state: we have computed the states of the outcome and checked for the presence of hazards. In most applications, a network started in a stable total state should move to another stable total state when its

Table 2.8. Septenary simulation

	x_1	x_2	x_3	y_1	y_2	y_3	y_4	y_5
initial state	0	1	1	1	0	1	1	0
	01	1	1	1	0	1	1	0
	01	1	1	10	01	1	1	01
	01	1	1	10	01	10	1	01
	01	1	1	10	01	10	101	01
result Ã	01	1	1	10	01	10	101	Φ
	1	1	1	10	01	10	101	Φ
	1	1	1	0	1	10	101	101
	1	1	1	0	1	0	1	101
result B	1	1	1	0	1	0	1	1

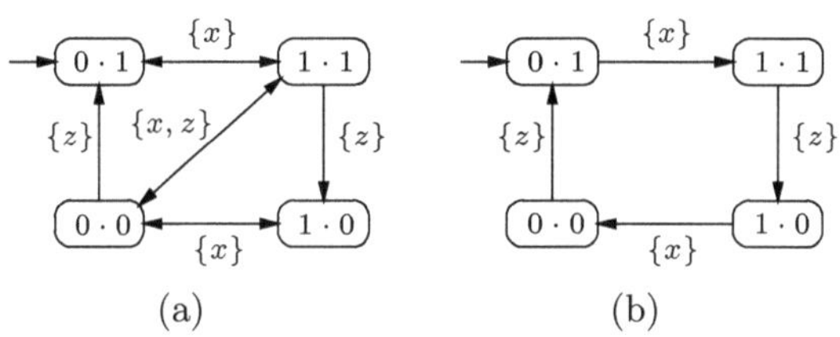

Fig. 2.8. Inverter behavior: (a) unrestricted, (b) fundamental-mode

inputs change, and its outputs should have no hazards. We now examine the response of a network to a *sequence* of input changes. We use three examples, an inverter, a latch and a flip-flop, to illustrate how sequential behaviors are modeled by automata, and how the operation of a circuit can be restricted to avoid unreliable responses. The flip-flop is also an example of a synchronous circuit analyzed by asynchronous methods to show the details of its operation.

Since we are modeling physical processes, we assume that only a finite number of input changes can take place in any finite time interval. If we do not put any other restrictions on the environment, the behaviors that result can be unreliable, and therefore not useful, as is shown below.

We represent network behaviors by digraphs with states as vertices. A directed edge from one vertex to another is a *transition*, and the set of input and output variables changing in each transition is indicated as a label.

Consider an inverter operating in an unrestricted environment. Suppose it has input x, internal state y, output $z = y$, and starts in stable state $x \cdot y = x \cdot z = 0 \cdot 1$. If the input changes to 1, the inverter changes to state $1 \cdot 1$, which is unstable; see Fig. 2.8(a). From state $1 \cdot 1$, the inverter can move to state $1 \cdot 0$. However, in state $1 \cdot 1$, the environment may change x back to 0 before the inverter has a chance to respond; then state $0 \cdot 1$ can again be reached. In state $1 \cdot 1$, it is also possible that y changes to 0 as x is changing, resulting in a transition to $0 \cdot 0$. Thus, if the input is withdrawn, we do not know which state will result. The rest of the behavior is similar.

It is clear that some restrictions must be imposed on the environment to avoid unreliable behavior. A restriction that has been used for many years is operation in *fundamental mode* [10]. In this mode, a circuit starts in a stable state, and an input can only be changed if the circuit is stable. For each input change from a stable total state we use binary analysis to find the outcome. It is tacitly assumed that the circuit always reaches a unique stable state, and that the input is held constant until this happens. The fundamental-mode behavior of the inverter is shown in Fig. 2.8(b).

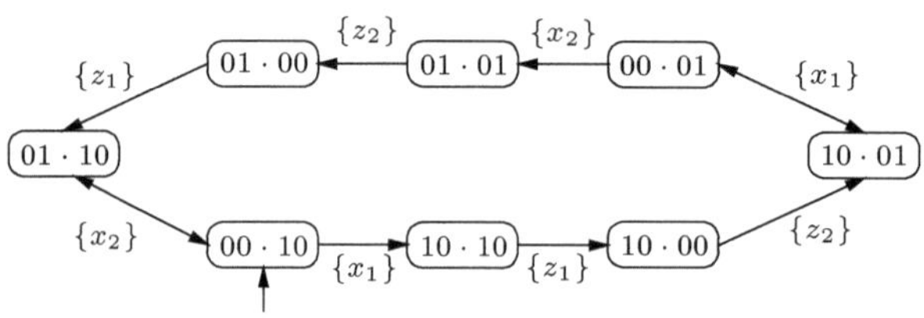

Fig. 2.9. Part of fundamental-mode behavior of latch

As a second example, consider the latch of Fig. 2.3 started in total state $00 \cdot 10$; see Fig. 2.9. Note that $z_1 = y_1$ and $z_2 = y_2$. State $00 \cdot 10$ is stable. If the input changes to 01, the latch remains stable. When the input is 10, the excitations are $Y_1 = 0$, $Y_2 = \overline{y_1}$. If the input changes to 10 from state $00 \cdot 10$, the new state $10 \cdot 10$ is unstable, and $y_1 y_2$ changes as follows: $\underline{1}\,0 \rightarrow 0\,\underline{0} \rightarrow 0\,1$. Thus the outcome is a unique stable state; there are no races, oscillations or hazards. The rest of the behavior is symmetric.

The latch behavior above is not only fundamental-mode, but is also restricted by the condition that both inputs should never be 1 at the same time. The latch is used as a memory device, storing a value, traditionally called Q, in variable $y_1 = z_1$, and its complement $\overline{Q}$ in variable $y_2 = z_2$. Input x_1 then corresponds to the *reset* input R, and x_2 is the *set* input S. When $RS = 00$, the latch has two stable states $Q\overline{Q} = 01$ and $Q\overline{Q} = 10$; it *remembers* the previously stored value. If $RS = 10$, the latch is *reset* to, or stays in, state $Q\overline{Q} = 01$, and, if $RS = 01$, the latch is *set* to, or stays in, state $Q\overline{Q} = 10$. If the input 11 is applied and held, both Q and $\overline{Q}$ become 0, violating the condition that $\overline{Q}$ should be the complement of Q. Furthermore, if both inputs then become 0 at the same time, the latch enters a match-dependent oscillation, as shown in Fig. 2.6(b). Since the latch's behavior is unpredictable after it leaves the oscillation, the input combination 11 is not used in practice.

The circuit of Fig. 2.10 is a *master/slave toggle flip-flop*. It has two inputs: the *toggle* input t, and the *clock* input c. It is normally considered a synchronous circuit, but we analyze it as an asynchronous circuit operating in fundamental mode and under the restriction that t should not change when $c = 1$. Variables y_3, y_4 form the *master* latch, and y_7, y_8 are the *slave* latch; these variable values will be shown in boldface. Suppose $ct = 00$, and

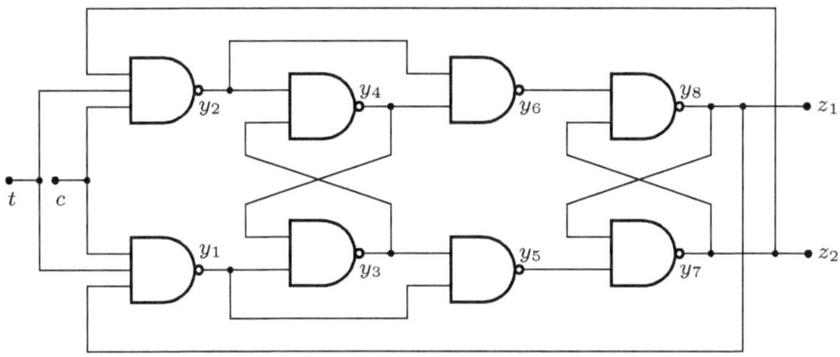

Fig. 2.10. Toggle flip-flop

$y = y_1 \dots y_8 = 11\mathbf{01}10\mathbf{01}$; this state is stable, both latches hold the values 01, and the *slave follows the master*. If t becomes 1, there is no change. If c then becomes 1, and the input is constant at $ct = 11$, the excitations are: $Y_1 = \overline{y_8}$, $Y_2 = \overline{y_7}$, $Y_3 = \overline{y_1 * y_4}$, $Y_4 = \overline{y_2 * y_3}$, $Y_5 = \overline{y_1 * y_3}$, $Y_6 = \overline{y_2 * y_4}$, $Y_7 = \overline{y_5 * y_8}$, $Y_8 = \overline{y_6 * y_7}$. The internal state changes as follows:

$$\underline{1}1\mathbf{01}10\mathbf{01} \to 01\mathbf{\underline{0}1}10\mathbf{01} \to 01\mathbf{1\underline{1}}10\mathbf{01} \to 01\mathbf{10}1\underline{0}\mathbf{01} \to 01\mathbf{10}11\mathbf{01}.$$

The new values 10 are now stored in the master, while the slave remembers the old values 01. When the clock again becomes 0, we have

$$\underline{0}1\mathbf{10}11\mathbf{01} \to 11\mathbf{10}\underline{1}1\mathbf{01} \to 11\mathbf{10}01\mathbf{\underline{0}1} \to 11\mathbf{10}01\mathbf{1\underline{1}} \to 11\mathbf{10}01\mathbf{1\underline{0}}.$$

The slave again follows the master, and the operation can be repeated when c next becomes 1. We can summarize the flip-flop behavior as follows: If $t = 0$ when the c becomes 1, the outputs of the flip-flop do not change. If $t = 1$ when c becomes 1, the flip-flop *toggles*, *i.e.*, its outputs are complemented. For more examples of this approach to analysis see [11].

In general, a *behavior* of a network $N = (x, y, z, Y, E_z)$ with m inputs $x_1, \dots, x_m$, n state variables $y_1, \dots, y_n$, and p outputs $z_1, \dots, z_p$ is a tuple

$$B = (x, y, z, Q, q_\epsilon, T, E_z),$$

where $Q = \{0,1\}^{m+n}$ is the set of *total states*, $q_\epsilon \in Q$ is the *initial state*, $T \subseteq Q \times Q - \{(q,q) \mid q \in Q\}$ is the set of *transitions*, to be defined.

Several types of behaviors have been studied [10]; here we consider only certain special ones. The *unrestricted transition relation* is defined as follows. In any total state $a \cdot b$, $a \in \{0,1\}^m$, $b \in \{0,1\}^n$, if $bR_a b'$, $b \neq b'$ or $a \neq a'$, we have $(a \cdot b, a' \cdot b') \in T$. The behavior of the inverter in Fig. 2.8(a) is an example of unrestricted behavior.

The *fundamental-mode transition relation* is defined as follows: $(a \cdot b, a' \cdot b') \in T$ if and only if $bR_a b'$ and either $b = b'$ or $a = a'$. Thus, an input can change only if $a \cdot b$ is stable; otherwise, the state follows the R_a relation. The latch behavior in Fig. 2.9 is a part of its fundamental-mode behavior.

A behavior is *direct* if it is fundamental-mode and every transition from an unstable state leads directly to a stable state. An example of a direct behavior is shown in Fig. 2.11 for the latch. Here the details of the two 2-step transitions of Fig. 2.9 are suppressed and only the final outcome is shown.

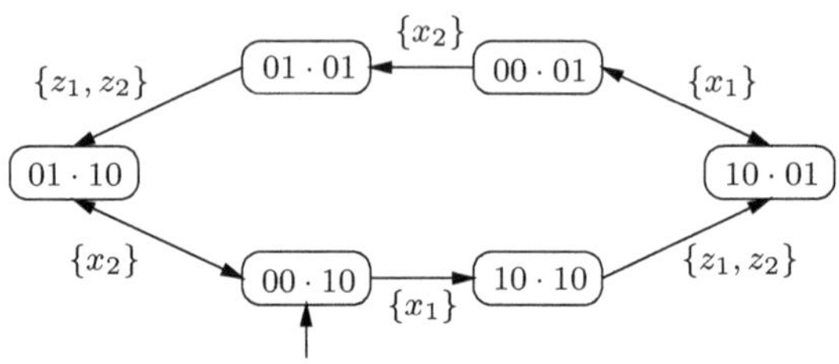

Fig. 2.11. Part of direct behavior of latch

With each behavior $B = (x, y, z, Q, q_\epsilon, T, E_z)$, we associate a finite nondeterministic *behavior automaton* $A = (\Sigma, Q, q_\epsilon, F, T)$, where $\Sigma = 2^{\mathcal{X} \cup \mathcal{Z}} \setminus \{\emptyset\}$ is the input alphabet of the automaton consisting of nonempty subsets of the set of input and output variables. Elements of Σ are used as labels on the transitions; the label is the empty word ϵ if there are no input or output changes accompanying a transition. Also, $F = Q$, *i.e.*, every state is final (accepting). The language $L = L(A)$ of all words accepted by A is always *prefix-closed, i.e.*, the prefix of every word in the language is also in the language.

2.9 Delay-Insensitivity

Behavior automata can be used as specifications of behaviors to be realized by networks. Ideally, a network realizing a given behavior should be *delay-insensitive*, in the sense that it should continue to realize the behavior in the presence of arbitrary gate and wire delays. As we shall see, very few behaviors can be realized delay-insensitively. This will be made more precise.

Fundamental-mode operation has the major drawback that the environment needs to know how long to wait for a stable state to be reached, and this is an unrealistic requirement. An alternative is the "input/output mode" of operation [3, 4, 10, 15, 29]. Here the environment is allowed to change an input after a response from the component has been received, or no response is expected. To illustrate this mode of operation, we present the circuit of Fig. 2.12(a) consisting of two delays and an OR gate.

Suppose the circuit is started in total state $0 \cdot 000$, which is stable. In fundamental mode, if the input changes to 1, the internal state changes as follows: $\underline{0}\,0\,0 \to 1\,\underline{0}\,0 \to 1\,1\,\underline{0} \to 1\,1\,1$. If the input then changes to 0 again, we have $\underline{1}\,1\,1 \to 0\,1\,1$. The direct behavior corresponding to these two input changes is shown in Fig. 2.12(b).

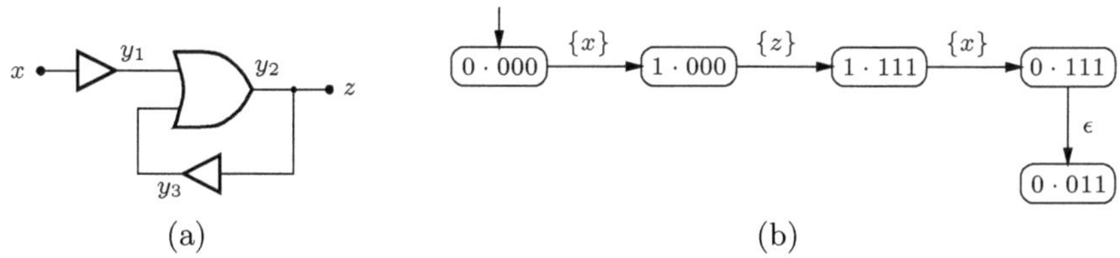

Fig. 2.12. Illustrating operating modes: (a) circuit, (b) direct behavior

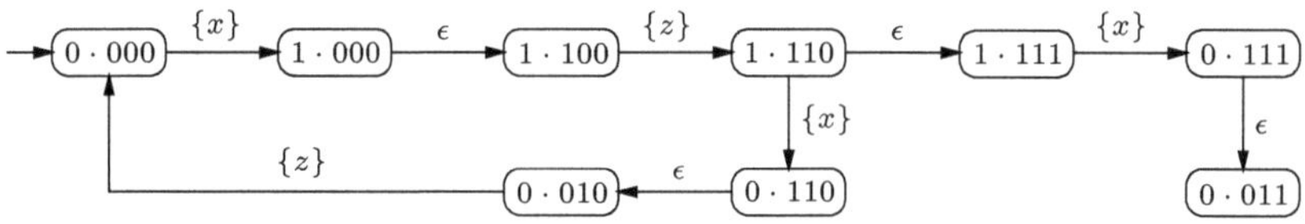

Fig. 2.13. Part of input/output-mode behavior

On the other hand, suppose the network operates in *input/output mode.* Then the environment is permitted to change the input after the output has changed, without waiting for the entire state to stabilize. A part of the input/output-mode behavior is shown in Fig. 2.13. While the correct input/output sequence xzx is still present, it is also possible for the circuit to have the sequence $xzxz$, which is *not* permitted in fundamental-mode.

Even in fundamental mode, very few behaviors are realizable by delay-insensitive networks. Suppose a behavior B has transitions $a \cdot b \to a' \cdot b \to a' \cdot b'$, $a' \cdot b' \to a \cdot b' \to a \cdot b''$, and $a \cdot b'' \to a' \cdot b'' \to a' \cdot b'''$, where $a \cdot b$, $a' \cdot b'$, $a \cdot b''$, and $a' \cdot b'''$ are stable. A result of Unger's [44, 45] states that, if $b' \neq b'''$, any network realizing B has an *essential hazard*, meaning that, if wire delays are taken into account, the network state reached from $a \cdot b$ when the input changes to a' may be either b' or b''', implying that the network is not delay-insensitive. Unger's result has been rephrased by Seger [10, 36, 37]:

Theorem 7. *Let N be any complete network, let $a \cdot b$ be a stable state of N, and assume that $out(R_{a'}(b)) = \{b'\}$, $out(R_a(b')) = \{b''\}$, and $out(R_{a'}(b'')) = \{b'''\}$. Then $b''' = b'$.*

This theorem shows that behavior B either has no delay-insensitive realization (b' and b''' can be confused by any network purporting to realize it), or is very restricted (cannot have $b''' \neq b'$). Seger's proof uses the results of Section 2.6 about the equivalence of ternary simulation and binary analysis.

Behaviors realizable by delay-insensitive networks operating in the input/output mode are even more restricted, as was shown in the example of Fig. 2.12(a). It has been shown by Brzozowski and Ebergen [4, 10] that a simple behavior similar to that of Fig. 2.12(b), where a first input change produces an output change, but the second input change produces no response, cannot be realized by any gate circuit operating in input/output mode. Consequently, most components used in asynchronous circuit design cannot be delay-insensitive. It is, therefore, necessary to design such components using some timing restrictions. However, it *is* possible to design *networks* of such components to operate independently of the component and wire delays.

2.10 Delay-Insensitivity in Networks

We introduce a very general definition of an asynchronous module, which includes not only gates but also more complex components used in asynchronous design. We suppress the details of the implementation of the module, and use only an abstract model [13].

Definition 1. *A* module *is a nondeterministic sequential machine* $M = (\mathcal{S}, \mathcal{X}, y, \mathcal{Z}, \delta, \lambda)$ *of the Moore type, where*

- $\mathcal{S}$ *is a finite set of* internal states*;*
- $\mathcal{X} = \{x_1, \ldots, x_m\}$ *is the set of* binary input variables, *where* $m \geq 0$*;*
- y *is the* internal state variable *taking values from* $\mathcal{S}$*;*
- $\mathcal{Z} = \{z_1, \ldots, z_p\}$ *is the set of* binary output variables, *where* $p \geq 0$*;*
- δ *is the* excitation function, $\delta : \{0,1\}^m \times \mathcal{S} \to 2^{\mathcal{S}} \setminus \{\emptyset\}$, *satisfying the restriction that for any* $a \in \{0,1\}^m$ *and* $b \in \mathcal{S}$, *either* $\delta(a,b) = \{b\}$ *(in which case (a,b) is stable) or* $b \notin \delta(a,b)$ *((a,b) is unstable);*
- $\lambda = (\lambda_1, \ldots, \lambda_p)$ *is the* output function, $\lambda : \mathcal{S} \to \{0,1\}^p$.

If $m = 0$, the module is a *source*, and if $p = 0$, then it is a *sink*. If the cardinality of $\delta(a,b)$ is 1 for all a, b, then the module is *deterministic*. For deterministic modules we write $\delta(a,b) = c$, instead of $\delta(a,b) = \{c\}$. If the module is deterministic, δ is independent of $\mathcal{S}$, $\mathcal{S} = \{0,1\}$, $\mathcal{Z} = \{z\}$, and $z = y$, then the module is equivalent to a gate.

Figure 2.14 shows six deterministic 2-state modules. In each of them, the state set is $\{0,1\}$, but we could also use any abstract 2-element set like $\{a,b\}$. Notation $y \cdot z$ means that the state is y and the associated output is z. The excitations and outputs are: (a) $\delta = x$, $z = y$, (b) $\delta = \overline{x}$, $z = y$, (c) $\delta = x$, $z_1 = z_2 = y$, (d) $\delta = x_1 \not\equiv x_2$, $z = y$, (e) $\delta = x_1 * x_2 + (x_1 + x_2)y$, $z = y$, and (f) $\delta = x_1 + \overline{x_2} * y$, $z_1 = y$, $z_2 = \overline{y}$. Modules (a), (b), and (d) are gates.

The module in Fig. 2.15 is nondeterministic; it is a primitive sort of *arbiter*. An input $x_i = 0$ indicates no request, and $x_i = 1$ is a request, for $i = 1, 2$. If the inputs are 00, there are no requests, the state is 0, and the outputs are 00. If $x = 10$, x_1 is requesting service; the arbiter grants this service by changing state to 1, and setting $z_1 = 1$. The arbiter remains in this state until x_1 becomes 0; the second input is ignored, since the request was granted to x_1. If the input becomes 01 in state 1, the arbiter first returns to state 0, to terminate the service to x_1, and then makes a grant to x_2. The situation is symmetric if $x = 01$ in state 00. If $x = 11$, two transitions possible: the arbiter nondeterministically decides to serve either x_1 or x_2.

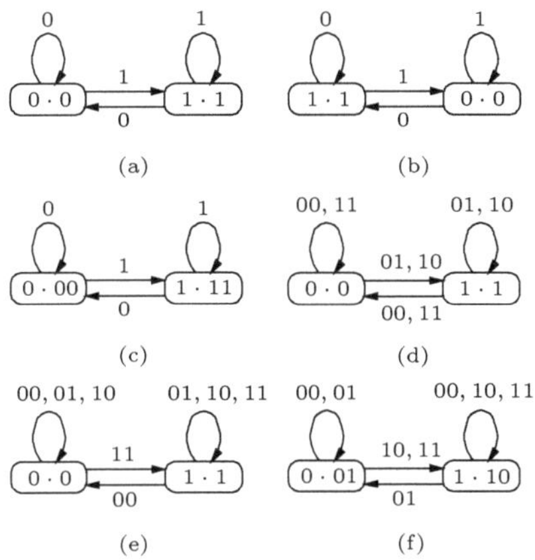

Fig. 2.14. 2-state modules: (a) delay, (b) inverter, (c) (isochronic) fork, (d) XOR, (e) C-element, (f) set-dominant latch

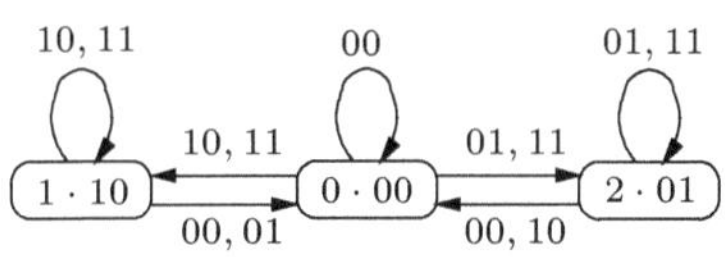

Fig. 2.15. Arbiter

Module outputs cannot change before the internal state changes, since the outputs are uniquely determined by the state. There are no delays in the output wires, since wire delays *are* often negligible, especially if the module takes up a small area of a chip. Also, designers sometimes use the *isochronic fork* assumption, that signals at two ends of a fork arrive at the same time, and we can model this. If needed, delay modules can be added to the wires.

We consider only *closed* or *autonomous* networks of modules; such networks are without external inputs or outputs. An open network can be made closed by adding a module representing the environment. A *network N of modules* consists of a set $\{M^1, \ldots, M^n\}$ of modules and a set of wires, each wire connecting an output z_k^h of module M^h to an input x_j^i of module M^i. Each module output must be connected to a module input, and *vice versa*. As in the case of gates, a wire can connect only two points.

The set of network state variables is $\mathcal{Y} = \{y^1, \ldots, y^n\}$, and the excitation of module i is δ_i. A state of N is an n-tuple in $\mathcal{S} = \mathcal{S}^1 \times \ldots \times \mathcal{S}^n$. The excitation of each module is a function of $y^1, \ldots, y^n$.

A network consisting of a 2-output C-element and two inverters is shown in Fig. 2.16. Here the C-element has excitation $\delta_1 = y^2 * y^3 + (y^2 + y^3) * y^1$, and two outputs, $z_1^1 = z_2^1 = y^1$. The inverter excitations are $\delta_2 = \delta_3 = \overline{y^1}$. Suppose the state of the network is 011. Since each inverter is in state 1, its output

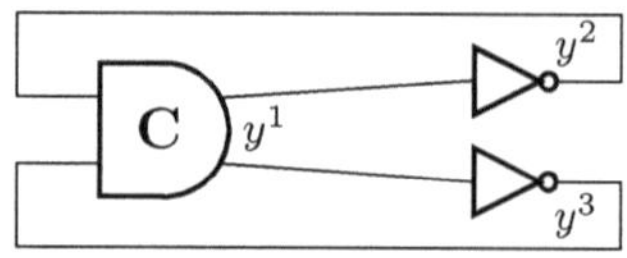

Fig. 2.16. Network

is 1, and the inputs of the C-element are both 1. The C-element is therefore unstable. At the same time, the outputs of the C-element are both 0, and both inverters are stable. This analysis leads to the behavior of Fig. 2.17.

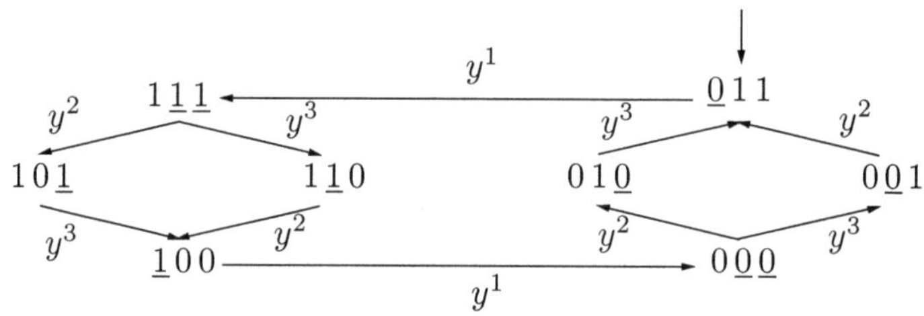

Fig. 2.17. Network behavior

In the analysis of networks of modules we use the *general single-winner* (GSW) model. It turns out that this simplification is justifiable, as we shall see later. Let $\mathcal{R}'$ be the GSW relation; then states s and t are related by $\mathcal{R}'$ if they differ in exactly one component i (that is $s_i \neq t_i$), y^i is unstable in state s, and $t_i \in \delta_i(s)$. In general, the GSW behavior of a network N is an initialized directed graph $B = (q, Q, \mathcal{R})$, where $q \in \mathcal{S}$ is the initial state, Q is the set of states reachable from q by the GSW relation, and $\mathcal{R}$ is the GSW relation $\mathcal{R}'$ restricted to Q. Moreover, we label each edge of this graph by the variable that changes during the corresponding transition.

A behavior $B = (q, Q, \mathcal{R})$ can be viewed as a nondeterministic finite automaton $\mathcal{B} = (\mathcal{Y}, Q, q, \mathcal{R}, F)$, where $\mathcal{Y}$ is the input alphabet, Q is the set of states, q is the initial state, $\mathcal{R}$ is the labeled transition relation, and $F = Q$ is the set of final states. The language $L(\mathcal{B})$ is the set of all possible sequences of module variable changes.

Let N be a network started in state q. A *delay extension* of N is a network $\hat{N}$ started in $\hat{q}$ and obtained from N by adding any number of delays in the wires; here $\hat{q}$ is an extension of q obtained by adding values for the new delays in such a way that they are initially stable. Behavior $\hat{B} = (\hat{q}, \hat{Q}, \hat{\mathcal{R}})$ of $\hat{N}$ is *livelock-free* if every sequence of changes of only the inserted delays is finite. Every delay-extension is livelock-free [13].

A network is *strongly delay-insensitive* with respect to a state q, if, for any delay extension $\hat{N}$, the behavior $\hat{B} = (\hat{q}, \hat{Q}, \hat{\mathcal{R}})$ of $\hat{N}$ is observationally equivalent to the behavior $B = (q, Q, \mathcal{R})$ of N. Roughly speaking, two behaviors are *observationally equivalent*, or *weakly bisimilar*, if they can simulate each other in a step-by-step fashion. Of course, we ignore the added delays in $\hat{N}$, and compare only that part of the state that corresponds to the original modules. Bisimilar states must agree in the observable variables, for any transition from one state there must be a corresponding transition from the other state, and these transitions must lead to bisimilar states.

We say that $\hat{B}$ is *safe* with respect to B, if, whenever $\hat{s} \in \hat{Q}$ and $s \in Q$ are bisimilar, then, any sequence

$$\hat{s} \xrightarrow{*} \hat{u} \xrightarrow{y^i} \hat{v} \xrightarrow{*} \hat{t}$$

from state $\hat{s}$ to $\hat{t}$ (that involves zero or more unobservable changes, followed by a change in y^i, followed by zero or more unobservable changes), implies the existence of a corresponding sequence from s to t, where $\hat{t}$ and t are bisimilar:

$$\hat{s} \xrightarrow{*} \hat{u} \xrightarrow{y^i} \hat{v} \xrightarrow{*} \hat{t} \quad \Rightarrow \quad s \xrightarrow{y^i} t.$$

Behavior $\hat{B}$ is *complete* with respect to B if the converse implication holds. One verifies that $\hat{B}$ is always complete with respect to B [13]. Therefore to verify strong delay-insensitivity, one needs to check only safety. This definition involves an infinite test, because there are infinitely many delay-extensions of a network. However, there is a finite test for the equivalent property of quasi semi-modularity.

Semi-modularity is a concept introduced by Muller [31, 32] for deterministic networks of gates. This notion has been generalized in [13] as follows. *Quasi semi-modularity* of a behavior B requires that, once a state variable y^i becomes unstable, and b is a possible next state for y^i, then b should remain a possible next state until y^i changes. In other words, no change in some other variable y^j can remove b from the excitation of y^i.

To illustrate these notions, we present a part of a network in Fig. 2.18. In the top left, the two delays have outputs 1, and excitations $\{0\}$ and $\{1\}$. The arbiter is in state 0, and has excitation $\{1, 2\}$. After the top delay changes to 0, the excitation of the arbiter becomes $\{2\}$. This is a violation of quasi semi-modularity, because a change in the delay has caused the arbiter state 1 to be removed from the excitation.

The bottom parts illustrate what happens if an initially stable delay is added. After the left delay is changed, the arbiter retains its original excitation, because of the added delay. Thus, in the extension of the network, the state of the arbiter can become 1, while this is not possible in the original network. This is a violation of safety, and this network is not strongly delay-insensitive.

The following results have been proved by Brzozowski and Zhang [13, 49]:

Theorem 8. *If a network is strongly delay-insensitive, then its behavior is quasi semi-modular.*

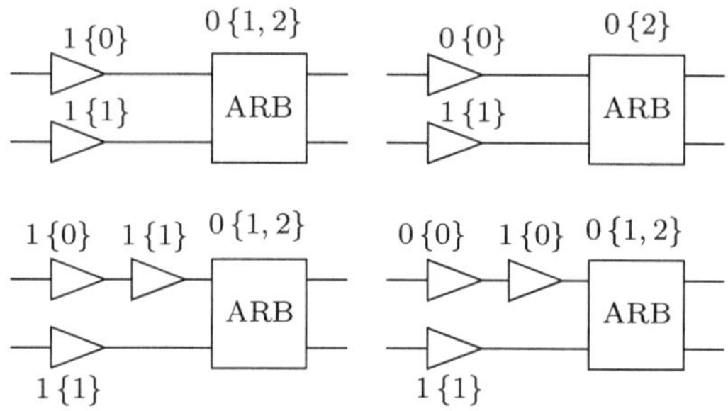

Fig. 2.18. Safety violation

The converse result holds under the condition that each wire has at least one delay. A network satisfying this condition is called *delay-dense*.

Theorem 9. *If the behavior of a delay-dense network is quasi semi-modular, then the network is strongly delay-insensitive.*

In our analysis we have used the GSW model, but strong delay-insensitivity and quasi semi-modularity can be generalized to the GMW model. The following results of Silver's [39, 40] justify the use of the GSW model.

Theorem 10. *A network is strongly delay-insensitive in the GSW model if and only if it is strongly delay-insensitive in the GMW model.*

Theorem 11. *The GSW behavior of a network is single-change quasi semi-modular if and only if its GMW behavior is multiple-change quasi semi-modular.*

2.11 Trace Structures

This section is an introduction to formal specifications of asynchronous circuits, and is based on the work of Ebergen [16, 17]. The design of modules, as we have seen, necessarily involves making some assumptions about relative sizes of delays. Assuming that the modules have been designed, Ebergen describes their behavior in a program notation inspired by Hoare's Communicating Sequential Processes [23]. Using this notation, one can then design networks of modules, and reason about their correctness. These correctness concerns are now separate from the timing concerns.

We begin by defining the behavior of several modules using a notation similar to regular expressions. Our view is now more abstract, and hides the implementation details. Thus we no longer need to know whether a signal is high or low, but interpret a signal change as a *communication* at a circuit terminal. Because this communication can be an input or an output, depending on whether we are talking about a module or its environment, we use the following notation: $a?$ denotes an input event, whereas $a!$ is an output. Note that a is not a variable, but a label of a terminal.

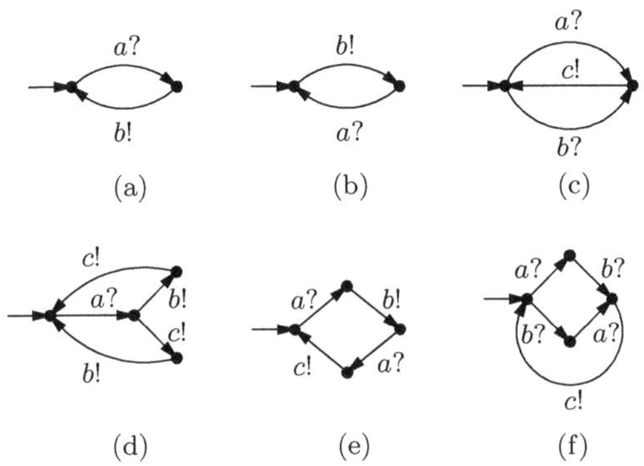

Fig. 2.19. Module state graphs: (a) WIRE, (b) IWIRE, (c) MERGE, (d) FORK, (e) TOGGLE, (f) JOIN

Figure 2.19 shows the state graphs of several modules. The specification of each module is a contract between the module and the environment. The WIRE receives an input $a?$ and responds by producing an output $b!$ Thus, the correct event sequences for the WIRE are: $\epsilon, a?, a?b!, a?b!a?, a?b!a?b!, \ldots$ A convenient way to describe this language is to denote all input/output alternations by the regular expression $(a?b!)^*$ and then close this set under the prefix operation. Thus the language of a WIRE is $\mathbf{pref}((a?b!)^*)$. It is not correct for the environment to send two consecutive inputs to the WIRE without receiving an output after the first signal. It is also incorrect for the WIRE to produce an output without receiving an input, or to produce two consecutive outputs, without an input between them. These rules are similar to the input/output mode discussed earlier.

The *initialized wire* or IWIRE first produces an output, and then acts like a WIRE; its language is $\mathbf{pref}((b!a?)^*)$. A MERGE module has two inputs $a?$ and $b?$ and one output $c!$ It receives a signal on either one of its input terminals, and then produces an output; its language is $\mathbf{pref}(((a? + b?)c!)^*)$. A FORK has one input $a?$ and two outputs $b!$ and $c!$, and its language is $\mathbf{pref}((a?b!c! + a?c!b!)^*)$. A TOGGLE has one input $a?$ and two outputs $b!$ and $c!$, and its language is $\mathbf{pref}((a?b!a?c!)^*)$. The JOIN is a component that detects whether both of its input events $a?$ and $b?$ have happened before responding with an output $c!$, and its language is $\mathbf{pref}(((a?b? + b?a?)c!)^*)$. Thus, this module provides synchronization. A similar specification of a 3-input JOIN is long and clumsy, as the reader can verify. To overcome this, a new operation $\|$, a type of parallel composition called *weave*, is introduced. Using $\|$, we can write $\mathbf{pref}(((a?\|b?)c!)^*)$ instead of $\mathbf{pref}(((a?b? + b?a?)c!)^*)$. In trace theory, parallel events are denoted by interleaving, as they are in the GSW model.

Expressions, like those above, describing module languages are called "commands." Formally, a *command* over an alphabet Σ is defined inductively as follows:

1. ABORT, SKIP, and $a?$ and $a!$, for all $a \in \Sigma$, are commands.
2. If E and F are commands, then so are: (EF), $(E+F)$, E^*, $\mathbf{pref}E$, $E\!\downarrow_\Gamma$, for $\Gamma \subseteq \Sigma$, and $(E\|F)$.
3. Any command is obtained by a finite number of applications of 1 and 2.

The following order of operator precedence is used to simplify notation: star, **pref**, concatenation, addition, and weave at the lowest level.

Next, we define the semantics of commands. A *trace structure* is a triple (I, O, T), where I is the *input alphabet*, O is the *output alphabet*, $I \cap O = \emptyset$, and $T \subseteq (I \cup O)^*$ is the *trace set*, a *trace* being another term for a word over the *alphabet* $\Sigma = I \cup O$ of the trace structure. The *meaning* of a command is a trace structure defined below, where we equate a command with the trace structure that it represents. The notation $t\!\downarrow_\Gamma$ for $\Gamma \subseteq \Sigma$ denotes the projection of the word t to the alphabet Γ obtained from t by replacing all the letters of t that are not in Γ by the empty word.

$$\text{ABORT} = (\emptyset, \emptyset, \emptyset) \tag{2.7}$$
$$\text{SKIP} = (\emptyset, \emptyset, \{\epsilon\}) \tag{2.8}$$
$$a? = (\{a\}, \emptyset, \{a\}) \tag{2.9}$$
$$a! = (\emptyset, \{a\}, \{a\}) \tag{2.10}$$

Next, if $E = (I, O, T)$ and $E' = (I', O', T')$, then

$$(EE') = (I \cup I', O \cup O', TT') \tag{2.11}$$
$$(E + E') = (I \cup I', O \cup O', T \cup T') \tag{2.12}$$
$$E^* = (I, O, T^*) \tag{2.13}$$
$$\mathbf{pref}E = (I, O, \{u \mid uv \in T\}) \tag{2.14}$$
$$E\!\downarrow_\Gamma = (I \cap \Gamma, O \cap \Gamma, \{t\!\downarrow_\Gamma \mid t \in T\}) \tag{2.15}$$
$$(E\|E') = (I \cup I', O \cup O', \{t \in (\Sigma \cup \Sigma')^* \mid t\!\downarrow_\Sigma \in T, t\!\downarrow_{\Sigma'} \in T'\}) \tag{2.16}$$

The operation $E\!\downarrow_\Gamma$ is that of *hiding* symbols. When an output $a!$ of one module becomes an input $a?$ of another, both these symbols can become internal, and may be removed from the description of the combined system. To illustrate the weave operation, let $E = (\{a\}, \{c\}, \{\epsilon, a, ac\})$ and $E' = (\{b\}, \{c\}, \{\epsilon, b, bc\})$. Then $(E\|E') = (\{a, b\}, \{c\}, \{\epsilon, a, b, ab, ba, abc, bac\})$.[4] As another example, reconsider the JOIN. We can also write its language as $\mathbf{pref}(a?c!\|b?c!)^*$; the concatenation operations enforce that an input precedes the output, and the weave provides synchronization on the common letter c,

[4] The weave is not the same as the *shuffle* operation in formal language theory; the shuffle of E and E' includes such words as ac and acb.

making sure that an output occurs only after both inputs have arrived. The C-element is similar to the JOIN, except that its inputs may be "withdrawn" after they are supplied. Another name for the C-element is the *rendezvous*.

As we have already stated, a trace structure specifies all possible sequences of communications that can occur between a component and its environment. An input or output produced when it is not allowed by the specification represents a violation of *safety*. If each specified trace can indeed occur, then a *progress* condition is satisfied. Note that the occurrence of all traces is not guaranteed, and the progress condition is not sufficiently strong to exclude *deadlock* (a path in the behavior where the desired events do not occur, though they may occur in another path) and *livelock*. In spite of these limitations, the theory has been used successfully.

We close this section with an example of the RGDA arbiter, where the letters stand for "request, grant, done, acknowledge." The arbiter communicates with two processes, each with four terminals called r_1, g_1, d_1, a_1 and r_2, g_2, d_2, a_2. The communications with each process taken separately must follow the RGDA pattern; this is captured by the commands $\mathbf{pref}(r_1?g_1!d_1?a_1!)^*$ and $\mathbf{pref}(r_2?g_2!d_2?a_2!)^*$. Thus the arbiter receives a "request," issues a "grant," receives a "done" signal and produces an "acknowledge" signal. Of course, this specification is not complete, because we must ensure that the arbiter serves only one process at a time. To specify that the process must be served exclusively, we insist that a "grant" can be made only after a "done" has been received. This is covered by the command $\mathbf{pref}(g_1!d_1? + g_2d_2!)^*$. Since all three conditions are required, the complete specification becomes

$$\mathbf{pref}(r_1?g_1!d_1?a_1!)^* \parallel \mathbf{pref}(r_2?g_2!d_2?a_2!)^* \parallel \mathbf{pref}(g_1!d_1? + g_2d_2!)^*.$$

2.12 Further Reading

The literature on asynchronous circuits is quite extensive; see [2] for a survey on delay-insensitivity and additional references. The term *delay-insensitivity* was introduced in connection with the Macromodules project [29]; Molnar used the term "foam-rubber wrapper" to describe a component with delays in its input and output wires [29]. Seitz developed a design methodology called *self-timed* systems (Chapter 7 in [27]). Van de Snepscheut introduced trace theory for modeling asynchronous circuits [47]. Udding [42, 43] was the first to formalize the concept of delay-insensitivity and of the foam-rubber wrapper postulate. A different formalization of the latter was done by Schols [34]. A modified trace theory was used by Dill [15]. Communicating sequential processes were studied by Hoare [23], and a similar model was proposed by He, Josephs, and Hoare [21]. Algebraic methods for asynchronous circuits were introduced by Josephs and Udding [22]. Van Berkel studied "handshake processes" [46]. Verhoeff extended Udding's rules for delay-insensitivity to include progress concerns [48]. Negulescu developed the algebra of "process spaces" for asynchronous circuits [33].

For a survey of asynchronous design methodologies see [8]. The annual proceedings of the International Symposium on Advanced Research in Asynchronous Circuits and Systems published by the IEEE Computer Society are an excellent source on information on the progress in asynchronous design. An extensive bibliography on asynchronous circuits is available at `http://www.win.tue.nl/async-bib/`.

Acknowledgment I am very grateful to Mihaela Gheorghiu and Jo Ebergen for their constructive comments.

References

1. J.A. Bondy and U.S.R. Murty. *Graph Theory with Applications,* American Elsevier, 1976
2. J.A. Brzozowski. Delay-Insensitivity and Ternary Simulation, *Theoretical Computer Science,* 245 (2000) 3-25.
3. J.A. Brzozowski and J.C. Ebergen. Recent Developments in the Design of Asynchronous Circuits, *Proc. Fundamentals of Computation Theory,* (J. Csirik and J. Demetrovics and F. Gécseg, eds.), Springer-Verlag, Berlin, 1989, 78–94.
4. J.A. Brzozowski and J.C. Ebergen. On the Delay-Sensitivity of Gate Networks, *IEEE Trans. on Computers,* 41, 11 (1992), 1349–1360.
5. J.A. Brzozowski and Z. Eśik. Hazard Algebras, *Formal Methods in System Design,* 23, 3 (2003), 223–256.
6. J.A. Brzozowski, Z. Ésik and Y. Iland. Algebras for Hazard Detection, *Beyond Two - Theory and Applications of Multiple-Valued Logic*, (M. Fitting, and E. Orłowska, eds.), Physica-Verlag, Heidelberg, 2003, 3–24.
7. J.A. Brzozowski and M. Gheorghiu. Gate Circuits in the Algebra of Transients, *Theoretical Informatics and Applications,* to appear.
8. J.A. Brzozowski, S. Hauck and C-J. Seger, Design of Asynchronous Circuits, Chapter 15 in [10]
9. J.A. Brzozowski and C-J. Seger. A Characterization of Ternary Simulation of Gate Networks, *IEEE Trans. Computers*, C-36, 11 (1987), 1318–1327.
10. J.A. Brzozowski and C-J. Seger. *Asynchronous Circuits,* Springer, Berlin, 1995.
11. J.A. Brzozowski and M. Yoeli. *Digital Networks*, Prentice-Hall, 1976
12. J.A. Brzozowski and M. Yoeli. On a Ternary Model of Gate Networks, *IEEE Trans. on Computers,* C-28, 3 (1979), 178–183.
13. J.A. Brzozowski and H. Zhang. Delay-Insensitivity and Semi-Modularity, *Formal Methods in System Design,* 16, 2 (2000), 187–214.
14. T.J. Chaney and C.E. Molnar. Anomalous Behavior of Synchronizer and Arbiter Circuits, *IEEE Trans. on Computers,* C-22, 4 (1973), 421–422.
15. D.L. Dill. *Trace Theory for Automatic Hierarchical Verification of Speed-Independent Circuits,* PhD Thesis, Computer Science Department, Carnegie Mellon University, February 1988. Also, The MIT Press, Cambridge, MA, 1989.
16. J.C. Ebergen. *Translating Programs into Delay-Insensitive Circuits,* PhD Thesis, Department of Mathematics and Computing Science, Eindhoven University of Technology, Eindhoven, The Netherlands, October 1987. Also, CWI Tract 56, Centre for Math. and Computer Science, Amsterdam, The Netherlands, 1989.

17. J.C. Ebergen. A Formal Approach to Designing Delay-Insensitive Circuits, *Distributed Computing,* 5, 3 (1991), 107–119.
18. E.B. Eichelberger. Hazard Detection in Combinational and Sequential Switching Circuits, *IBM J. Res. and Dev.,* 9 (1965), 90–99.
19. M. Gheorghiu. *Circuit Simulation Using a Hazard Algebra*, MMath Thesis, Dept. of Computer Science, University of Waterloo, Waterloo, ON, Canada, 2001. `http://maveric.uwaterloo.ca/publication.html`
20. M. Gheorghiu and J.A. Brzozowski. Simulation of Feedback-Free Circuits in the Algebra of Transients, *Int. J. Foundations of Computer Science*, 14, 6 (2003), 1033-1054.
21. J. He, M.B. Josephs and C.A.R. Hoare. A Theory of Synchrony and Asynchrony, *Programming Concepts and Methods,* (M. Broy and C.B. Jones, eds.), North-Holland, Amsterdam, 1990, 459–478.
22. M.B. Josephs and J.T. Udding. An Algebra for Delay-Insensitive Circuits, *Computer-Aided Verification,* (E.M. Clarke and R.P. Kurshan, eds.), AMS-ACM, Providence, RI, 1990, 147–175.
23. C.A.R. Hoare. *Communicating Sequential Processes,* Prentice-Hall, Inc., Englewood Cliffs, NJ, 1985.
24. D.A. Huffman. The Design and Use of Hazard-Free Switching Circuits, *J. ACM,* 4 (1957), 47–62.
25. A.J. Martin. Compiling Communicating Processes into Delay-Insensitive VLSI Circuits, *Distributed Computing,* 1 (1986), 226–234.
26. E.J. McCluskey. Transient Behavior of Combinational Logic Circuits, *Redundancy Techniques for Computing Systems,* (R. H. Wilcox and W. C. Mann, eds.), Spartan Books, Washington, DC, 1962, 9–46.
27. C. Mead and L. Conway. *Introduction to VLSI Systems,* Addison-Wesley, Reading, MA, 1980
28. R.E. Miller. *Switching Theory, Volume II: Sequential Circuits and Machines,* Wiley, New York, 1965.
29. C.E. Molnar, T.P. Fang and F.U. Rosenberger. Synthesis of Delay-Insensitive Modules, *Proc. 1985 Chapel Hill Conference on VLSI,* (H. Fuchs, ed.), Computer Science Press, Rockville, Maryland, 1985, 67–86.
30. M. Mukaidono. Regular Ternary Logic Functions—Ternary Logic Functions Suitable for Treating Ambiguity, *Proc. 13th Ann. Symp. on Multiple-Valued Logic,* 1983, 286–291.
31. D.E. Muller. *A Theory of Asynchronous Circuits,* Tech. Report 66, Digital Computer Laboratory, University of Illinois, Urbana-Champaign, Illinois, USA, 1955.
32. D.E. Muller and W.S. Bartky. A Theory of Asynchronous Circuits, *Proc. Int. Symp. on the Theory of Switching,* Annals of the Computation Laboratory of Harvard University, Harvard University Press, 1959, 204–243.
33. R. Negulescu. *Process Spaces and Formal Verification of Asynchronous Circuits,* PhD Thesis, Dept. of Computer Science, University of Waterloo, Waterloo, ON, Canada, 1998.
34. H. Schols. *A Formalisation of the Foam Rubber Wrapper Principle,* Master's Thesis, Department of Mathematics and Computing Science, Eindhoven University of Technology, Eindhoven, The Netherlands, February 1985.
35. C-J.H. Seger. *Ternary Simulation of Asynchronous Gate Networks,* MMath Thesis, Dept. of Comp. Science, University of Waterloo, Waterloo, ON, Canada, 1986.

36. C-J.H. Seger. *Models and Algorithms for Race Analysis in Asynchronous Circuits,* PhD Thesis, Dept. of Comp. Science, University of Waterloo, Waterloo, ON, Canada, 1988.
37. C-J.H. Seger. On the Existence of Speed-Independent Circuits, *Theoretical Computer Science,* 86, 2 (1991), 343–364.
38. C.E. Shannon. A Symbolic Analysis of Relay and Switching Circuits, *Trans. AIEE,* 57 (1938), 713–723.
39. S. Silver. *True Concurrency in Models of Asynchronous Circuit Behaviors,* MMath Thesis, Dept. of Computer Science, University of Waterloo, Waterloo, ON, Canada, 1998.
40. S. Silver and J.A. Brzozowski. True Concurrency in Models of Asynchronous Circuit Behavior, *Formal Methods in System Design,* 22, 3 (2003), 183—203.
41. I.E. Sutherland and J. Ebergen. Computers without Clocks, *Scientific American,* August 2002, 62–69.
42. J.T. Udding. *Classification and Composition of Delay-Insensitive Circuits,* PhD Thesis, Department of Mathematics and Computing Science, Eindhoven University of Technology, Eindhoven, The Netherlands, September 1984.
43. J.T. Udding. A Formal Model for Defining and Classifying Delay-Insensitive Circuits and Systems, *Distributed Computing,* 1, 4 (1986), 197–204.
44. S.H. Unger. Hazards and Delays in Asynchronous Sequential Switching Circuits, *IRE Trans. on Circuit Theory,* CT–6 (1959), 12–25.
45. S.H. Unger. *Asynchronous Sequential Switching Circuits,* Wiley-Interscience, New York, 1969.
46. K. van Berkel. *Handshake Circuits,* Cambridge University Press, Cambridge, England, 1993.
47. J.L.A. van de Snepscheut. *Trace Theory and VLSI Design,* PhD Thesis, Department of Computing Science, Eindhoven University of Technology, Eindhoven, The Netherlands, May 1983. Also, *Lecture Notes in Computer Science,* vol. 200, Springer-Verlag, Berlin, 1985
48. T. Verhoeff. *A Theory of Delay-Insensitive Systems,* PhD Thesis, Department of Mathematics and Computing Science, Eindhoven University of Technology, Eindhoven, The Netherlands, May 1994.
49. H. Zhang. *Delay-Insensitive Networks,* MMath Thesis, Dept. of Computer Science, University of Waterloo, Waterloo, ON, Canada, 1997.

3

Text Searching and Indexing

Maxime Crochemore[1] and Thierry Lecroq[2]

[1] Institut Gaspard-Monge, University of Marne-la-Valle
Cit Descartes, 5 Bd Descartes, Champs-sur-Marne, F-77454 Marne-la-Valle CEDEX 2, France
E-mail: Maxime.Crochemore@univ-mlv

[2] Computer Science Department and ABISS
Faculty of Science, University of Rouen, 76821 Mont-Saint-Aignan, France
E-mail: Thierry.Lecroq@univ-rouen.fr

Although data is stored in various ways, text remains the main form of exchanging information. This is particularly evident in literature or linguistics where data is composed of huge corpora and dictionaries. This applies as well to computer science, where a large amount of data is stored in linear files. And this is also the case in molecular biology where biological molecules can often be approximated as sequences of nucleotides or amino acids. Moreover, the quantity of available data in this fields tends to double every 18 months. This is the reason why algorithms should be efficient even if the speed of computers increases at a steady pace.

Pattern matching is the problem of locating a specific pattern inside raw data. The pattern is usually a collection of strings described in some formal language. In this chapter we present several algorithms for solving the problem when the pattern is composed of a single string.

In several applications, texts need to be structured before being searched. Even if no further information is known about their syntactic structure, it is possible and indeed extremely efficient to build a data structure that support searches. In this chapter we present suffix arrays, suffix trees, suffix automata and compact suffix automata.

3.1 Pattern Matching

String-matching consists in finding all the occurrences of a pattern x of length m in a text y of length n $(m, n > 0)$. Both strings x and y are built on a finite alphabet V.

Applications require two kinds of solution depending on which string, the pattern or the text, is given first. Algorithms based on the use of automata or combinatorial properties of strings are commonly implemented to preprocess

M. Crochemore and T. Lecroq: *Text Searching and Indexing*, Studies in Computational Intelligence (SCI) **25**, 43–80 (2006)
www.springerlink.com

the pattern in time $O(m)$ and solve the first kind of problem in time $O(n)$. The notion of indexes realized by trees or automata is used in the second kind of solutions to preprocess the text in time $O(n)$. The search of the pattern can then be done in time $O(m)$. This section only investigates algorithms of the first kind.

String-matching is a very important subject in the wider domain of text processing. String-matching algorithms are basic components used in implementations of practical software existing under most operating systems. Moreover, they emphasize programming methods that serve as paradigms in other fields of computer science (system or software design). Finally, they also play an important role in theoretical computer science by providing challenging problems.

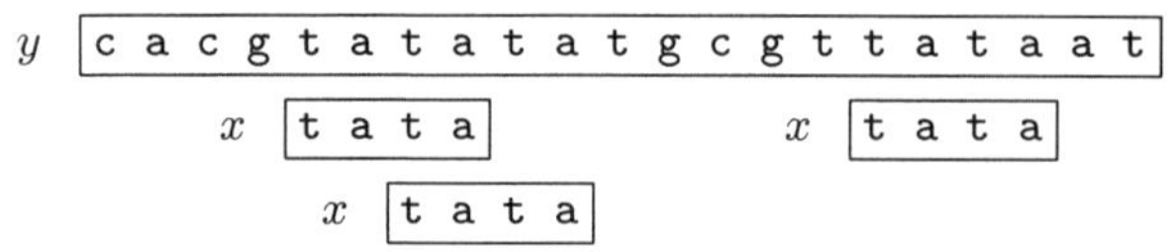

Fig. 3.1. There are three occurrences of $x =$ `tata` in $y =$ `cacgtatatatgcgttataat`.

Figure 3.1 shows the occurrences of the pattern $x =$ `tata` in the text $y =$ `cacgtatatatgcgttataat`. The basic operations allowed for comparing symbols are equality and inequality ($=$ and $\neq$).

String-matching algorithms of the present section work as follows. They scan the text through a window which size is generally equal to m. They first align the left ends of the window and the text, then compare the symbols of the window with the symbols of the pattern — this specific work is called an *attempt* — and after a whole match of the pattern or after a mismatch they *shift* the window to the right. They repeat the same procedure again until the right end of the window goes beyond the right end of the text. This mechanism is usually called the *sliding window mechanism.* A string-matching algorithm is thus a succession of scan and shift. Figure 3.2 illustrates this notion.

We associate each attempt with the positions j and $j + m - 1$ in the text when the window is positioned on $y[j \mathinner{..} j + m - 1]$: we say that the attempt is at the left position j and at the right position $j + m - 1$.

The naive algorithm consists in checking, at all positions in the text between 0 and $n - m$, whether an occurrence of the pattern starts there or not. Then, after each attempt, it shifts the pattern by exactly one position to the right. It memorizes no information (see Figure 3.3). It requires no preprocessing phase, and a constant extra space in addition to the pattern and the text. During the searching phase the symbol comparisons can be done in any order. The time complexity of this searching phase is $O(m \times n)$ (the bound is met when searching for $\mathtt{a}^{m-1}\mathtt{b}$ in $\mathtt{a}^n$ for instance). The expected number of

```
STRING-MATCHING(x, m, y, n)
put window at the beginning of y
while window on y do
    ▷ scan
    if window = x then
        report it
    ▷ shift
    shift window to the right and
    memorize some information for use during next scans and shifts
```

Fig. 3.2. Scan and shift mechanism for string-matching.

symbol comparisons is $2n$ on a two-symbol alphabet, with equiprobability and independence conditions.

```
NAIVE-SEARCH(x, m, y, n)
j ← 0
while j ≤ n − m do
    i ← 0
    while i < m and x[i] = y[i + j] do
        i ← i + 1
    if i = m then
        OUTPUT(x occurs in y at position j)
    j ← j + 1
```

Fig. 3.3. The naive string-matching algorithm.

3.1.1 Complexities of the Problem

The following theorems state some known results on the problem.

Theorem 1 ([18]). *The search can be done optimally in time $O(n)$ and space $O(1)$.*

Theorem 2 ([41]). *The search can be done in optimal expected time $O(\frac{\log m}{m} \times n)$.*

Theorem 3 ([7]). *The maximal number of comparisons done during the search is $\geq n + \frac{9}{4m}(n-m)$, and can be made $\leq n + \frac{8}{3(m+1)}(n-m)$.*

We now give lower and upper bounds on symbol comparisons with different strategies depending on the access to the text:
Access to the whole text:

- upper: $2n - 1$ [36];
- lower: $\frac{4}{3}n$ [19].

Search with a sliding window of size m:

- lower: $n + \frac{9}{4m}(n - m)$ [7];
- upper: $n + \frac{8}{3(m+1)}(n - m)$ [7].

Search with a sliding window of size 1:

- lower and upper: $(2 - \frac{1}{m})n$ [24, 5];

The delay is defined as the maximum number of comparisons on each text symbol:

- lower: $\min\{1 + \log_2 m, card(V)\}$ [24];
- upper: $\min\{\log_\Phi(m + 1), card(V)\}$ [38], $\min\{1 + \log_2 m, card(V)\}$ [24] and $\log \min\{1 + \log_2 m, card(V)\}$ [25].

3.1.2 Methods

Actually, searching for the occurrences of a pattern x in a text y consists in identifying all the prefixes of y that are elements of the language V^*x (see Figure 3.4).

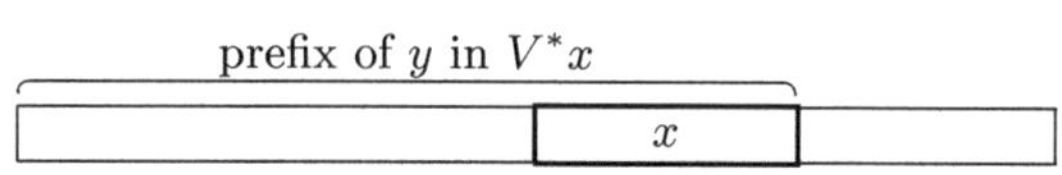

Fig. 3.4. An occurrence of the pattern x in the text y corresponds to a prefix of y in V^*x.

To solve this problem there exist several methods of different types:

- **Sequential searches**: methods in this category adopt a window of size exactly one symbol. They are well adapted to applications in telecommunication. They are based on efficient implementations of automata [33, 38, 24, 5].
- **Time-space optimal searches**: these methods are mainly of theoretical interest and are based on combinatorial properties of strings [18, 9, 10, 20, 13].
- **Practically-fast searches**: these methods are typically used in text editors or data retrieval software. They are based on combinatorics on words and theory of automata and often use heuristics [4, 17, 2, 11].

3.1.3 Morris and Pratt Algorithm

Periods and Borders

For a non-empty string u, an integer p such that $0 < p \leq |u|$ is a **period** of u if any of these equivalent conditions is satisfied:

1. $u[i] = u[i+p]$, for $0 \leq i < |u| - p$;
2. u is a prefix of some y^k, $k > 0$, $|y| = p$;
3. $u = yw = wz$, for some strings y, z, w with $|y| = |z| = p$. The string w is called a **border** of u: it occurs both as a prefix and a suffix of u.

The period of u, denoted by $period(u)$, is its smallest period (it can be $|u|$). **The border** of u, denoted by $border(u)$, is its longest border (it can be empty).

Example 1. $u =$ `abacabacaba`

periods	borders of u
4	`abacaba`
8	`aba`
10	`a`
11	empty string

The Searching Algorithm

The notions of period and border naturally lead to a simple on-line search algorithm where the length of the shift is given by the period of the matched prefix of the pattern. Furthermore the algorithm implements a memorization of the border of the matched prefix of the pattern.

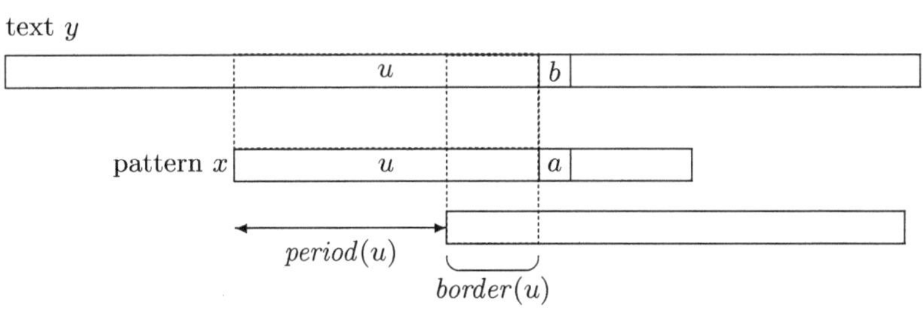

Fig. 3.5. A typical situation during a sequential search.

A typical situation during a sequential search is the following: a prefix u of the pattern has been matched, a mismatch occurs between a symbol a in the pattern and a symbol b in the text ($a \neq b$). Then a shift of length $period(u) = |u| - |border(u)|$ can be applied (see Figure 3.5). The comparisons are then resumed between symbols $x[|border(u)|]$ of x and b in y (no backtrack is necessary on the text y). The corresponding algorithm MP, due to Morris and Pratt [36], is shown Figure 3.6. It uses a table *MPnext* defined by $MPnext[i] = |border(x[0\,..\,i-1])|$ for $i = 0, \ldots, m$.

```
MP(x, m, y, n)
i ← 0
j ← 0
while j < n do
   while i = m or (i ≥ 0 and x[i] ≠ y[j]) do
      i ← MPnext[i]
   i ← i + 1
   j ← j + 1
   if i = m then
      OUTPUT(x occurs in y at position j − i)
```

Fig. 3.6. The Morris and Pratt string-matching algorithm.

Computing Borders of Prefixes

The table *MPnext* is defined by $MPnext[i] = |border(x[0\,..\,i-1])|$ for $i = 0, \ldots, m$. It can be computed by using the following remark: a border of a border of u is a border of u. A border of u is either $border(u)$ or a border of it. It can be linearly computed by the algorithm presented in Figure 3.7. This algorithm uses an index j that runs through decreasing lengths of borders. The computation of the table *MPnext* proceeds as the searching algorithm, as if $y = x[1\,..\,m-1]$.

```
COMPUTE-MP-NEXT(x, m)
MPnext[0] ← −1
for i ← 0 to m − 1 do
   j ← MPnext[i]
   while j ≥ 0 and x[i] ≠ x[j] do
      j ← MPnext[j]
   MPnext[i + 1] ← j + 1
return MPnext
```

Fig. 3.7. A linear time algorithm for computing the table *MPnext* for a string x of length m.

3.1.4 Knuth-Morris-Pratt Algorithm

Interrupted Periods and Strict Borders

For a fixed string x and a non-empty prefix u of x, w is a strict border of u if both:

- w is a border of u;
- wb is a prefix of x, but ub is not.

An integer p is an interrupted period of u if $p = |u| - |w|$ for some strict border $|w|$ of u.

Example 2. Prefix `abacabacaba` of `abacabacabacc`

interrupted periods	strict borders of `abacabacaba`
10	a
11	empty string

The Searching Algorithm

The Morris-Pratt algorithm can be further improved. Consider a typical situation (Figure 3.5) where a prefix u of x has been matched and a mismatch occurs between the symbol a in x and the symbol b in y. Then the shift in the Morris-Pratt algorithm consists in aligning the prefix occurrence of the border of u in x with its suffix occurrence in y. But if this prefix occurrence in x is followed by the symbol a then another mismatched will occur with the symbol b in y. An alternative solution consists in precomputing for each prefix $x[0..i-1]$ of x the longest border followed by a symbol different from $x[i]$ for $i = 1, \ldots, m$. Those borders are called **strict borders** and then the length of the shifts are given by **interrupted periods**. It changes only the preprocessing of the string-matching algorithm KMP which is due to Knuth, Morris and Pratt [33].

Computing Strict Borders of Prefixes

The preprocessing of the algorithm KMP consists in computing the table *KMPnext*. *KMPnext*[0] is set to -1. Then for $i = 1, \ldots, m-1$, $k = \mathit{MPnext}[i]$

$$\mathit{KMPnext}[i] = \begin{cases} k & \text{if } x[i] \neq x[k] \text{ or if } i = m, \\ \mathit{KMPnext}[k] & \text{if } x[i] = x[k]. \end{cases}$$

The table *KMPnext* can be computed with the algorithm presented Figure 3.8.

```
COMPUTE-KMP-NEXT(x, m)
KMPnext[0] ← −1
j ← 0
for i ← 1 to m − 1 do
   if x[i] = x[j] then
      KMPnext[i] ← KMPnext[j]
   else KMPnext[i] ← j
      j ← KMPnext[j]
      while j ≥ 0 and x[i] ≠ x[j] do
         j ← KMPnext[j]
   j ← j + 1
KMPnext[m] ← j
return KMPnext
```

Fig. 3.8. Preprocessing of the Knuth-Morris-Pratt algorithm.

3.1.5 Complexities of MP and KMP Algorithms

Let us consider the algorithm given in Figure 3.6. Every positive comparisons increase the value of j. The value of j runs from 0 to $n-1$. Thus there are at most n positive comparisons. Negative comparisons increase the value of $j-i$ because such comparisons imply a shift. The value of $j-i$ runs from 0 to $n-1$. Thus there are at most n negative comparisons. Altogether, the algorithm makes no more than $2n$ symbol comparisons. This gives the following theorem.

Theorem 4. *On a text of length n, MP and KMP string-searching algorithms run in time $O(n)$. They make less than $2n$ symbol comparisons.*

The **delay** is defined as the maximum number of comparisons on one text symbol.

Theorem 5. *With a pattern of length m, the delay for* MP *algorithm is no more than m. The delay for* KMP *algorithm is no more than $\log_{\Phi}(m+1)$, where Φ is the golden ratio, $(1+\sqrt{5})/2$.*

Theorem 5 shows the advantage of KMP algorithm over MP algorithm. Its proof relies on combinatorial properties of strings. The next section sketches a further improvement.

3.1.6 Searching With an Automaton

The MP and KMP algorithms simulate a finite automaton. It is possible to build and use the string-matching automaton $SMA(x)$ which is the smallest deterministic automaton accepting the language V^*x.

Example 3. SMA(abaa)

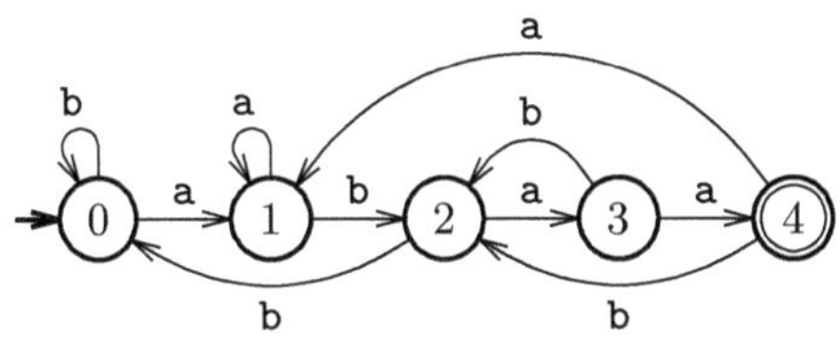

Search for abaa in $y =$ babbaabaabaabba:

y	b	a	b	b	a	a	b	a	a	b	a	a	b	b	a
state 0	0	1	2	0	1	1	2	3	4	2	3	4	2	0	1

Two occurrences of x occur in y at (right) positions 8 and 11. This is given by the fact that at these positions the search reaches the only terminal state of the string-matching automaton (state 4).

Searching Algorithm

The searching algorithm consists in a simple parsing of the text y with the string-matching automaton $SMA(x)$ (see Figure 3.9).

```
SEARCH-WITH-AN-AUTOMATON(x, y)
(Q, V, initial, {terminal}, δ) is the automaton SMA(x)
q ← initial
while not end of y do
    a ← next symbol of y
    q ← δ(q, a)
    if q = terminal then
        report an occurrence of x in y
```

Fig. 3.9. Searching with an automaton.

Construction of $SMA(x)$

The on-line construction of the smallest deterministic automaton accepting the language V^*x actually consists in unwinding appropriate arcs. The following example presents one step of the construction.

Example 4. From *SMA*(abaa)

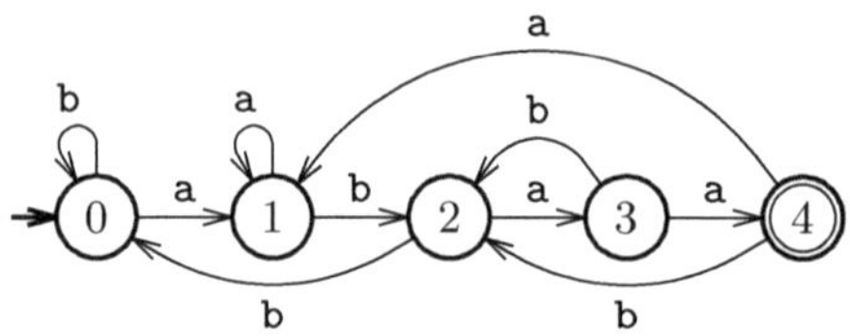

to *SMA*(abaab)

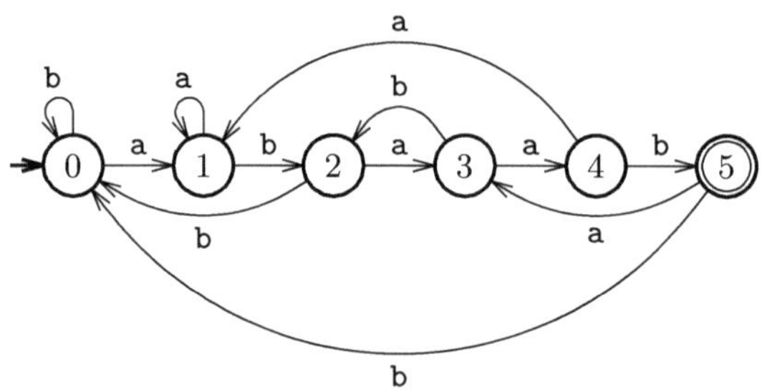

Updating *SMA*(abaa) to *SMA*(abaab) consists in storing the target state of the transition from the terminal state 4 in *SMA*(abaa) labeled by b (the new symbol): this state is 2. Then a new terminal state 5 is added and the transition from 4 by b is redirected to 5 and transitions for all the symbols of the alphabet from 5 go as all the transitions from 2: 5 by a leads to 3 since 2 by a leads to 3 (same for 5 by b leading to 0 since 2 by b leads to 0).

```
AUTOMATON-SMA(x)
let initial be a new state
Q ← {initial}
for each a ∈ V do
    δ(initial, a) ← initial
terminal ← initial
while not end of x do
    b ← next symbol of x
    r ← δ(terminal, b)
    add new state s to Q
    δ(terminal, b) ← s
    for each a ∈ V do
        δ(s, a) ← δ(r, a)
    terminal ← s
return (Q, V, initial, {terminal}, δ)
```

Fig. 3.10. The construction of the automaton *SMA*(x).

The complete construction can be achieved by the algorithm given in Figure 3.10.

Significant Arcs

We now characterize the number of significant arcs in the string-matching automaton $SMA(x)$.

Example 5. Complete automaton $SMA(\texttt{ananas})$:

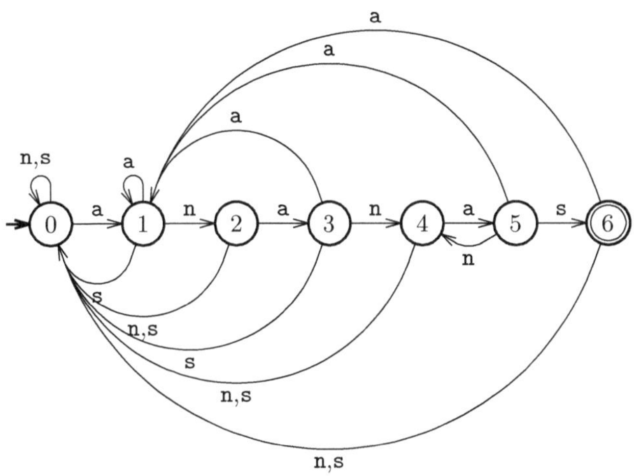

In such an automaton we distinguish two kinds of arcs:

- **Forward arcs**: arcs that spell the pattern;
- **Backward arcs**: other arcs which do not reach the initial state.

Example 6. $SMA(\texttt{ananas})$ represented with only forward and backward arcs:

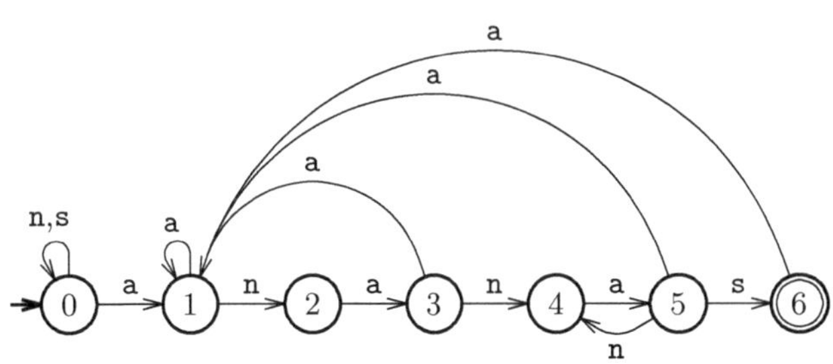

Backward Arcs in $SMA(x)$

The different states of $SMA(x)$ are identified with prefixes of x. A backward arc is of the form (u, b, vb) with u, v prefixes of x and $b \in V$ a symbol where vb is the longest suffix of ub that is a prefix of x, and $u \neq v$. Note that ub is not a prefix of x. Let $p(u, b) = |u| - |v|$ (a period of u).

Let (u, b, vb) and $(u', b', v'b')$ be two backward arcs. If $p(u, b) = p(u', b') = p$, then $vb = v'b'$. Otherwise, if, for instance, vb is a proper prefix of $v'b'$, vb

occurs at position p like v' does, ub is a prefix of x, which is a contradiction. Thus $v = v'$, $b = b'$, and then $u = u'$. Each period p, $1 \le p \le |x|$, corresponds to at most one backward arc, thus there are at most $|x|$ such arcs. This gives the following lemma.

Lemma 1. *The automaton SMA(x) contains at most $|x|$ backward arcs.*

The bound of the previous lemma is tight: *SMA*(ab^{m-1}) has m backward arcs ($a \neq b$) and thus constitutes a worst case for the number of backward arcs.

A fairly immediate consequence is that the implementation of *SMA*(x) and its construction can be done in $O(|x|)$ time and space, independently of the alphabet size.

Complexity of Searching With SMA

The complexities of the search with the string-matching automaton depend upon the implementation chosen for the automaton.

With a complete SMA implemented by transition matrix, the preprocessing on the pattern x can be done in time $O(m \times card(V))$ using a space in $O(m \times card(V))$. Then the search on the text y can be done in time $O(n)$ using a space in $O(m \times card(V))$. The delay is then constant.

With a SMA implemented by lists of forward and backward arcs. The preprocessing on the pattern x can be done in time $O(m)$ using a space in $O(m)$. Then the search on the text y can be done in time $O(n)$ using a space in $O(m)$. The delay becomes $\min\{card(V), \log_2 m\}$ comparisons. This constitutes an improvement on KMP algorithm.

3.1.7 Boyer-Moore Algorithm

The Boyer-Moore string-matching algorithm [4] performs the scanning operations from right to left inside the window on the text.

Example 7. $x =$ `cgctagc` and $y =$ `cgctcgcgctatcg`

```
y  c g c t c g c g c t a t c g
x  c g c t a g c
        x  c g c t a g c
            x  c g c t a g c
```

It uses two rules:

- the matching shift: good-suffix rule;
- the occurrence heuristics: bad-character rule;

to compute the length of the shift after each attempt. Extra rules can be used if some memorization are done from one attempt to the next.

The Matching Shift

A typical situation during the searching phase of the Boyer-Moore algorithm is depicted in Figure 3.11. During an attempt where the window is positioned on $y[j..j+m-1]$, a suffix $u = x[i+1..m-1]$ of x has been matched (from right to left) in y. A mismatch has occurred between symbol $x[i] = a$ in x and $y[j] = b$ in y.

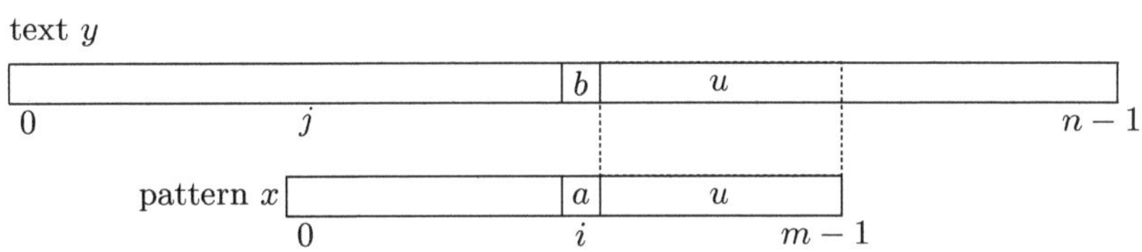

Fig. 3.11. A typical situation during the Boyer-Moore algorithm.

Then a valid shift consists in aligning the occurrence of u in y with a reoccurrence of u in x preceded by a symbol $c \neq a$ (see Figure 3.12). If no such reoccurrence exists, the shift consists in aligning the longest suffix of u in y which is a prefix of x (see Figure 3.13).

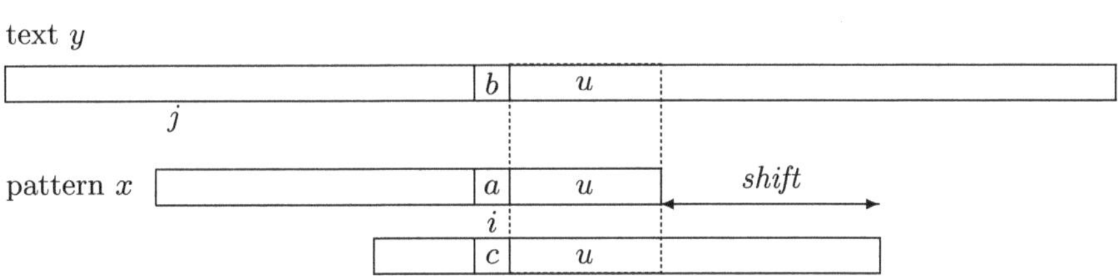

Fig. 3.12. The matching shift: a reoccurrence of u exists in x with $c \neq a$.

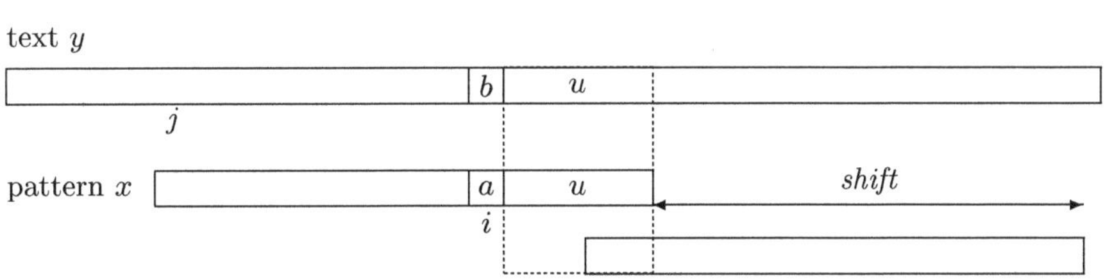

Fig. 3.13. The matching shift: no reoccurrence of u exists in x.

The first case for the matching shift which consists in the computation of the rightmost reoccurrences of each suffix u of x can be done in $O(m)$ time and space. The second case which basically corresponds to the computation of the period of x can also be performed in $O(m)$ time and space.

A table D implements the good-suffix rule: for $i = 0, \ldots, m-1$,

$$D[i] = \min\{|z| > 0 \mid (x \text{ suffix of } x[i \,..\, m-1]z) \text{ or } (bx[i \,..\, m-1]z \text{ suffix of } x \text{ and } bx[i \,..\, m-1] \text{ not suffix of } x, \text{ for } b \in V)\}$$

and $D[m] = 1$.

The Occurrence Heuristics

During an attempt, of the searching phase of the Boyer-Moore algorithm, where the window is positioned on $y[j..j+m-1]$, a suffix u of x has been matched (from right to left) in y. A mismatch has occurred between symbol $x[i] = a$ in x and $y[i+j] = b$ in y. The occurrence shift consists in aligning the symbol b in y with its rightmost occurrence in x (possibly leading to a negative shift) (see Figure 3.14).

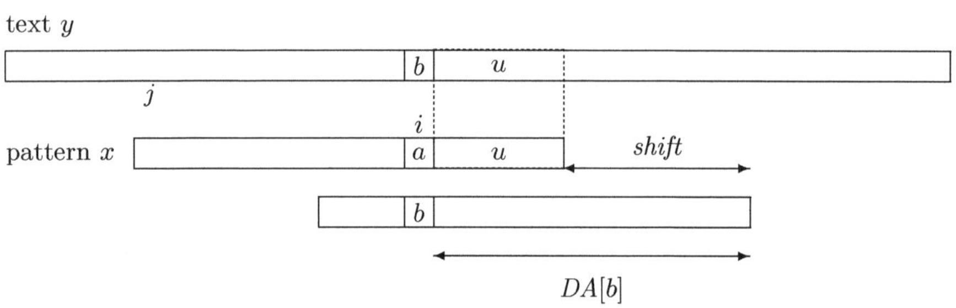

Fig. 3.14. The occurrence shift.

A table DA implements the bad-character rule: $DA[a] = \min(\{|z| > 0 \mid az \text{ suffix of } x\} \cup \{m\})$ for any symbol $a \in V$.

Then the length of the shift to apply is given by $DA[b] - |u| = DA[b] - m + i$.

BM Algorithm

The Boyer-Moore string-matching algorithm performs no memorization of previous matches. It applies the maximum between the two shifts. It is presented in Figure 3.15.

Suffix Displacement

For $0 \le i \le m-1$ we denote by $suf[i]$ the length of the longest suffix of x ending at position i in x. Let us denote by $lcsuf(u, v)$ the longest common suffix of two words u and v.

The computation of the table suf is done by the algorithm SUFFIXES presented in Figure 3.16. Figure 3.17 depicts the variables and the invariants

```
BM(x, m, y, n)
j ← 0
while j ≤ n − m do
    i ← m − 1
    while i ≥ 0 and x[i] = y[i + j] do
        i ← i − 1
    if i = −1 then
        OUTPUT(j)
    j ← j + max{D[i + 1], DA[y[i + j]] − m + i + 1}
```

Fig. 3.15. The Boyer-Moore algorithm.

```
SUFFIXES(x, m)
suf[m − 1] ← m
g ← m − 1
for i ← m − 2 downto 0 do
    if i > g and suf[i + m − 1 − f] < i − g then
        suf[i] ← suf[i + m − 1 − f]
    else g ← min{g, i}
        f ← i
        while g ≥ 0 and x[g] = x[g + m − 1 − f] do
            g ← g − 1
        suf[i] ← f − g
return suf
```

Fig. 3.16. Algorithm SUFFIXES.

of the main loop of algorithm SUFFIXES. The values of *suf* are computed for each position i in x in decreasing order. The algorithm uses two variables f and g which satisfy:

- $g = \min\{j - suf[j] \mid i < j < m - 1\}$;
- f is such that $i < f < m - 1$ and $f - suf[f] = g$.

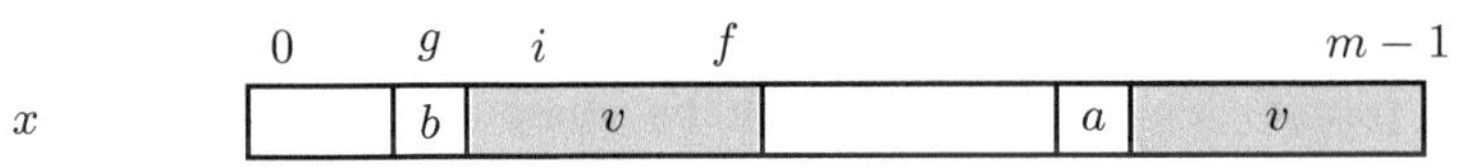

Fig. 3.17. Variables i, f, g of algorithm SUFFIXES. The main loop has invariants: $v = x[g + 1 \mathinner{..} f] = lcsuf(x, x[0 \mathinner{..} f])$ and $a \neq b$ $(a, b \in V)$ and $i < f$. The picture corresponds to the case where $g < i$.

We are now able to give, in Figure 3.18, the algorithm COMPUTE-D that computes the table D using the table *suf*. The invariants of the second loop of algorithm COMPUTE-D are presented in Fig. 3.19.

```
Compute-D(x, m)
1 j ← 0
2 for i ← m − 1 downto −1 do
3     if i = −1 or suf[i] = i + 1 then
4         while j < m − 1 − i do
5             D[j] ← m − 1 − i
6             j ← j + 1
7 for i ← 0 to m − 2 do
8     D[m − 1 − suf[i]] ← m − 1 − i
9 return D
```

Fig. 3.18. Computation of the matching shift.

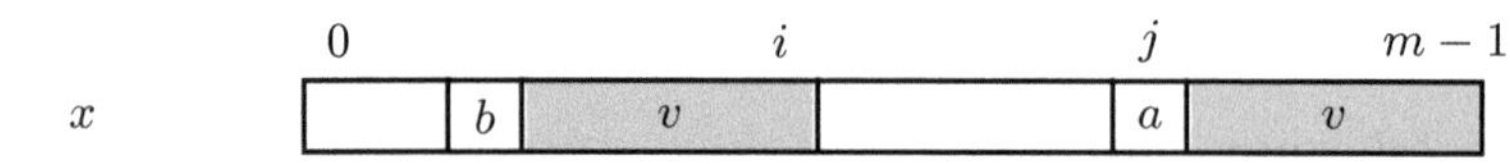

Fig. 3.19. Variables i and j of algorithm Compute-D. Situation where $suf[i] < i+1$. The loop of lines 7-8 has the following invariants: $v = lcsuf(x, x[0\,..\,i])$ and $a \neq b$ $(a, b \in V)$ and $suf[i] = |v|$. Thus $D[j] \leq m-1-i$ with $j = m-1-suf[i]$.

The algorithm of Figure 3.20 computes the table DA.

```
Compute-DA(x, m)
for each a ∈ V do
    DA[a] ← m
for i ← 0 to m − 2 do
    DA[x[i]] ← m − i − 1
return DA
```

Fig. 3.20. Computation of the occurrence shift.

Complexity of BM Algorithm

Preprocessing phase: The match shift can be computed in $O(m)$ time while the occurrence shift can be computed in $O(m + card(V))$ time.

Searching phase: When one wants to find all the occurrences of the pattern int the text, the worst case running time of the Boyer-Moore string-matching algorithm is $O(n \times m)$. The minimum number of symbol comparisons is n/m and the maximum number of symbol comparisons $n \times m$.

Extra space: The extra space needed for the two shift functions is $O(m + card(V))$ and it can be reduced to $O(m)$.

Symbol Comparisons in Variants of BM

In [33], it is proved that for finding the first occurrence of a pattern x of length m in a text y of length n, the BM algorithm performs no more than $7n$ comparisons between symbols of the text and symbols of the pattern. The bound is lowered to $4n$ comparisons in [23]. R. Cole [6] gives a tight bound of $3n - m$ comparisons for non-periodic patterns (*i.e.* $period(x) > m/2$)

For finding all the occurrences of a pattern x of length m in a text y of length n, linear variants of the BM algorithm have been designed. The Galil algorithm [17] implements a prefix memorization technique when an occurrence of the pattern is located in the text. It gives a linear number of comparisons between symbols of the pattern and symbols of the text and requires a constant extra space.

The Turbo-BM algorithm [11] implements a last-suffix memorization technique which leads to a maximal of $2n$ comparisons. It also requires a constant extra space. Actually it stores the last match in the text when a matching shift is applied (the memorized factor is called *memory*) (see Figure 3.21). This enables it to perform jumps, in subsequent attempts, on these memorized factors of the text, saving thus some symbol comparisons. It can also perform, in some cases, larger shifts by using turbo-shifts. Its preprocessing is the same as the BM algorithm. The searching phase need an $O(1)$ extra space to store *memory* as a pair (length, right position).

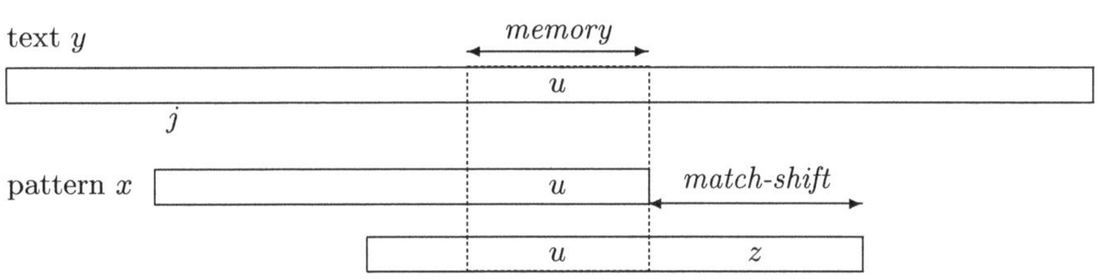

Fig. 3.21. When a match shift is applied the Turbo-BM algorithm memorizes the factor u of y.

The Apostolico-Giancarlo [2] implements an all-suffix memorization technique that gives a maximal number of comparisons equal to $1.5n$ [12]. It requires an $O(m)$ extra space. The Apostolico and Giancarlo remembers the length of the longest suffix of the pattern ending at the right position of the window at the end of each attempt (see Figure 3.22). These information are stored in a table *skip*. Let us assume that during an attempt at a position less than j the algorithm has matched a suffix of x of length k at position $j+i$ with $0 < i < m$ then $skip[j+i]$ is equal to k. Let $suf[i]$, for $0 \leq i < m$ be equal to the length of the longest suffix of x ending at the position i in x (see Section 3.1.7). During the attempt at position j, if the algorithm compares successfully the factor of the text $y[j+i+1..j+m-1]$ only in the case where $k = suf[i]$, a "jump" has to be done over the text factor $y[j+i-k+1..j+i]$ in order to resume the comparisons between the symbols $y[j+i-k]$ and $x[i-k]$. In all the other cases, no more comparisons have to be done to conclude the attempt and a shift can be performed.

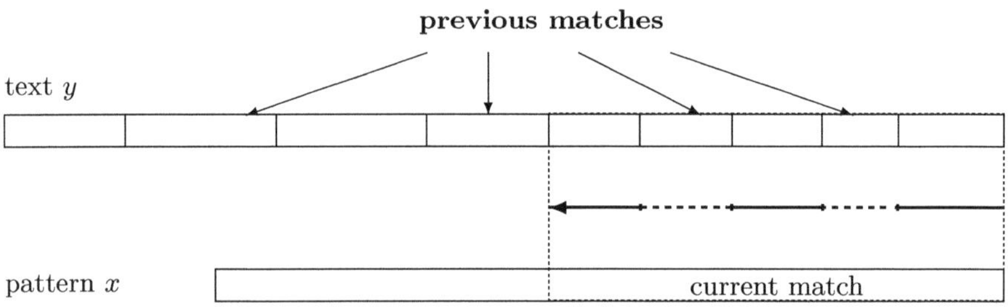

Fig. 3.22. The Apostolico-Giancarlo algorithm stores all the matches that occur between suffixes of x and subwords of y.

3.2 Searching a List of Strings — Suffix Arrays

In this section we consider two main questions that are related by the technique used to solve them. The first question on word list searching is treated in the first subsection, and the second one, indexing a text, is treated in Subsection 3.2.3.

3.2.1 Searching a List of Words

Input a list L of n strings of V^* stored in increasing lexicographic order in a table: $L_0 \leq L_1 \leq \cdots \leq L_{n-1}$ and a string $x \in V^*$ of length m.
Problem find either i, $-1 < i < n$, with $x = L_i$ if x occurs in L, or d, $-1 \leq d \leq n$, that satisfy $L_d < x < L_{d+1}$ otherwise.

Example 8. **List** L

$$\begin{aligned} L_0 &= \texttt{a a a b a a} \\ L_1 &= \texttt{a a a b b} \\ L_2 &= \texttt{a a b b b b} \\ L_3 &= \texttt{a b} \\ L_4 &= \texttt{b a a a} \\ L_5 &= \texttt{b b} \end{aligned}$$

The search for `aaabb` outputs 1 as does the search for `aaba`.

3.2.2 Searching Algorithm

A standard way of solving the problem is to use a binary search because the list of strings is sorted. Its presentation below makes use of the function *lcp* that computes the longest common prefix (LCP) of two strings.

```
SIMPLE-SEARCH(L, n, x, m)
d ← −1
f ← n
while d + 1 < f do
    ▷ Invariant: L_d < x < L_f
    i ← ⌊(d + f)/2⌋
    ℓ ← |lcp(x, L_i)|
    if ℓ = m and ℓ = |L_i| then
        return i
    else if (ℓ = |L_i|) or (ℓ ≠ m and L_i[ℓ] < x[ℓ]) then
        d ← i
    else f ← i
return d
```

The running time of the binary search is $O(m \times \log n)$ if we assume that the LCP computation of two string takes a linear time, doing it by pairwise symbol comparisons. The worst case is met with the list $L = (\mathtt{a}^{m-1}\mathtt{b}, \mathtt{a}^{m-1}\mathtt{c}, \mathtt{a}^{m-1}\mathtt{d}, \dots)$ and the string $x = \mathtt{a}^m$.

Indeed, it is possible to reduce the running time of the binary search to $O(m + \log n)$ by storing the LCPs of some pairs of strings of the list. These pairs are of the form (L_d, L_f) where (d, f) is a pair of possible values of d and f in the binary search algorithm. Since there are $2n + 1$ such pairs, the extra space required by the new algorithm SEARCH is $O(n)$.

The design of the algorithm is based on properties arising in three cases (plus symmetric cases) that are described below. The algorithm maintains three variables defined as: $ld = |lcp(x, L_d)|$, $lf = |lcp(x, L_f)|$, $i = \lfloor (d + f)/2 \rfloor$. In addition, the main invariant of the loop of the algorithm is $L_d < x < L_f$.

Case one

If $ld \leq |lcp(L_i, L_f)| < lf$, then $L_i < x < L_f$ and $|lcp(x, L_i)| = |lcp(L_i, L_f)|$.

Case two

If $ld \leq lf < |lcp(L_i, L_f)|$, then $L_d < x < L_i$ and $|lcp(x, L_i)| = |lcp(x, L_f)|$.

Case three

If $ld \leq lf = |lcp(L_i, L_f)|$, then we have to compare x and L_i to discover if they match or which one is the smallest. But this comparison symbol by symbol is to start at position lf because the strings have a common prefix of length lf.

The resulting algorithm including the symmetric cases where $lf \leq ld$ is given in Figure 3.23 and it satisfies the next proposition because LCP can be implemented to run in constant time after preprocessing the list (in time linear in the input size).

```
SEARCH(L, n, Lcp, x, m)
(d, ld) ← (−1, 0)
(f, lf) ← (n, 0)
while d + 1 < f do
    ▷ Invariant : L_d < x < L_f
    i ← ⌊(d + f)/2⌋
    if ld ≤ Lcp(i, f) < lf then
        (d, ld) ← (i, Lcp(i, f))
    else if ld ≤ lf < Lcp(i, f) then
        f ← i
    else if lf ≤ Lcp(d, i) < ld then
        (f, lf) ← (i, Lcp(d, i))
    else if lf < ld < Lcp(d, i) then
        d ← i
    else ℓ ← max{ld, lf}
        ℓ ← ℓ + |lcp(x[ℓ . . m − 1], L_i[ℓ . . |L_i| − 1])|
        if ℓ = m and ℓ = |L_i| then
            return i
        else if (ℓ = |L_i|) or (ℓ ≠ m and L_i[ℓ] < x[ℓ]) then
            (d, ld) ← (i, ℓ)
        else (f, lf) ← (i, ℓ)
return d
```

Fig. 3.23. Search for x in L in time $O(m + \log n)$.

Proposition 1. *Algorithm* SEARCH *finds a string x of length m in a sorted list of n strings in time $O(m + \log n)$.*

It makes no more than $m+\lceil\log(n+1)\rceil$ comparisons of symbols and requires $O(n)$ extra space.

A straightforward extension of the algorithm SEARCH used for suffix arrays in the rest of the section computes the pair (d, f), $-1 \leq d < f \leq n$, that satisfies: $d < i < f$ if and only if x prefix of L_i.

Preprocessing the list is a classical matter.

Sorting can be done by repetitive applications of bin sorting and takes time $O(||L||)$, where $||L|| = \sum_{i=0}^{n-1} |L_i|$.

Computing LCPs of strings consecutive in the sorted list takes the same time by mere symbol comparisons. Computing other LCPs is based on next lemma and takes time $O(n)$.

Lemma 2. *Let $L_0 \leq L_1 \leq \ldots \leq L_{n-1}$. Let d, i and f, $-1 < d < i < f < n$. Then $|lcp(L_d, L_f)| = \min\{|lcp(L_d, L_i)|, |lcp(L_i, L_f)|\}$.*

So, the complete preprocessing time is $O(||L||)$.

3.2.3 Suffix Array

A suffix array is a structure for indexing texts. It is used for the implementation of indexes supporting operations of searching for patterns, their number of occurrences, or their list of positions. Contrary to suffix trees or suffix automata whose efficiency relies on the design of a data structure, suffix arrays are grounded on efficient algorithms, one of them being the search algorithm of the previous section.

The suffix array of a text $y \in V^*$ of length n is composed of the elements described for the list of strings, applied to the list of suffixes of the text. So, it consists of both the permutation of positions on the text that gives the sorted list of suffixes and the corresponding array of lengths of their LCPs. They are denoted by p and LCP and defined by:

$$y[p[0]\mathinner{..}n-1] < y[p[1]\mathinner{..}n-1] < \ldots < y[p[n-1]\mathinner{..}n-1]$$

and

$$LCP[i] = |lcp(y[p[i-1]\mathinner{..}n-1], y[p[i]\mathinner{..}n-1])|.$$

Example 9. $y =$ `aabaabaabba`

i	$p[i]$	$LCP[i]$	
0	10	0	`a`
1	0	1	`a a b a a b a a b b a`
2	3	6	`a a b a a b b a`
3	6	3	`a a b b a`
4	1	1	`a b a a b a a b b a`
5	4	5	`a b a a b b a`
6	7	2	`a b b a`
7	9	0	`b a`
8	2	2	`b a a b a a b b a`
9	5	4	`b a a b b a`
10	8	1	`b b a`

There are several algorithms for computing a suffix array efficiently, two of them running in linear time are presented here as a sample. Note that the solutions of Section 3.2.1 would lead to algorithms running in time $O(n^2)$ because $||\mathrm{Suf}(y)|| = O(n^2)$. But they would not exploit the dependencies between the suffixes of the text.

We consider that the alphabet is a bounded segment of integers, as it can be considered in most real applications. The schema for sorting the suffixes is as follows.

1. bucket sort positions i according to $First_3(y[i\,..\,n-1])$, ($First_3(x)$ is either the first three symbols of x if $|x| \geq 3$ or x if $|x| < 3$ for a string $x \in V^*$) for $i = 3q$ or $i = 3q+1$;
   let $t[i]$ be the rank of i in the sorted list.
2. recursively sort the suffixes of the 2/3-shorter word
   $t[0]t[3]\cdots t[3q]\cdots t[1]t[4]\cdots t[3q+1]\cdots$
   let $s[i]$ be the rank of suffix i in the sorted list ($i = 3q$ or $i = 3q+1$)
3. sort suffixes $y[j\,..\,n-1]$, for j of the form $3q+2$, by bucket sorting pairs $(y[j], s[j+1])$.
4. merge lists obtained at steps 2 and 3
   Note: comparing suffixes i (first list) and j (second list) remains to compare:
   $(x[i], s[i+1])$ and $(x[j], s[j+1])$ if $i = 3q$
   $(x[i]x[i+1], s[i+2])$ and $(x[j]x[j+1], s[j+2])$ if $i = 3q+1$

The recursivity of the algorithm yields the recurrence relation $T(n) = T(2n/3) + O(n)$ for its running time, which gives $T(n) = O(n)$.

Example 10. $y =$ `aabaabaabba`

i	0	1	2	3	4	5	6	7	8	9	10
$y[i]$	a	a	b	a	a	b	a	a	b	b	a

Rank t	
0	a
1	a a b
2	a b a
3	a b b
4	b a

Rank s	i	Suf(11142230)
0	10	0
1	0	1 1 1 4 2 2 3 0
2	3	1 1 4 2 2 3 0
3	6	1 4 2 2 3 0
4	1	2 2 3 0
5	4	2 3 0
6	7	3 0
7	9	4 2 2 3 0

Rank	j	$(y[j], s[j+1])$
0	2	(b, 2)
1	5	(b, 3)
2	8	(b, 7)

i	0	1	2	3	4	5	6	7	8	9	10
$y[i]$	a	a	b	a	a	b	a	a	b	b	a
$r[i]$	1	4	8	2	5	9	3	6	10	7	0
$p[i]$	10	0	3	6	1	4	7	9	2	5	8

Table r is defined by: $r[j] =$ rank of suffix at position j in the sorted list of all suffixes. It is the inverse of p.

There is a second linear-time algorithm for computing LCPs (see Figure 3.24) of consecutive suffixes in the sorted list (other LCPs are computed as in Section 3.2.1). Its running time analysis is straightforward. The next example illustrates the following lemma that is the clue of algorithm LCP.

Example 11. $y =$ `aabaabaabba`

i	0	1	2	3	4	5	6	7	8	9	10	11
$y[i]$	a	a	b	a	a	b	a	a	b	b	a	
$p[i]$	10	0	3	6	1	4	7	9	2	5	8	
$LCP[i]$	0	1	6	3	1	5	2	0	2	4	1	0

j	$r[j]$	
0	1	a a b a a b a a b b a
3	2	a a b a a b b a

j	$r[j]$	
1	4	a b a a b a a b b a
4	5	a b a a b b a

Lemma 3. *Let* $j \in (1, 2, \ldots, n-1)$ *with* $r[j] > 0$. *Then* $LCP[r[j-1]] - 1 \leq LCP[r[j]]$.

```
LCP(y, n, p, r)
 ℓ ← 0
 for j ← 0 to n − 1 do
     ℓ ← max{0, ℓ − 1}
     if r[j] > 0 then
         i ← p[r[j] − 1]
         while y[i + ℓ] = y[j + ℓ] do
             ℓ ← ℓ + 1
         LCP[r[j]] ← ℓ
 LCP[0] ← 0
 LCP[n] ← 0
 return LCP
```

Fig. 3.24. Computation of the LCPs

The next statement summarizes the elements of the present section.

Proposition 2. *Computing the suffix array of a text of length* n *can be done in time* $O(n)$ *with* $O(n)$ *memory space.*

3.3 Indexes

Indexes are data structures that are used to solve the pattern matching problem in static texts. An index for a text y of is a structure that contains all the factors of y. It must enable to deal with the following basic operations:

String-matching: computing the existence of a pattern x of length m in the text y;

All occurrences: computing the list of positions of occurrences of a pattern x of length m in y;

Repetitions: computing a longest subword of y occurring at least k times;

Marker: computing a shortest subword of y occurring exactly once.

Other possible applications includes:

- finding all the repetitions in texts;
- finding regularities in texts;
- approximate matchings.

3.3.1 Implementation of Indexes

Indexes are implemented by suffix arrays or by suffix trees or suffix automata in $O(n)$ space. Such structures represent all the subwords of y since every subword of y is a prefix of a suffix of y. Table 3.1 summarizes the complexities of different operations on indexes with these structures.

	suffix array	suffix tree or suffix automaton
Construction	$O(n)$	$O(n \times \log card(V))$
String-matching	$O(m + \log n)$	$O(m \times \log card(V))$
All occurrences	$O(m + \log n + \lvert output \rvert)$	$O(m \times \log card(V)) + \lvert output \rvert)$
Repetitions	$O(n)$	$O(n)$
Marker	$O(n)$	$O(n)$

Table 3.1. Complexities of different operations on indexes.

3.3.2 Efficient Constructions

The notion of position tree is due to Weiner [40], who presents an algorithm for computing its compact version. An off-line computation of suffix trees is given by McCreight [35]. Ukkonen [39] gives an on-line algorithm and Farach [15] designs an alphabet independent algorithm for the suffix tree construction. Other implementations of suffix trees are given in [1, 29, 28, 37, 22].

The suffix automaton is also kwown as the DAWG for Directed Acyclic Word Graph. Its linearity was discovered by Blumer *et al.* [3]. The minimality of the structure as an automaton is due to Crochemore [8] who shown how to construct the factor automaton with the same complexity.

PAT arrays were designed by Gonnet [21]. Suffix arrays were first designed by Manber and Myers [34], for recent results see [30, 31, 32].

SB-trees are used to store this structures in external memory [16].

Crochemore and Vérin [14] first introduced compact suffix automata. An on-line algorithm for its construction is given in [27].

3.3.3 Trie of Suffixes

The trie of suffixes of y, denoted by $\mathcal{T}(y)$ is a digital tree which branches are labeled by suffixes of y. Actually it is a tree-like deterministic automaton accepting the language Suf(y).

Nodes of $\mathcal{T}(y)$ are identified with subwords of y. **Terminal nodes** of $\mathcal{T}(y)$ are identified with suffixes of y. An output is defined, for each terminal node, which is the starting position of the suffix in y.

Example 12. Suffix trie of ababbb

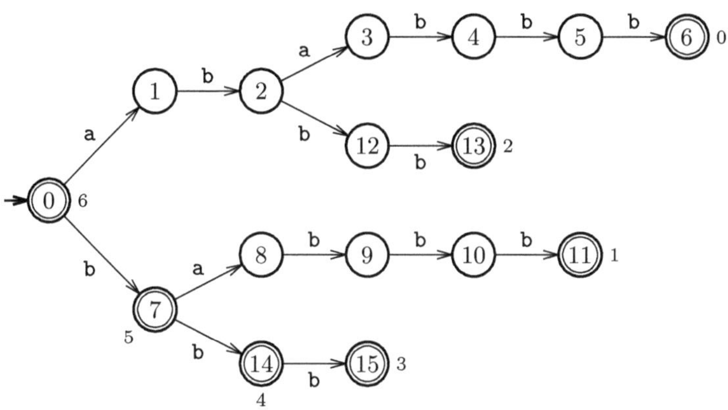

Starting with an empty tree, the trie $\mathcal{T}(y)$ is build by successively inserting the suffixes of y from the longest one (y itself) to the shortest one (the empty word).

Forks

Let us examine the insertion of $u = y[i\,..\,n-1]$ in the structure accepting longer suffixes $(y, y[1\,..\,n-1], \ldots, y[i-1\,..\,n-1])$. The **head** of u is the longest prefix $y[i\,..\,k-1]$ of u occurring before i. The **tail** of u is the the rest $y[k\,..\,n-1]$ of suffix u.

Example 13. With $y =$ ababbb, the head of abbb is ab and its tail is bb.

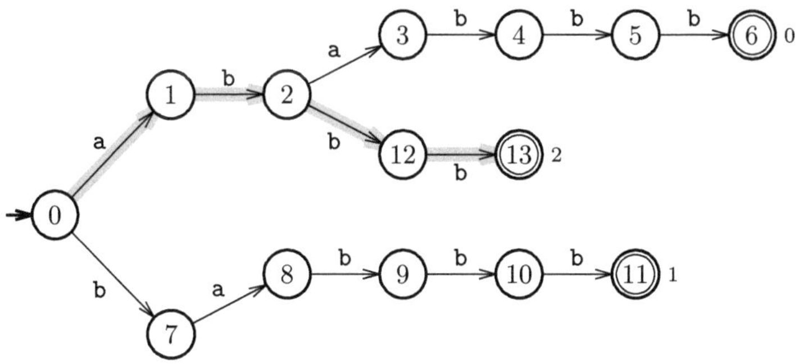

A **fork** is any node that has out-degree 2 at least, or that has both out-degree 1 and is terminal. The head of a prefix of y is a fork. The initial node is a fork if and only if y is non empty.

The insertion of a suffix $u = y[i \mathinner{..} n-1]$ consists first in finding the fork corresponding to the head of u and then in inserting the tail of u from this fork.

Suffix Link

A **function** s_y, called *suffix link* is defined as follows: if node p is identified with subword av, $a \in V, v \in V^*$ then $s_y(p) = q$ where node q is identified with v.

Example 14. Suffix links are represented by dotted arrows.

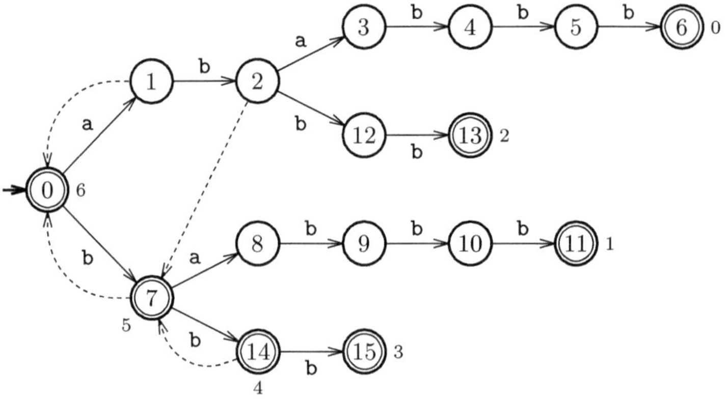

The suffix links create shortcuts that are used to accelerate heads computations. It is useful for forks only. If node p is a fork, so is $s_y(p)$. If the head of $y[i-1 \mathinner{..} n-1]$ is of the form au ($a \in V, u \in V^*$) then u is a prefix of the head of $y[i \mathinner{..} n-1]$. Then, using suffix links, the insertion of the suffix $y[i \mathinner{..} n-1]$ consists first in finding the fork corresponding to the head of $y[i \mathinner{..} n-1]$ (starting from suffix link of the fork associated with au) and then in inserting the tail of $y[i \mathinner{..} n-1]$ from this fork.

3.3.4 Suffix Tree

The suffix tree of y, denoted by $\mathcal{S}(y)$, is a compact trie accepting the language $\mathrm{Suf}(y)$. It is obtained from the suffix trie of y by deleting all nodes having out-degree 1 that are not terminal. Edges are then labeled by subwords of y instead of symbols.

Example 15. $\mathcal{S}$(ababbb)

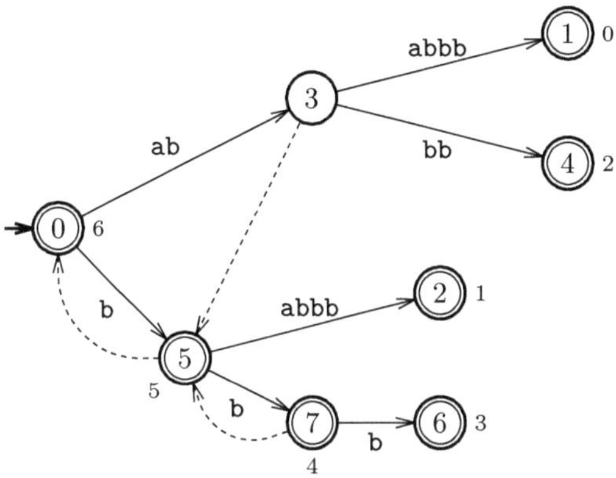

The number of nodes of $\mathcal{S}(y)$ is not more than $2n$ (if $n > 0$) since all internal nodes either have two children at least or are terminal and there are at most n terminal nodes.

Labels of Edges

The edge labels are represented by pairs (j, ℓ) representing subwords $y[j\,..\,j+\ell-1]$ of y.

Example 16. $\mathcal{S}$(ababbb)

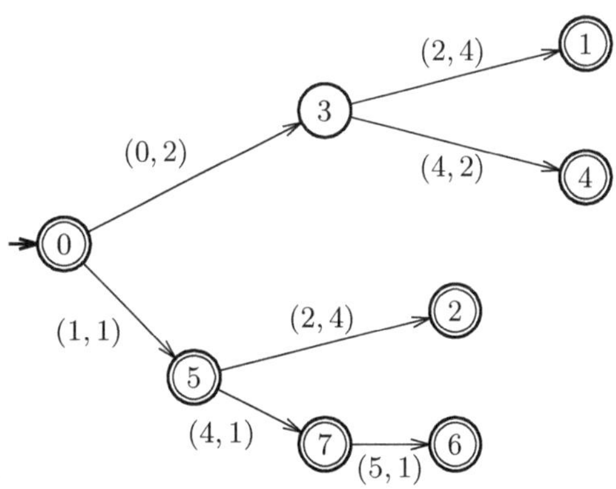

i	0	1	2	3	4	5
$y[i]$	a	b	a	b	b	b

This technique requires to have y residing in main memory. Thus the size of $\mathcal{S}(y)$ is $O(n)$.

Scheme of Suffix Tree Construction

The algorithm for building the suffix tree of y is given in Figure 3.25. It uses algorithms FAST-FIND and SLOW-FIND that are described next. Starting with an empty tree, $\mathcal{S}(y)$ is built by successively inserting the suffixes of y from the longest one (y itself) to the shortest one (the empty word). Using suffix links, the insertion of the suffix $y[i\,..\,n-1]$ consists first in finding the fork corresponding to the head of $y[i\,..\,n-1]$ (starting from suffix link of the fork associated with the head of $y[i-1\,..\,n-1]$) and then in inserting the tail of $y[i\,..\,n-1]$ from this fork.

```
SUFFIX-TREE(y, n)
T ← NEW-AUTOMATON()
for i ← 0 to n − 1 do
   find fork of head of y[i..n − 1] using
   FAST-FIND from node s_y(fork), and then SLOW-FIND
   k ← position of tail of y[i..n − 1]
   if k < n then
      q ← NEW-STATE()
      Adj[fork] ← Adj[fork] ∪ {((k, n − k), q)}
      output[q] ← i
   else output[fork] ← i
return T
```

Fig. 3.25. Scheme of the construction of the suffix tree of string y of length n.

This scheme requires an adjacency-list representation of labeled arcs.

Let us examine more closely the insertion of the suffix $y[i\,..\,n-1]$ in the tree. The search for the node associated with the head of $y[i\,..\,n-1]$ proceeds in two steps:

1. Assume that the head of $y[i-1\,..\,n-1]$ is $auv = y[i-1\,..\,k-1]$ ($a \in V, u, v \in V^*$) and let *fork* be the associated node. If the suffix link of *fork* is defined, it leads to node s, then the second step starts from this node. Otherwise, the suffix link from *fork* is found by rescanning as follows. Let r be the parent node of *fork* and let (j, ℓ) be the label of edge (r, fork). For the ease of description, assume that $auv = au(y[k-\ell\,..\,k])$ (it may happened that $auv = y[k-\ell\,..\,k]$). There is a suffix link from node r to node p associated with v. The crucial observation here is that $y[k-\ell\,..\,k]$ is the prefix of the label of some branch starting at node p. Then the algorithm rescans $y[k-\ell\,..\,k]$ in the tree: let q be the child of p along that branch and let (h, m) be the label of the edge (p, q). If $m < \ell$ then a recursive scan of $y[k-\ell+m\,..\,k]$ starts from node q. If $m > \ell$ then the edge (p, q) is broken to insert a new node s; labels are updating correspondingly. If $m = \ell$, s is simply set to q. This search is performed by the algorithm FAST-FIND given in Figure 3.26. The suffix link of *fork* is then set to s.
2. A downward search starts from node s to find the fork associated with the head of $y[i\,..\,n-1]$. This search is dictated by the symbols of the tail

of $y[i \mathinner{.\,.} n-1]$, one by one from left to right. If necessary a new internal node is created at the end of this scanning (see Figure 3.27).

```
FAST-FIND(r, j, k)
if j ≥ k then
   return r
else q ← TARGET(r, y[j])
   (j′, ℓ) ← label(r, q)
   if j + ℓ ≤ k then
      return FAST-FIND(q, j + ℓ, k)
   else Adj[r] ← Adj[r] \ {((j′, ℓ), q)}
      p ← NEW-STATE()
      Adj[r] ← Adj[r] ∪ {((j, k − j), p)}
      Adj[p] ← Adj[p] ∪ {((j′ + k − j, ℓ − k + j), q)}
      return p
```

Fig. 3.26. Search for $y[j \mathinner{.\,.} k]$ from node r.

```
SLOW-FIND(p, k)
while k < n and TARGET(p, y[k]) ≠ NIL do
   q ← TARGET(p, y[k])
   (j, ℓ) ← label(p, q)
   i ← j
   do
      i ← i + 1
      k ← k + 1
   while i < j + ℓ and k < n and y[i] = y[k]
   if i < j + ℓ then
      Adj[p] ← Adj[p] \ {((j, ℓ), q)}
      r ← NEW-STATE()
      Adj[p] ← Adj[p] ∪ {((j, i − j), r)}
      Adj[r] ← Adj[r] ∪ {((j + i − j, ℓ − i + j), q)}
      return (r, k)
   p ← q
return (p, k)
```

Fig. 3.27. Search of the longest prefix of $y[k \mathinner{.\,.} n-1]$ from node p. A new node is created when the target lies in the middle of an arc.

The insertion of the tail from the fork associated to the head of $y[i \mathinner{.\,.} n-1]$ is done by adding a new edge labeled by the tail leading to a new node. It is done in constant time.

Example 17. $\mathcal{S}$(abababbb)

End of insertion of suffix babbb

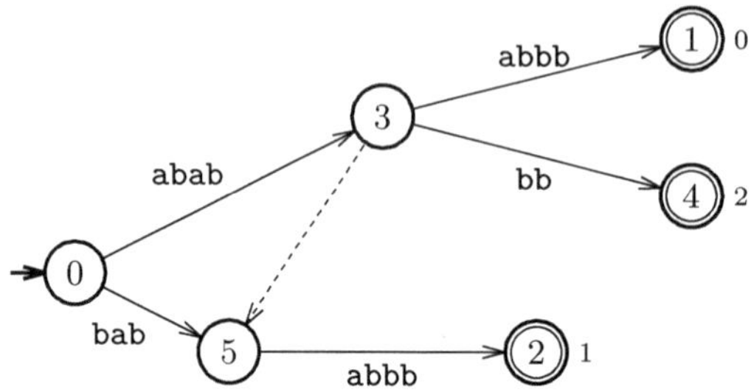

i	0	1	2	3	4	5	6	7
$y[i]$	a	b	a	b	a	b	b	b

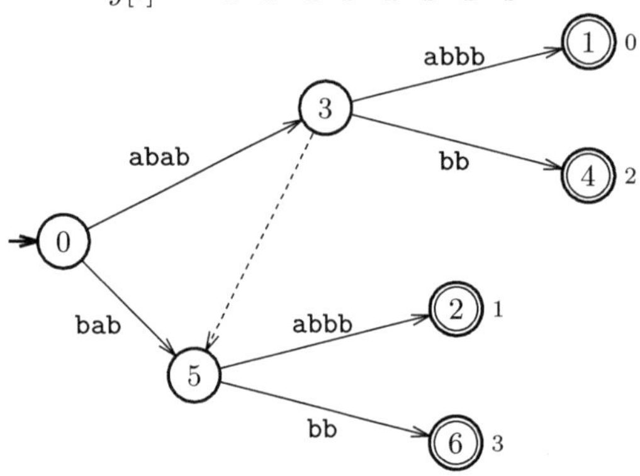

The head of babbb is bab so its tail is bb.

Complete Algorithm

We are now able to give the complete algorithm for building the suffix tree of a text y of length n (see Figure 3.28). A table s implements the suffix links.

Complexity

The execution of SUFFIX-TREE(y) takes $O(|y| \times \log card(V))$ time in the comparison model. Indeed the main iteration increments i, which never decreases, iterations in FAST-FIND increment j, which never decreases, iterations in SLOW-FIND increment k, which never decreases and basic operations run in constant time or in time $O(\log card(V))$ time in the comparison model.

3.3.5 Suffix Automaton

The minimal deterministic automaton accepting Suf(y) is denoted by $\mathcal{A}(y)$. It can be seen as the minimization of the trie $\mathcal{T}(y)$ of suffixes of y.

```
SUFFIX-TREE(y, n)
T ← NEW-AUTOMATON()
s[initial[T]] ← initial[T]
(fork, k) ← (initial[T], 0)
for i ← 0 to n − 1 do
   k ← max{k, i}
   if s[fork] = NIL then
      r ← parent of fork
      (j, ℓ) ← label(r, fork)
      if r = initial[T] then
         ℓ ← ℓ − 1
      s[fork] ← FAST-FIND(s[r], k − ℓ, k)
   (fork, k) ← SLOW-FIND(s[fork], k)
   if k < n then
      q ← NEW-STATE()
      Adj[fork] ← Adj[fork] ∪ {((k, n − k), q)}
      output[q] ← i
   else output[fork] ← i
return T
```

Fig. 3.28. The complete construction of the suffix tree of y of length n.

Example 18. $\mathcal{A}$(ababbb)

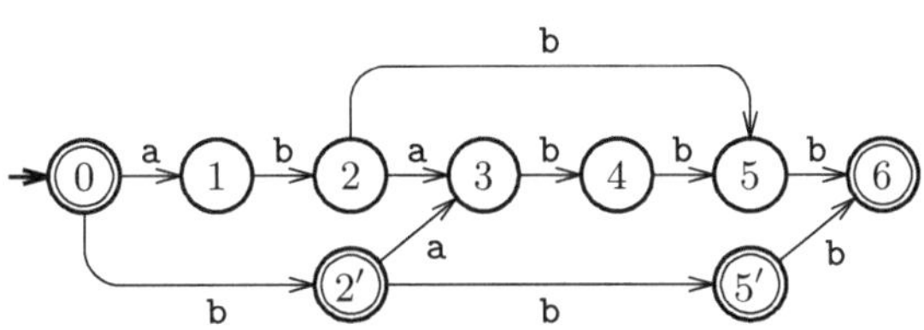

The states of $\mathcal{A}(y)$ are classes of factors (subwords) of y. Two subwords u and v of y are in the same equivalence class if they have the same right context in y. Formally $u \equiv_y v$ iff $u^{-1}\mathrm{Suf}(y) = v^{-1}\mathrm{Suf}(y)$.

The suffix automaton $\mathcal{A}(y)$ has a linear size:

- it has between $n + 1$ and $2n - 1$ states;
- it has between n and $3n - 4$ arcs.

Suffix Link

A function f_y, also called *suffix link* is defined as follows:
let $p = $TARGET$(initial[\mathcal{A}(y)], v)$, $v \in V^+$, $f_y(p) = $TARGET$(initial[\mathcal{A}(y)], u)$, where u is the longest suffix of v occurring in a different right context ($u \not\equiv_y v$).

Example 19. $\mathcal{A}$(aabbabb)

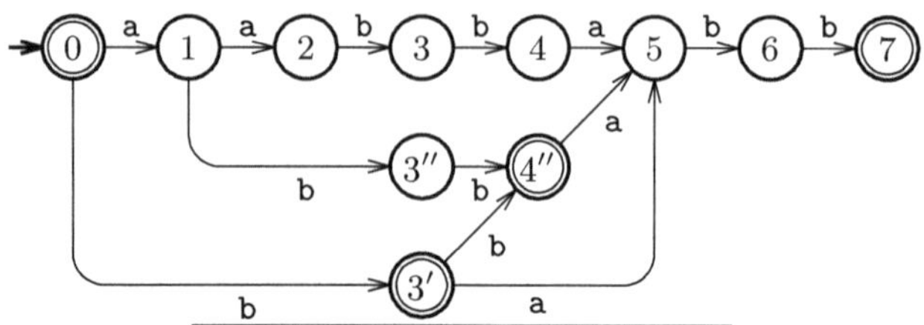

i	1	2	3	3′	3″	4	4″	5	6	7
$f_y(i)$	0	1	3″	0	3′	4″	3′	1	3″	4″

Suffix Path

For a state p of $\mathcal{A}(y)$, the suffix path of p denoted by $SP_y(p)$ is defined as follows:

$$SP_y(p) = \langle p, f_y(p), f_y^2(p), \ldots \rangle.$$

Solid Arc

For a state p of $\mathcal{A}(y)$, we denote by $L_y(p)$ the length of the longest string u in the class of p. It also corresponds to the length of the longest path from the initial state to state p (this path is labeled by u). An arc (p, a, q) of $\mathcal{A}(y)$ is solid iff $L_y(q) = L_y(p) + 1$.

Construction

Starting with a single state, the automaton $\mathcal{A}(y)$ is build by successively inserting the symbols of y from $y[0]$ to $y[n-1]$. The algorithm is presented in Figure 3.29. Tables f and L implements functions f_y and L_y respectively. Let us assume that $\mathcal{A}(w)$ is correctly build for a prefix w of y and let *last* be the state of $\mathcal{A}(w)$ corresponding to the class of w. The algorithm Extension(a) builds $\mathcal{A}(wa)$ from $\mathcal{A}(w)$ (see Figure 3.30). This algorithm creates a new state *new*. Then in the first while loop, transitions (p, a, new) are created for the first states p of $SP_y(last)$ that do not already have a defined transition for the symbol a. Let q be the first state of $SP_y(last)$ for which a transition is defined for the symbol a, if such a state exists. When the first while loop of Extension(a) ends three cases can arise:

1. p is not defined;
2. (p, a, q) is a solid arc;
3. (p, a, q) is not a solid arc.

Case 1: This situation arises when a does not occur in w. We have then $f_y(new) = initial[\mathcal{A}(w)]$.

Case 2: Let u be the longest string recognized in state p ($|u| = L_y(p)$). Then ua is the longest suffix of wa that is a subword of w. Thus $f_y(new) = q$.

Case 3: Let u be the longest string recognized in state p ($|u| = L_y(p)$). Then ua is the longest suffix of wa that is a subword of w. Since the arc (p, a, q) is not solid, ua is not the longest string recognized in state q. Then state q is splitted into two states: the old state q and a new state *clone*. The state *clone* has the same transitions than q. The strings v (of the form $v'a$) shorter than ua that were recognized in state q are now recognized in state *clone*.

```
SUFFIX-AUTOMATON(y, n)
T ← NEW-AUTOMATON()
L[initial[T]] ← 0
f[initial[T]] ← NIL
last ← initial[T]
for j ← 0 to n − 1 do
    ▷ Extension of T with the symbol y[j]
    last ← EXTENSION(y[j])
p ← last
do
    terminal[p] ← TRUE
    p ← f[p]
while p ≠ NIL
return T
```

Fig. 3.29. Construction of $\mathcal{A}(y)$.

Example 20. One step: from $\mathcal{A}$(ccccbbccc) to $\mathcal{A}$(ccccbbcccd)

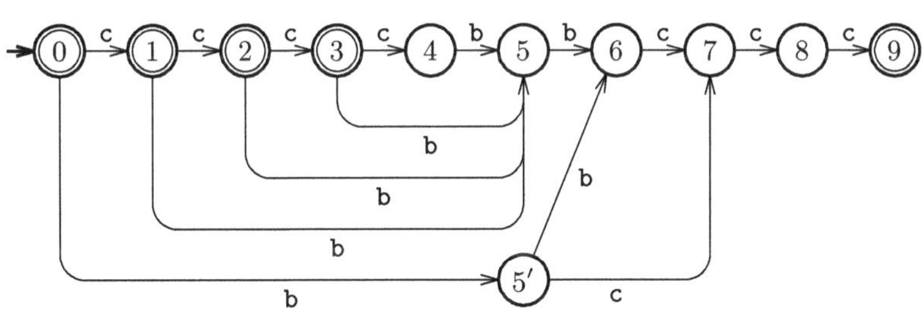

```
Extension(a)
new ← New-state()
L[new] ← L[last] + 1
p ← last
do
   Adj[p] ← Adj[p] ∪ {(a, new)}
   p ← f[p]
while p ≠ nil and Target(p, a) = nil
if p = nil then
   f[new] ← initial[T]
else q ← Target(p, a)
   if (p, a, q) is solid, i.e. L[p] + 1 = L[q] then
      f[new] ← q
   else clone ← New-state()
      L[clone] ← L[p] + 1
      for each pair (b, q') ∈ Succ[q] do
         Adj[clone] ← Adj[clone] ∪ {(b, q')}
      f[new] ← clone
      f[clone] ← f[q]
      f[q] ← clone
      do
         Adj[p] ← Adj[p] \ {(a, q)}
         Adj[p] ← Adj[p] ∪ {(a, clone)}
         p ← f[p]
      while p ≠ nil and Target(p, a) = q
return new
```

Fig. 3.30. Construction of $\mathcal{A}(wa)$ from $\mathcal{A}(w)$ for w a prefix of y.

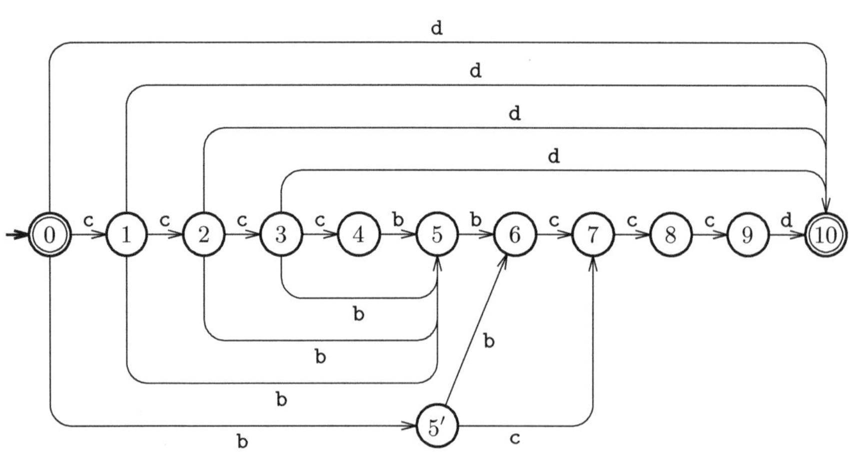

New arcs are created from states of the suffix path $\langle 9, 3, 2, 1, 0 \rangle$.
From $\mathcal{A}$(ccccbbccc) to $\mathcal{A}$(ccccbbcccc)

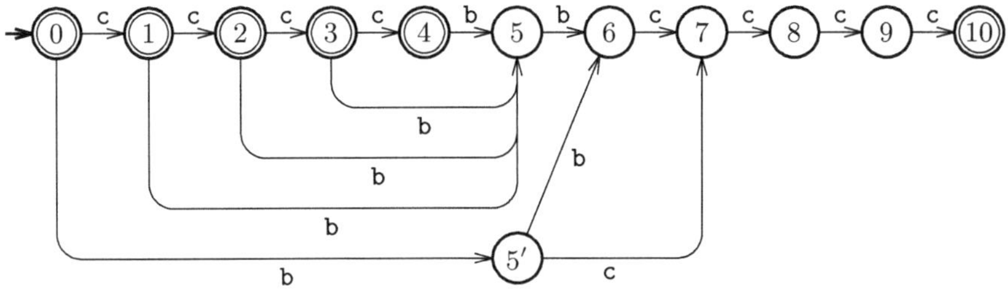

$f[9] = 3$ and $(3, \mathtt{c}, 4)$ is a solid arc (not a shortcut) then, $f[10] = $ TARGET$(3, \mathtt{c}) = 4$.

From $\mathcal{A}(\mathtt{ccccbbccc})$ to $\mathcal{A}(\mathtt{ccccbbcccb})$

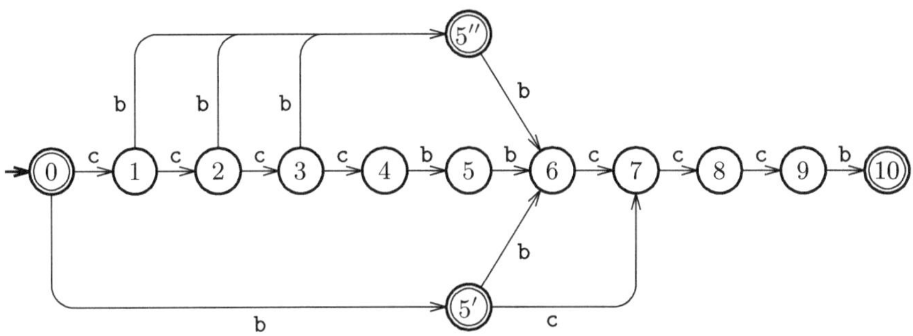

$f[9] = 3$, $(3, \mathtt{b}, 5)$is a non-solid arc , ccccb is a suffix but ccccb is not: state 5 is cloned into $5'' = f[10] = f[5]$, $f[5''] = 5'$. Arcs $(3, \mathtt{b}, 5)$, $(2, \mathtt{b}, 5)$ et $(1, \mathtt{b}, 5)$ are redirected onto $5''$.

3.3.6 Compact Suffix Automaton

The suffix tree results from a compaction of the suffix trie while the minimal suffix automaton results from a minimization of the suffix trie. Minimizing the suffix tree or compacting the minimal suffix automaton results in the same structure called the compact suffix automaton.

Example 21. Compact suffix automaton of ababbb

i	0	1	2	3	4	5
$y[i]$	a	b	a	b	b	b

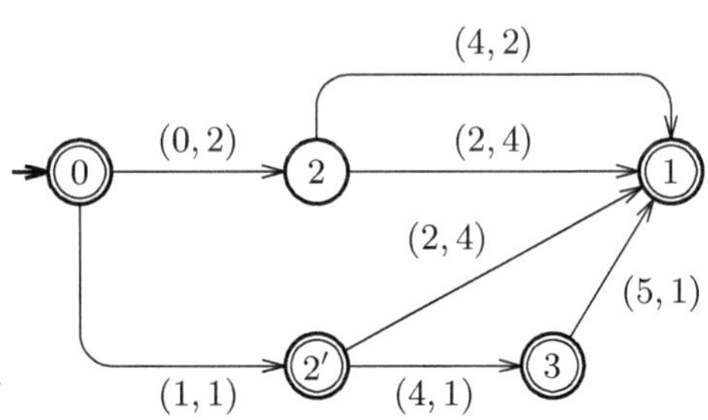

The size of the compact suffix automaton of a string y is linear in the length of y.

Direct Construction of the Compact Suffix Automaton

The direct construction of the compact suffix automaton is similar to both the suffix tree construction or the suffix automaton construction [14]. It consists of the sequential addition of suffixes in the structure from the longest one (y) to the shortest one (λ).

It uses the following features:

- "slow-find" and "fast-find" procedures;
- suffix links;
- solid and non-solid arcs;
- state splitting;
- re-directions of arcs.

The compact suffix automaton can be built in $O(n \log card(V))$ time using $O(n)$ space. In practice it can save up to 50% space on the suffix automaton [26].

References

1. A. Andersson and S. Nilsson. Improved behavior of tries by adaptive branching, *Inf. Process. Lett.*, 46, 6 (1993), 295–300.
2. A. Apostolico and R. Giancarlo. The Boyer-Moore-Galil string searching strategies revisited, *SIAM J. Comput.*, 15, 1 (1986), 98–105.
3. A. Blumer, J. Blumer, A. Ehrenfeucht, D. Haussler and R. McConnel. Linear size finite automata for the set of all subwords of a word: an outline of results, *Bull. Eur. Assoc. Theor. Comput. Sci.*, 21 (1983), 12–20.
4. R. S. Boyer and J. S. Moore. A fast string searching algorithm, *Commun. ACM*, 20, 10 (1977), 762–772.
5. D. Breslauer, L. Colussi and L. Toniolo. Tight comparison bounds for the string prefix-matching problem, *Inf. Process. Lett.*, 47, 1 (1993), 51–57.
6. R. Cole. Tight bounds on the complexity of the Boyer-Moore string matching algorithm, *SIAM J. Comput.*, 23, 5 (1994), 1075–1091.
7. R. Cole, R. Hariharan, M. Paterson, and U. Zwick. Tighter lower bounds on the exact complexity of string matching, *SIAM J. Comput.*, 24, 1 (1995), 30–45.
8. M. Crochemore. Linear searching for a square in a word, *Bull. Eur. Assoc. Theor. Comput. Sci.*, 24 (1984), 66–72.
9. M. Crochemore and D. Perrin. Two-way string-matching, *J. Assoc. Comput. Mach.*, 38, 3 (1991), 651–675.
10. M. Crochemore. String-matching on ordered alphabets, *Theor. Comput. Sci.*, 92, 1 (1992), 33–47.
11. M. Crochemore, A. Czumaj, L. Gąsieniec, S. Jarominek, T. Lecroq, W. Plandowski and W. Rytter. Speeding up two string matching algorithms, *Algorithmica*, 12, 4/5 (1994), 247–267.
12. M. Crochemore and T. Lecroq. Tight bounds on the complexity of the Apostolico-Giancarlo algorithm, *Inf. Process. Lett.*, 63, 4 (1997), 195–203.
13. M. Crochemore, L. Gąsieniec, and W. Rytter. Constant-space string-matching in sublinear average time, *Theor. Comput. Sci.*, 218, 1 (1999), 197–203.

14. M. Crochemore and R. Vérin. Direct construction of compact directed acyclic word graphs, *Proceedings of the 8th Annual Symposium on Combinatorial Pattern Matching*, (A. Apostolico and J. Hein, eds.), LNCS 1264, Springer-Verlag, Berlin, 1997, 116–129.
15. M. Farach. Optimal suffix tree construction with large alphabets, *Proceedings of the 38th IEEE Annual Symposium on Foundations of Computer Science*, Miami Beach, FL, 1997, 137–143.
16. P. Ferragina and R. Grossi. The string B-tree: A new data structure for string search in external memory and its applications, *J. Assoc. Comput. Mach.*, 46, 2 (1999), 236–280.
17. Z. Galil. On improving the worst case running time of the Boyer-Moore string searching algorithm, *Commun. ACM*, 22, 9 (1979), 505–508.
18. Z. Galil and J. Seiferas. Time-space optimal string matching, *J. Comput. Syst. Sci.*, 26, 3 (1983), 280–294.
19. Z. Galil and R. Giancarlo. On the exact complexity of string matching: lower bounds, *SIAM J. Comput.*, 20, 6 (1991), 1008–1020.
20. L. Gąsieniec, W. Plandowski and W. Rytter. Constant-space string matching with smaller number of comparisons: sequential sampling, *Proceedings of the 6th Annual Symposium on Combinatorial Pattern Matching*, (Z. Galil and E. Ukkonen, eds.), LNCS 937, Springer-Verlag, Berlin, 1995, 78–89.
21. G. H. Gonnet. The PAT text searching system, Report, Department of Computer Science, University of Waterloo, Ontario, 1987.
22. R. Grossi, A. Gupta and J. S. Vitter. When indexing equals compression: experiments with compressing suffix array and applications, *Proceedings of the 15th ACM-SIAM Annual Symposium on Discrete Algorithms*, (J. I. Munro, ed.), New Orleans, LO, 2004, 636–645.
23. L. J. Guibas and A. M. Odlyzko. A new proof of the linearity of the Boyer-Moore string searching algorithm, *SIAM J. Comput.*, 9, 4 (1980, 672–682.
24. C. Hancart. On Simon's string searching algorithm, *Inf. Process. Lett.*, 47, 2 (1993), 95–99.
25. C. Hancart, 1996. Personal communication.
26. J. Holub and M. Crochemore. On the implementation of compact dawg's, *Proceedings of the 7th Conference on Implementation and Application of Automata*, (J.-M. Champarnaud and D. Morel, eds.), LNCS 2608, Springer-Verlag, Berlin, 2002, 289–294.
27. S. Inenaga, H. Hoshino, A. Shinohara, M. Takeda, S. Arikawa, G. Mauri and G. Pavesi. On-line construction of compact directed acyclic word graphs, *Proceedings of the 12th Annual Symposium on Combinatorial Pattern Matching*, (A. Amir and G. M. Landau, eds.), LNCS 2089, Springer-Verlag, Berlin, 2001, 169–180.
28. R. W. Irving. Suffix binary search trees, Technical report, University of Glasgow, Computing Science Department, 1996.
29. J. Kärkkäinen. Suffix cactus: a cross between suffix tree and suffix array, *Proceedings of the 6th Annual Symposium on Combinatorial Pattern Matching*, (Z. Galil and E. Ukkonen, eds.), LNCS937, Springer-Verlag, Berlin, 1995, 191–204.
30. J. Kärkkäinen and P. Sanders. Simple linear work suffix array construction, *Proceedings of the 30th International Colloquium on Automata, Languages and Programming*, LNCS 2719, Springer-Verlag, Berlin, 2003, 943–955.

31. D. K. Kim, J. S. Sim, H. Park and K. Park. Linear-time construction of suffix arrays, *Proceedings of the 14th Annual Symposium on Combinatorial Pattern Matching*, (R. A. Baeza-Yates, E. Chávez, and M. Crochemore, eds.), LNCS 2676, Springer-Verlag, Berlin, 2003, 186–199.
32. P. Ko and S. Aluru. Space efficient linear time construction of suffix arrays, *Proceedings of the 14th Annual Symposium on Combinatorial Pattern Matching*, (R. A. Baeza-Yates, E. Chávez, and M. Crochemore, eds.), LNCS 2676, Springer-Verlag, Berlin, 2003, 200–210.
33. D. E. Knuth, J. H. Morris, Jr and V. R. Pratt. Fast pattern matching in strings, *SIAM J. Comput.*, 6, 1 (1977), 323–350.
34. U. Manber and G. Myers. Suffix arrays: a new method for on-line string searches, *SIAM J. Comput.*, 22, 5 (1993), 935–948.
35. E. M. McCreight. A space-economical suffix tree construction algorithm, *J. Algorithms*, 23, 2 (1976), 262–272.
36. J. H. Morris, Jr and V. R. Pratt. A linear pattern-matching algorithm, Report 40, University of California, Berkeley, 1970.
37. J. I. Munro, V. Raman and S. S. Rao. Space efficient suffix trees, *J. Algorithms*, 39, 2 (2001), 205–222.
38. I. Simon. Sequence comparison: some theory and some practice, *Electronic, Dictionaries and Automata in Computational Linguistics*, (M. Gross and D. Perrin, eds.), LNCS 377, Springer-Verlag, Berlin, 1989, 79–92.
39. E. Ukkonen. Approximate string matching with q-grams and maximal matches, *Theor. Comput. Sci.*, 92, 1 (1992), 191–212.
40. P. Weiner. Linear pattern matching algorithm, *Proceedings of the 14th Annual IEEE Symposium on Switching and Automata Theory*, Washington, DC, 1973, 1–11.
41. A. C. Yao. The complexity of pattern matching for a random string, *SIAM J. Comput.*, 8, 3 (1979), 368–387.

4

Quantum (Finite) Automata: An Invitation

Jozef Gruska*

Faculty of Informatics
Masaryk University
Botanická 68a, 602 00 Brno, Czech Republic
E-mail: gruska@fi.muni.cz

Summary. Quantum informatics is one of the most important new directions in informatics, especially in theoretical informatics, results of which have impacts far beyond the usual framework of (theoretical) informatics. Quantum automata form an interesting subarea of quantum informatics. Main models of quantum automata can be seen as quantum versions of the models of classical automata. However, also brand new quantum models of automata have been introduced and investigated. The main goal od this paper is to present basics of the basic models of quantum (mainly finite) automata, problems that have been and should be investigated and some of the basic results and methods in this area as well as a presentation of open problems and potential areas of research in quantum automata. All that is done in Section 3. Since the basic mathematical framework in which quantum automata theory is developed is so different from the one for models of classical automata, in Section 2 several fundamentals elements and primitives of quantum information processing are briefly introduced as well as some of the basic and important features and outcomes: quantum parallelism, quantum entanglement, quantum teleportation, quantum no-cloning theorem and quantum one-time pad cryptosystem. Moreover, a very brief general introduction to quantum information processing is given in Section 1 that includes discussion of the main current motivations and goals of the field, some of the very basic physical experiments that motivated development of quantum mechanics and the corresponding mathematical framework of Hilbert spaces. All that is to provide an invitation to this interesting area that is a natural and inspiring generalization of so powerful and diverse classical automata theory.

4.1 An Introduction to Quantum Information Processing

At the beginning of the quantum information processing and communication (QIPC) developments, one of the major motivations for the field was an observation that in quantum information processing we witness a merge of two of the most important areas of science of 20th century: quantum physics and

* Support of the GAČR grant 201/04/1153 and of the VEGA grant 1/7654/20 is to be acknowledged.

J. Gruska: *Quantum (Finite) Automata: An Invitation*, Studies in Computational Intelligence (SCI) **25**, 81–117 (2006)
www.springerlink.com

informatics, and therefore that we can expect in this new area results that will significantly influence science and technology of the 21th century.

Currently, there are five main reasons why QIPC is increasingly considered as of (very) large importance for science and technology:

- QIPC is believed to lead to new Quantum Information Processing Technology that could have deep and broad impacts.
- Several sciences and technology are approaching in their development the point at which they badly need expertise with isolation, manipulating and transmission of particles.
- It is increasingly believed that new, quantum information processing based, understanding of (complex) quantum phenomena and systems can (and should) be developed.
- Quantum cryptography seems to offer a new level of security, so called unconditional security, based not on unproven assumptions of complexity theory, as the classical public-key cryptography does, but on the laws and limitations of quantum physics, and can be soon feasible.
- QIPC has been proven to be more efficient in interesting/important cases.
- Theoretical computer science and Information theory got from QIPC a new dimension, importance and impulses.

The roots of QIPC are both in quantum physics and informatics, what is not so surprising because when properly seen these two areas have very close scientific goals.

Indeed, main scientific goals of physics can be seen as to study laws and limitations of the physical world and main scientific goals of informatics can be seen as to study laws and limitations of the information world.

How close are these two words is to be seen. In this context it is of interest the following view of the physicist John Wheeler: *I think of my lifetime in physics as divided into three periods. In the first period ...I was convinced that "everything is particle". I call my second period as "everything is field". Now I have a new vision, namely that "everything is information".*

In order to get a bit of insight into the quantum world, and into the mathematical framework we use to deal with it, we discuss in the rest of this section two basic quantum experiments and elements of the Hilbert space framework that correspond, in some way, to the physical concepts of quantum system. For more details see either the overview papers (Gruska, 1999a, Gruska, 2001) or the book Gruska (1999).

4.1.1 Physical Background - Basic Experiments and Motivations

So called *two-slit experiment*, due to T. Young in 1801, demonstrated a dual particle/wave role quantum particles exhibit and demonstrated also the existence of the *superposition principle* so fundamental for quantum world.

Let us have a source that shots electrons randomly in a wide range of directions and two walls, see Figure 4.1. Let the first wall has two tiny slits, each just large enough for one electron to get through at a time, perhaps changing its direction behind the slit, due to a reflection in the slit. The second wall has a detector that can be moved up and down and can count the number of electrons reaching a given position at the second wall. In this way we can experimentally determine probabilities that electrons reach given positions on the second wall. The results are shown in Figure 4.1b, by the curve $P_1(x)$ ($P_2(x)$) for the case that electrons reach the position x at the second wall and that only one slit, namely H_1 (H_2), is open. The results are as expected, the maximal values are exactly at points where the straight lines from the source through the slits reach the second wall. However, contrary to our intuition, in the case that both slits are open we get the curve $P_{12}(x)$, shown in Figure 4.1c. Very surprisingly, contrary to our intuition, at some places one observes fewer electrons when both slits are open than in the case only one slit is open! The resulting curve correspond to the one known from the classical physics in case a source produces waves and we witness a superposition of waves.

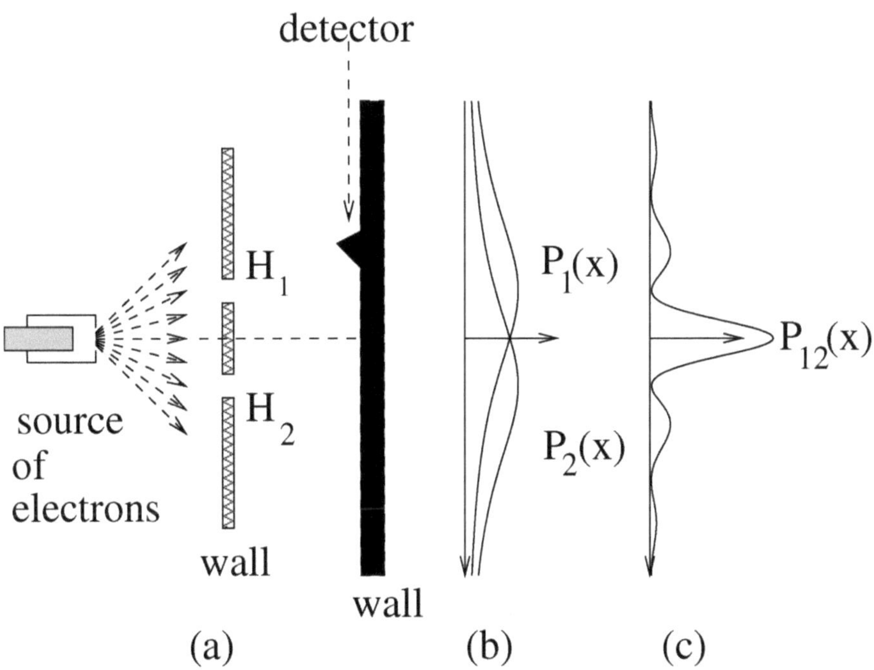

Fig. 4.1. Two slit experiment

There seem to be two surprising conclusions one can draw from these experimental results. By opening the second slit, it suddenly seems that electrons are somehow prevented from doing what they could do before! It seems that by allowing an electron to take an alternative route we have actually stopped it from traversing either route.

Electrons are particles, but they seem to have a wave-like behaviour as they pass through the slits! Each particle seems to behave as if it would be going through both holes at once and afterwards it would create waves that interfere, as in the classical wave experiment. (It is also important to emphasize that the results of this experiment do not depend on the frequency with which electrons are shot.)

In order to illustrate peculiarities of another surprising quantum phenomenon, quantum measurement, to be discussed next, we consider a modification of the previous experiment—with an observation (measurement) during the experiment.

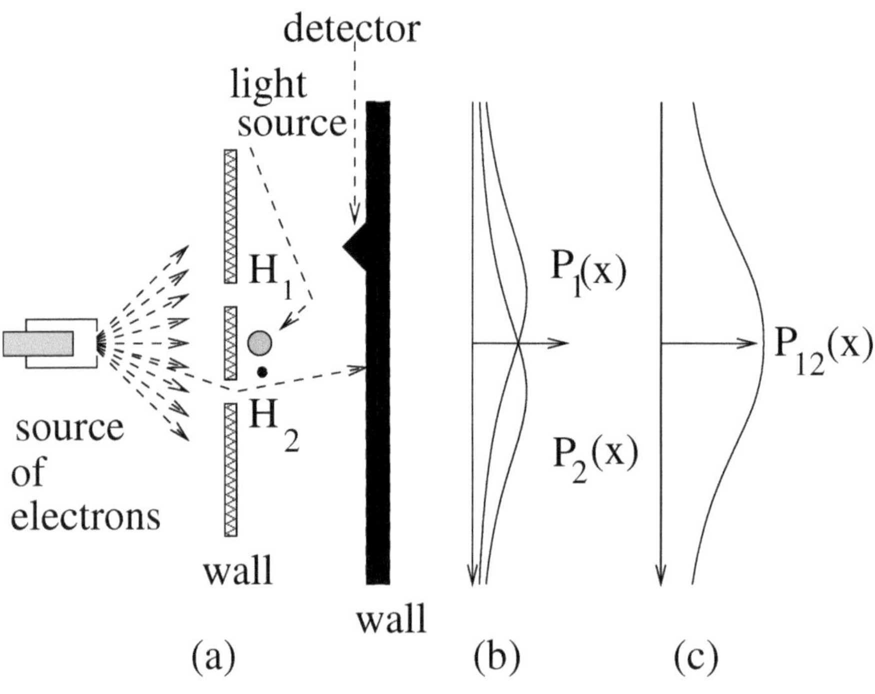

Fig. 4.2. Two-slit experiment with an observation

In the experiment depicted in Figure 4.2a, the basic setting is similar to that of the previous experiment. However, this time we have also, in addition, a source of light on the right hand side of the first wall, just in the middle between the two slits. If we now watch the experiment, we can detect (as indicated by the small black circle in the figure), through which slit a particular electron passes when going through the first wall. If an electron goes through the slit H_1, some light appears for a moment in the neighborhood of H_1, as a reflection; if an electron passes trough the slit H_2, we see some light near the slit H_2.

Also in this experiment we can determine probabilities that electrons reach positions on the second wall for the case where only one slit is open, Figure 4.2b, and for the case where both slits are open, Figure 4.2c. The curves in Figure 4.2b are as expected. However, the curve for the case where both slits are open is different from that in Figure 4.1c. This is again a counter-intuitive phenomenon. The explanation is that the resulting behaviour of electrons is due to the fact that we have been observing (or at least could observe) their behaviour by putting a light source next to the slits. In this case an **observation** or a **measurement** results in the lost of interference.

This way we have a particular case of a very well known phenomenon in the quantum world. A quantum system behaves differently if it is observed (measured) from when it is not! Moreover, the interference pattern disappears if we change our original electron experiment in some other way in order to find out which way electrons go. This can also be seen as another example of the uncertainty principle of quantum mechanics. Either you observe an interference pattern or you can find out which way the electron went, but you cannot do both. Seeing the interference pattern and detecting an electron are both measurements that cannot be performed in the same experiment—one has to choose one or the other. Observe also that detecting which slit an electron went through is sort of a "particle measurement"; recording the interference pattern is a "wave measurement". One can do one or the other, but not both.

As the last quantum experiment, let us consider, in an idealized form, one of the other famous experiments of quantum physics which demonstrates that some quantum phenomena are not determined except when they are measured.

The experiment was performed in 1921 by Otto Stern and Walter Gerlach in Berlin (see Figure 4.3). They shot a beam of atoms with random magnetic orientation (thought of, for this experiment, as little bar magnets, with North and South Poles) through a magnetic field, graded by intensity from top to bottom. The magnetic field is created by two vertically positioned magnets, to sort atoms according to their magnetic orientation, and the result are recorded on a screen. It was expected that the atoms emerging from the magnetic field would be smeared out, when they hit the screen, into a range of all possible deflections, with all possible orientations of their magnetic axes (as it would be the case with real magnets). Instead of that, it was discovered that atoms emerging from the magnetic field struck the screen in exactly two places, each with only one orientation, say "up" or "down", each with equal probability, so they came up in a "half-up and half-down manner". Moreover, the same phenomenon appeared when the magnets themselves were turned ninety degrees so that their magnetic field was horizontal and not vertical. Again, the atoms hit the screen in exactly two spots, to the left and right of the original line of the beam and again with the same probability. We can say they came out in a "half-left and half-right" manner.

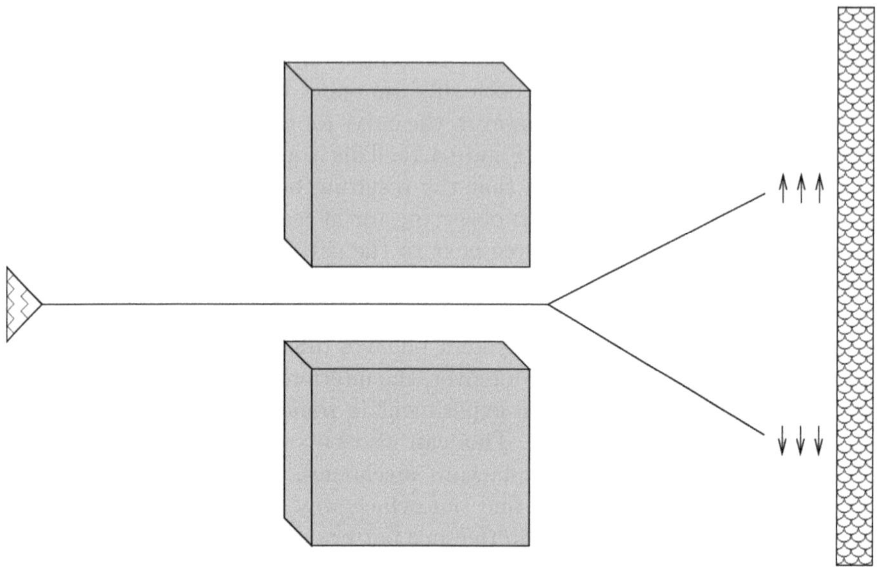

Fig. 4.3. Stern-Gerlach experiment with spin-$\frac{1}{2}$ particles

Actually, no matter how the magnetic field was lined up, it always split the beam of atoms into two. As if each atom was forced somehow to take up either one or the other of just two possible orientations, dictated by the alignment of the magnets.

It can be demonstrated that magnets do not physically 'sort' atoms passing through by directly manipulating their magnetic axes. The quantum theory explanation is the following one: Passing an atom through a magnetic field amounts to a measurement of its magnetic alignment, and until you make such a measurement there is no sense in saying what the atom's magnetic alignment might be; only when you make a measurement do you obtain one of only two possible outcomes, with equal probability, and those two possibilities are defined by the direction of the magnetic field that you use to make the measurement.

Conclusions From Experiments

It is easy to explain mathematically our first superposition/interference experiment if we accept the following three basic principles concerning transfers of quantum states.

P1 To each transfer from a quantum state ϕ to a state ψ a complex number

$$\langle\psi|\phi\rangle$$

is associated, which is called the **probability amplitude** of the transfer, such that

$$|\langle\psi|\phi\rangle|^2$$

is the **probability** of the transfer.

P2 If a transfer from a quantum state ϕ to a quantum state ψ can be decomposed into two subsequent transfers

$$\psi \leftarrow \phi' \leftarrow \phi,$$

then the resulting amplitude of the transfer is the **product** of amplitudes of sub-transfers: $\langle\psi|\phi\rangle = \langle\psi|\phi'\rangle\langle\phi'|\phi\rangle$

P3 If the transfer from a state ϕ to a state ψ has two independent alternatives with amplitudes α and β

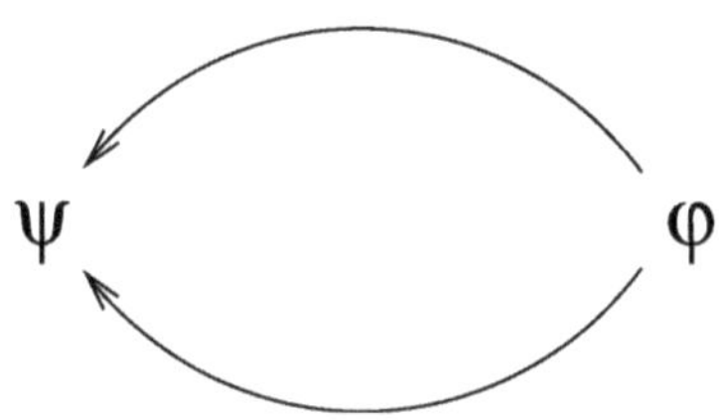

then the resulting amplitude of the whole transfer is $\alpha + \beta$ what is the sum of amplitudes of two alternative sub-transfers.

For example, in case of the first two-split experiment an electron can reach each position on the second wall in two independent ways - going through first or second slit. To each such transfer a probability amplitude is assigned and in the case these two probability amplitudes are some α and $-\alpha$, the probability that the electron reaches that position is zero! That is why at the two-split experiment there are, at some positions on the second wall, zero probabilities that electrons reach that positions.

The second two-split experiments demonstrates another important fact. Namely, that undetectable eavesdropping is impossible. This is behind one of the most famous and important results of quantum cryptography: there are unconditionally secure quantum protocols to generate secret shared classical keys at which undetectable eavesdropping is impossible.

Stern-Gerlach experiment shows that at quantum measurements states that are being measured collapse in a random way into one of the fixed number of states and to each one with the probability that can be precomputed, and one gets into the classical world only information which of the possible collapses occurred.

4.1.2 Mathematical Background – Hilbert Space Framework

Surprisingly, apparently very complex quantum physical world has, in principle, a simple framework in the mathematical world, that of the Hilbert space. However, relation between these two worlds, mathematical and physical, is far from simple and this problem, so called interpretation problem of the mathematical concepts of Hilbert space framework in the physical world, make busy, already for about 60 years, some of the best minds in the field.

A Hilbert space will be for us a complex vector space, finitely or infinitely dimensional. Mostly we will be dealing with finitely dimensional Hilbert spaces. An n-dimensional Hilbert space will be denoted $\mathcal{H}_n$. In case a set D is finite, by $\mathcal{H}_D$ we will denote Hilbert space of the dimension $|D|$ (cardinality of D), whose vector's elements will be indexed by elements of D.

So called Dirac's handy "bra-ket notation", so much used in QIPC, has three basic elements: for any states $\phi = (a_1, \ldots, a_n)$ and $\psi = (b_1, \ldots, b_n)$.

$|\phi\rangle$ — **ket-vector** – an equivalent to ϕ
$\langle\psi|\phi\rangle = \sum_{i=1}^{n} b_i^* a_i$ – **scalar product** (**inner-product**) of ψ and ϕ (an amplitude of going from the state ϕ to ψ).[2]
$\langle\psi|$ – **bra-vector** – a linear functional on H such that $\langle\psi|(|\phi\rangle) = \langle\psi|\phi\rangle$.

Using scalar products we can define on H_n the norm of a vector $|\phi\rangle$ as

$$||\phi|| = \sqrt{|\langle\phi|\phi\rangle|}$$

and the **metric**

$$\mathrm{dist}(\phi, \psi) = ||\phi - \psi||.$$

This allows to introduce on H a topology and such concepts as continuity.

Mathematical concept of n-dimensional Hilbert space corresponds to the physical concept of n-level quantum system. Vectors of norm 1 correspond to the physical concept of quantum state and will often be called quantum states.

Two quantum states $|\phi\rangle$ and $|\psi\rangle$ are called orthogonal if $\langle\phi|\psi\rangle = 0$. Orthogonality is a very important concept because two quantum states are physically perfectly distinguishable if and only if they are orthogonal.

It can be shown that each n-dimensional Hilbert space has (infinitely many) bases such that all vectors of each of such bases have norm 1 and are mutually orthogonal. In the following we will consider only such bases.

Of importance are also orthogonal decompositions of Hilbert spaces $\mathcal{H}_n$ to several mutually orthogonal subspaces $S_1, \ldots, S_k$, notation $H_n = S_1 \oplus S_2 \oplus \ldots \oplus S_k$. This refers to the case that each vector of H_n can be uniquely expressed as a union of vectors from mutually orthogonal subspaces $S_1, \ldots, S_k$.

---

[2] b_i^* is complex conjugate to b_i.

Quantum Evolution

Evolution in a quantum system $\mathcal{S}$ is described by Schrödinger linear equation

$$i\hbar\frac{\partial\psi(t)}{\partial t} = H(t)\psi(t),$$

where $\hbar$ is the Planck constant, $H(t)$ is a Hamiltonian of the system (which is a quantum analogue of a Hamiltonian in the classical mechanics—the total energy of the system—and can be represented by a Hermitian matrix) at the time t, and $|\psi(t)\rangle$ is the state of $\mathcal{S}$ at the time t.

In case that the Hamiltonian is time independent, the formal solution of the Schrödinger equation has the form

$$|\psi(t)\rangle = \mathrm{U}(t)|\psi(0)\rangle$$

where

$$\mathrm{U}(t) = e^{-Ht/\hbar}$$

is the evolution operator which can be represented by a unitary matrix. (A matrix U is unitary if $UU^\dagger = U^\dagger U = I$.) From that it follows that evolution (computation) of a quantum system is performed by a **unitary operator** and a (discrete) step of such an evolution we can see as a multiplication of a **unitary matrix** A with a vector $|\psi\rangle$, i.e.

$$A|\psi\rangle$$

Unitary operators map quantum states into quantum states in a reversible way. Indeed, for any unitary operation U it holds $U^\dagger U|\phi\rangle = |\phi\rangle$ for any state $|\phi\rangle$.

Another important class of operators that map quantum states into quantum states are **projectors**. They are always related to some subspaces of the given Hilbert space. Given a subspace S_0 of a quantum system $\mathcal{S}$ each state $|\phi\rangle$ can be uniquely decomposed into a state $|\phi_0\rangle$ from $\mathcal{S}_0$ and another state from the orthogonal subspace. The projector related to the subspace $\mathcal{S}_0$ projects then an arbitrary state $|\phi\rangle$ into its unique component $|\phi_0\rangle \in S + 0$. All projectors P have an important property $P^2 = P$. Projectors indeed map states into states, but, in general, in an irreversible way. Projectors play an important role in quantum measurements, to be discussed next.

Quantum Measurement

One of the most puzzling quantum phenomena is quantum measurement. It can be defined in several ways and three of them, discussed next, will play important role in our main chapter.

Measurement with respect to a basis. In case an orthonormal basis $\{\beta_i\}_{i=1}^n$ is chosen in a Hilbert space $\mathcal{H}_n$, any state $|\phi\rangle \in \mathcal{H}_n$ can be expressed in the form

$$|\phi\rangle = \sum_{i=1}^{n} a_i|\beta_i\rangle, \quad \sum_{i=1}^{n} |a_i|^2 = 1,$$

where $a_i = \langle\beta_i|\phi\rangle$ are probability amplitudes and their squares provide *probabilities* that if the state $|\phi\rangle$ is measured with respect to the basis $\{\beta_i\}_{i=1}^n$, then the state $|\phi\rangle$ collapses into the state $|\beta_i\rangle$ with probability $|a_i|^2$. The classical "outcome" of a measurement of the state $|\phi\rangle$ with respect to the basis $\{\beta_i\}_{i=1}^n$ is then the index i of that state $|\beta_i\rangle$ into which the state $|\phi\rangle$ collapses.

Measurement with respect to an orthogonal decomposition of $\mathcal{H}_n$ Let $\mathcal{H}_n = S_1 \oplus \ldots \oplus S_k$, where $S_1, \ldots, S_k$ are mutually orthogonal subspaces of $\mathcal{H}_n$. Any state $|\phi\rangle$ can then be uniquely expressed in the form

$$|\phi\rangle = \sum_{i=1}^{k} \psi_i,$$

where $\psi_i \in S_i$ is a vector, not necessarily of norm one. In such a case measurement of $|\phi\rangle$ with respect to the decomposition $S_1 \oplus \ldots \oplus S_k$ results, on the quantum level, in the collapse (projection) of $|\phi\rangle \in S_i$ to one of the states ψ_i, after normalization, and that happens with the probability $\langle|\psi_i|P_i|\psi_i\rangle$, where P_i is the projector to the subspace S_i, and to the classical world one gets information ("i") which of such collapses happened. In a special case that subspaces S_i are one dimensional and generated by vectors of an orthonormal bases, the measurement just discussed is reduced to the one considered above.

Measurement with respect to a Hermitian matrix A.[3] Each Hermitian matrix A has a unique (so called spectral) decomposition

$$A = \sum_{i=1}^{k} \lambda_k P_k,$$

where λ_i, $1 \leq i \leq k$, are all different eigenvalues of A and P_i is projector to the subspace generated by eigenvectors corresponding to the eigenvalue λ_i. A measurement of a state $|\phi\rangle$, with respect to such a Hermitian matrix (called also an observable) A is then a random projection of $|\phi\rangle$ by one of the projectors P_i, what happens with probability $\langle\phi|P_i|\phi\rangle$ and to the classical world information is delivered which of the projections took place (what is usually interpreted that the classical outcome of the measurement is the eigenvalue λ_i).

Compound Quantum Systems

have essentially different properties than classical compound systems. The existence of so called entangled states that exhibits non-locality is perhaps the most special feature of such states.

In order to introduce mathematical framework for compound systems the concept of the tensor product is of the key importance.

Tensor product of vectors is defined by

$$(x_1,\ldots,x_n)\otimes(y_1,\ldots,y_m)=(x_1y_1,\ldots,x_1y_m,x_2y_1,\ldots,x_2y_m,\ldots,x_ny_1,\ldots,x_ny_m)$$

Tensor product of matrices is defined by

$$A\otimes B=\begin{pmatrix} a_{11}B \ldots a_{1n}B \\ \vdots \qquad \vdots \\ a_{n1}B \ldots a_{nn}B \end{pmatrix}$$

where

$$A=\begin{pmatrix} a_{11} \ldots a_{1n} \\ \ldots \qquad \ldots \\ a_{n1} \ldots a_{nn} \end{pmatrix}$$

Example 1

$$\begin{pmatrix} 1 & 0 \\ 0 & 1 \end{pmatrix} \otimes \begin{pmatrix} a_{11} & a_{12} \\ a_{21} & a_{22} \end{pmatrix} = \begin{pmatrix} a_{11} & a_{12} & 0 & 0 \\ a_{21} & a_{22} & 0 & 0 \\ 0 & 0 & a_{11} & a_{12} \\ 0 & 0 & a_{21} & a_{22} \end{pmatrix}$$

$$\begin{pmatrix} a_{11} & a_{12} \\ a_{21} & a_{22} \end{pmatrix} \otimes \begin{pmatrix} 1 & 0 \\ 0 & 1 \end{pmatrix} = \begin{pmatrix} a_{11} & 0 & a_{12} & 0 \\ 0 & a_{11} & 0 & a_{12} \\ a_{21} & 0 & a_{22} & 0 \\ 0 & a_{21} & 0 & a_{22} \end{pmatrix}$$

Tensor product of two Hilbert spaces $H_n \otimes H_m$ is the complex vector space of dimension mn spanned by tensor products of vectors (states) from H_n and H_m. $\mathcal{H}_n \otimes H_m$ corresponds to the quantum system composed of the quantum systems corresponding to Hilbert spaces H_n and H_m.

An important difference between classical and quantum systems is the fact that a state of a compound classical (quantum) system can be (cannot be) always composed from the states of the subsystems. This will be discussed later.

4.2 Basics of Quantum Information Processing

In this section several basic concepts of quantum information processing as well as several very basic but very important results and methods are introduced. For more see Gruska (1999).

4.2.1 Qubits

A qubit is an element of the set of states

$$|\phi\rangle = \alpha|0\rangle + \beta|1\rangle$$

where α, β are complex numbers such that $|\alpha|^2 + |\beta|^2 = 1$, that lie in two dimensional Hilbert space $\mathcal{H}_2$ in which $\{|0\rangle, |1\rangle\}$ is a **(standard) basis.**

Vector notation of the standard basis is

$$|0\rangle = \begin{pmatrix} 1 \\ 0 \end{pmatrix}, \quad |1\rangle = \begin{pmatrix} 0 \\ 1 \end{pmatrix}.$$

Another important basis is *dual basis* with vectors

$$|0'\rangle = \frac{1}{\sqrt{2}}(|0\rangle + |1\rangle), \quad |1'\rangle = \frac{1}{\sqrt{2}}(|0\rangle - |1\rangle).$$

A very important unitary operation on qubits is Hadamard operation represented in the standard basis by the matrix

$$H = \frac{1}{\sqrt{2}} \begin{pmatrix} 1 & 1 \\ 1 & -1 \end{pmatrix}$$

Hadamard operation maps standard basis into dual and vice versa. Indeed,

$$H|0\rangle = |0'\rangle \qquad H|0'\rangle = |0\rangle$$

$$H|1\rangle = |1'\rangle \qquad H|1'\rangle = |1\rangle$$

A general form of a unitary matrix of degree 2, that is a unitary one-qubit operation, is

$$U = e^{i\gamma} \begin{pmatrix} e^{i\alpha} & 0 \\ 0 & e^{-i\alpha} \end{pmatrix} \begin{pmatrix} \cos\theta & i\sin\theta \\ i\sin\theta & \cos\theta \end{pmatrix} \begin{pmatrix} e^{i\beta} & 0 \\ 0 & e^{-i\beta} \end{pmatrix}.$$

Measurement of Qubits

A qubit state $\alpha|0\rangle + \beta|1\rangle$ can "contain" unbounded large amount of classical information in α and β. However, **an unknown quantum state** cannot be identified, what means that there is no way to determine α and β and to get classical information hidden in α and β. Indeed, by a **measurement** of the qubit state

$$\alpha|0\rangle + \beta|1\rangle,$$

with respect to the basis

$$\{|0\rangle, |1\rangle\},$$

we can obtain only one bit of classical information and only in the following random way:

$$0 \text{ with probability } |\alpha|^2$$
$$1 \text{ with probability } |\beta|^2$$

In the first (second) case the state collapses into the state $|0\rangle$ ($|1\rangle$).

On the other side, if the state

$$|\phi\rangle = \alpha|0\rangle + \beta|1\rangle$$

is measured with respect to the dual basis, then, since

$$|0'\rangle = \frac{1}{\sqrt{2}}(|0\rangle + |1\rangle) \qquad |1'\rangle = \frac{1}{\sqrt{2}}(|0\rangle - |1\rangle)$$

and therefore

$$|0\rangle = \frac{1}{\sqrt{2}}(|0'\rangle + |1'\rangle) \qquad |1\rangle = \frac{1}{\sqrt{2}}(|0'\rangle - |1'\rangle),$$

we have

$$|\phi\rangle = \frac{1}{\sqrt{2}}((\alpha + \beta)|0'\rangle + (\alpha - \beta)|1'\rangle)$$

what implies that measurement of $|\phi\rangle$ with respect to the dual basis provides

$$0 \text{ with probability } \tfrac{1}{2}|\alpha + \beta|^2$$

or

$$1 \text{ with probability } \tfrac{1}{2}|\alpha - \beta|^2.$$

Pauli Operators

Very important one-qubit unary operators are *Pauli operators*, that are expressed in the standard basis as follows;

$$\sigma_x = \begin{pmatrix} 0 & 1 \\ 1 & 0 \end{pmatrix}, \quad \sigma_y = \begin{pmatrix} 0 & -i \\ i & 0 \end{pmatrix}, \quad \sigma_z = \begin{pmatrix} 1 & 0 \\ 0 & -1 \end{pmatrix}$$

Observe that Pauli operators transform a qubit state $|\phi\rangle = \alpha|0\rangle + \beta|1\rangle$ as follows

$$\sigma_x(\alpha|0\rangle + \beta|1\rangle) = \beta|0\rangle + \alpha|1\rangle \quad \sigma_z(\alpha|0\rangle + \beta|1\rangle) = \alpha|0\rangle - \beta|1\rangle$$

and for $\sigma'_y = \sigma_x\sigma_z$

$$\sigma'_y(|\alpha|0\rangle + \beta|1\rangle) = \beta|0\rangle - \alpha|1\rangle.$$

Operators σ_x, σ_z and σ'_y represent therefore a *bit error*, a *sign error* and a *bit-sign error*.

4.2.2 Mixed States - Density Matrices

A probability distribution $\{(p_i, |\phi_i\rangle)\}_{i=1}^k$ on pure states is called a *mixed state* to which one assigns a so called *density operator*

$$\rho = \sum_{i=1}^{k} p_i |\phi_i\rangle\langle\phi_i|.$$

One interpretation of a mixed state $\{(p_i, |\phi_i\rangle)\}_{i=1}^k$ is that a source X produces the state $|\phi_i\rangle$ with probability p_i. Any matrix representing a density operator is called density matrix. Very important are facts that the same density matrix can correspond to two different mixed states and that two mixed states with the same density matrix are physically undistinguishable.

Example 2 To the (maximally) mixed state

$$(\frac{1}{2}, |0\rangle), (\frac{1}{2}, |1\rangle)$$

which represents a random bit, corresponds the density matrix

$$\frac{1}{2}\begin{pmatrix}1\\0\end{pmatrix}(1,0) + \frac{1}{2}\begin{pmatrix}0\\1\end{pmatrix}(0,1) = \frac{1}{2}\begin{pmatrix}1\,0\\0\,1\end{pmatrix} = \frac{1}{2}I_2.$$

Surprisingly, as it will be seen below, many other important mixed states have as their density matrix that of the maximally mixed state.

4.2.3 A Sase Study – Quantum One-time-pad Cryptosystem

It is quite surprising that using only the concepts of qubits and of mixed states we can already demonstrate an important and elegant application of quantum tools and laws in quantum cryptography. Namely, we show that a natural quantum version of the classically perfectly secure one-time pad cryptosystem results in an absolutely secure quantum one-time pad cryptosystem.

Classical one-time-pad cryptosystem has the following elements: plaintext: an n-bit string p; shared key: an n-bit string k; cryptotext: an n-bit string c and the following encoding/decoding rules:

$$\text{encoding:} \quad c = p \oplus k$$
$$\text{decoding:} \quad p = c \oplus k$$

Quantum one-time-pad cryptosystem has the following elements: plaintext: an n-qubit string $|p\rangle = |p_1\rangle \dots |p_n\rangle$; shared key: two n-bit strings k, k'; cryptotext: an n-qubit string $|c\rangle = |c_1\rangle \dots |c_n\rangle$

and the following encoding/decoding rules:

$$\text{encoding:} \quad |c_i\rangle = \sigma_x^{k_i}\sigma_z^{k'_i}|p_i\rangle$$
$$\text{decoding:} \quad |p_i\rangle = \sigma_z^{k'_i}\sigma_x^{k_i}|c_i\rangle$$

This encoding is absolutely secure. Indeed, in the case of encryption of a qubit

$$|\phi\rangle = \alpha|0\rangle + \beta|1\rangle$$

by the just introduced quantum one-time pad cryptosystem, what is being transmitted is the mixed state

$$(\frac{1}{4}, |\phi\rangle), (\frac{1}{4}, \sigma_x|\phi\rangle).(\frac{1}{4}, \sigma_z|\phi\rangle), (\frac{1}{4}, \sigma_x\sigma_z|\phi\rangle)$$

the density matrix of which is

$$\frac{1}{2}I_2.$$

which is identical to the density matrix corresponding to that of a random bit, that is to the mixed state $(\frac{1}{2}, |0\rangle), (\frac{1}{2}, |1\rangle)$.

This allows us to formulate a surprising quantum version of the classical "Shannon theorem": *n bits are necessary and sufficient to encrypt securely n bits.* Quantum version has the following form: *$2n$ classical bits are necessary and sufficient to encrypt securely n qubits.*

The above result is very surprising. Since a quantum bit $\alpha|0\rangle + \beta|1\rangle$ may contain infinitely large amount of the classical bits in α and β, it would seem that to hide perfectly one qubit we may need a lot of bits.

4.2.4 Quantum Registers

Hilbert space $\mathcal{H}_4$, called also *a two-qubit Hilbert space or register*, can be seen as tensor product of two one-qubit Hilbert spaces

$$\mathcal{H}_2 \otimes \mathcal{H}_2$$

and therefore one of its basis consists of the states

$$|0\rangle \otimes |0\rangle, \quad |0\rangle \otimes |1\rangle, \quad |1\rangle \otimes |0\rangle, \quad |1\rangle \otimes |1\rangle$$

that are usually denoted shortly as the states

$$|00\rangle, \quad |01\rangle, \quad |10\rangle, \quad |11\rangle,$$

Similarly, the states of the standard/computational basis of an n-qubit Hilbert space $\mathcal{H}_{2^n}$ are all states

$$|i_1 i_2 \dots i_n\rangle = |i_1\rangle \otimes \dots \otimes |i_n\rangle$$

where $i_k \in \{0, 1\}$. A general state of a 2-qubit register is:

$$|\phi\rangle = \alpha_{00}|00\rangle + \alpha_{01}|01\rangle + \alpha_{10}|10\rangle + \alpha_{11}|11\rangle$$

where

$$|\alpha_{00}|^2 + |\alpha_{01}|^2 + |\alpha_{10}|^2 + |\alpha_{11}|^2 = 1$$

and $|00\rangle, |01\rangle, |10\rangle, |11\rangle$ are vectors of the "standard" basis of H_4, i.e.

$$|00\rangle = \begin{pmatrix}1\\0\\0\\0\end{pmatrix} \quad |01\rangle = \begin{pmatrix}0\\1\\0\\0\end{pmatrix} \quad |10\rangle = \begin{pmatrix}0\\0\\1\\0\end{pmatrix} \quad |11\rangle = \begin{pmatrix}0\\0\\0\\1\end{pmatrix}$$

An important unitary matrix of the degree 4, to transform states of 2-qubit registers, is

$$CNOT = XOR = \begin{pmatrix}1\,0\,0\,0\\0\,1\,0\,0\\0\,0\,0\,1\\0\,0\,1\,0\end{pmatrix}.$$

It holds: CNOR$|x,y\rangle = |x, x \oplus y\rangle$, when $x, y \in \{0,1\}$.

The following *Bell states* are among the most important states of $\mathcal{H}_4$.

$$|\Phi^+\rangle = \frac{1}{\sqrt{2}}(|00\rangle + |11\rangle, \quad |\Phi^-\rangle = \frac{1}{\sqrt{2}}(|00\rangle - |11\rangle)$$

$$|\Psi^+\rangle = \frac{1}{\sqrt{2}}(|01\rangle + |10\rangle, \quad |\Psi^-\rangle = \frac{1}{\sqrt{2}}(|01\rangle - |10\rangle).$$

They form an orthogonal basis in $\mathcal{H}_4$.

A general form of a two-qubit state is

$$|\phi\rangle = \alpha_{00}|00\rangle + \alpha_{01}|01\rangle + \alpha_{10}|10\rangle + \alpha_{11}|11\rangle, \tag{4.1}$$

where $|a_{00}|^2 + |a_{01}|^2 + |a_{10}|^2 + |a_{11}|^2$.

We shall consider two different measurements of the above general state. The first one is measurement with respect to the basis $\{|00\rangle, |01\rangle, |10\rangle, |11\rangle\}$ that provides as the results:

$$\begin{array}{lll} |00\rangle \text{ and } 00 & \text{with probability} & |\alpha_{00}|^2 \\ |01\rangle \text{ and } 01 & \text{with probability} & |\alpha_{01}|^2 \\ |10\rangle \text{ and } 10 & \text{with probability} & |\alpha_{10}|^2 \\ |11\rangle \text{ and } 11 & \text{with probability} & |\alpha_{11}|^2. \end{array}$$

On the other hand, we can measure the state 4.1 with respect to the first or second qubit, that is with respect to the decomposition of $\mathcal{H}_4$ into orthogonal subspaces generated by states $\{|00\rangle, |01\rangle\}$ and $\{|10\rangle, |11\rangle\}$. As the result of the measurement of the first qubit one gets

$$0 \text{ with probability } |\alpha_{00}|^2 + |\alpha_{01}|^2$$

and $|\phi\rangle$ is reduced to the vector $\dfrac{\alpha_{00}|00\rangle + \alpha_{01}|01\rangle}{\sqrt{|\alpha_{00}|^2 + |\alpha_{01}|^2}}$

or 1 with probability $|\alpha_{10}|^2| + |\alpha_{11}|^2$

and $|\phi\rangle$ is reduced to the vector $\dfrac{\alpha_{10}|10\rangle + \alpha_{11}|11\rangle}{\sqrt{|\alpha_{10}|^2 + |\alpha_{11}|^2}}$

4.2.5 No-cloning Theorem

One of the most important, and at the same time very elementary, results of the quantum information theory says, informally only, that unknown quantum state cannot be cloned. More formally, no-cloning theorem says that there is no unitary transformation U such that for any qubit state $|\psi\rangle$

$$U(|\psi\rangle|0\rangle) = |\psi\rangle|\psi\rangle$$

Proof is very simple. Indeed, assume that such a U exists and for two different states $|\alpha\rangle$ and $|\beta\rangle$

$$U(|\alpha\rangle|0\rangle) = |\alpha\rangle|\alpha\rangle, \quad U(|\beta\rangle|0\rangle) = |\beta\rangle|\beta\rangle$$

If

$$|\gamma\rangle = \frac{1}{\sqrt{2}}(|\alpha\rangle + |\beta\rangle),$$

then

$$U(|\gamma\rangle|0\rangle) = \frac{1}{\sqrt{2}}(|\alpha\rangle|\alpha\rangle+|\beta\rangle|\beta\rangle) \neq |\gamma\rangle|\gamma\rangle = \frac{1}{2}(|\alpha\rangle|\alpha\rangle+|\beta\rangle|\beta\rangle+|\alpha\rangle|\beta\rangle+|\beta\rangle|\alpha\rangle).$$

4.2.6 Quantum Entanglement

Perhaps the most important "non-classical" quantum concept is that of *entangled states*: This concept concerns states of multipartite systems.

For a bipartite quantum system $\mathcal{H} = \mathcal{H}_A \otimes \mathcal{H}_B$, we say that its state $|\phi\rangle$ is *entangled* if it cannot be decomposed into a tensor product of a state from $\mathcal{H}_A$ and a state from $\mathcal{H}_B$.

For example, it is easy to verify that a two-qubit state

$$|\phi\rangle = a|00\rangle + b|01\rangle + c|10\rangle + d|11\rangle,$$

is not entangled, that is

$$|\phi\rangle = (x_1|0\rangle + y_1|1\rangle) \otimes (x_2|0\rangle + y_2|1\rangle)$$

if and only if $\frac{a}{b} = \frac{x_2}{y_2} = \frac{c}{d}$, that is if

$$ad - bc = 0.$$

Therefore, all Bell states are (important) examples of entangled states.

Importance of such entangled states as Bell states comes from the following observations.

Assume that two parties, Alice and Bob again, share two particles in the Bell state $|\Phi^+\rangle$, and they are (very) far apart.

If one of the parties measures his/her particle in the standard basis the state "collapses", with the same probability, either into the state $|00\rangle$ and the party gets as the classical outcome 0 (and so does the other party at his/her subsequent measurement, no matter when this happens), or into the state $|11\rangle$ and the party gets as the classical outcome 1 (and so does the other party at his/her measurement, no matter when that happens).

A measurement of one party therefore instantaneously predetermines the result of the subsequent measurement of the other party. The existence of such entangled states as $|\Phi^+\rangle$ allows therefore to create simultaneously the same random events by separated parties.

Moreover, at such a measurement we have therefore an example of a non-local impact of one physical phenomena on the other, what is against our common sense, formed by our education in the classical physics.

After its discovery, entanglement and its non-locality impacts have been seen as a peculiarity of the existing quantum theory that needs some modification to get rid of them, and as a source of all kind mysteries and counter-intuitive consequences.

Currently, after the discovery of quantum teleportation and of such powerful quantum algorithms as Shor's factorization algorithm, entanglement is seen and explored as a new and powerful quantum resource that allows

- to perform tasks that are not possible otherwise;
- to speed-up much some computations and to economize (even exponentially) some communications;
- to increase capacity of (quantum) communication channels;
- to implement perfectly secure information transmissions;
- to develop more general and more powerful theories of computations and communications than the framework of classical physics allows;
- to develop a new, better, information based, understanding of the key quantum phenomena and by that, a deeper, information processing based, understanding of Nature.

4.2.7 Quantum Gates and Circuits

Unitarity is the main new requirement quantum gates have to satisfy.

Definition 1 *A* **quantum gate** *with n inputs and n outputs is specified by a unitary operator $U : \mathcal{H}_{2^n} \rightarrow \mathcal{H}_{2^n}$, and it is represented by a unitary matrix A_U of degree 2^n.*

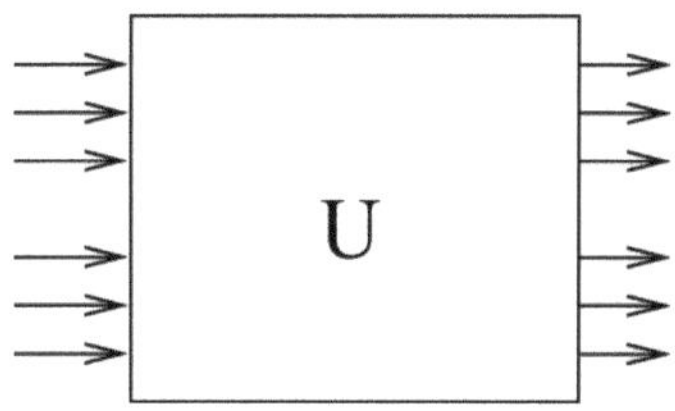

Example 3 The CNOT-gate (or XOR-gate) with CNOT operation is usually denoted as follows:

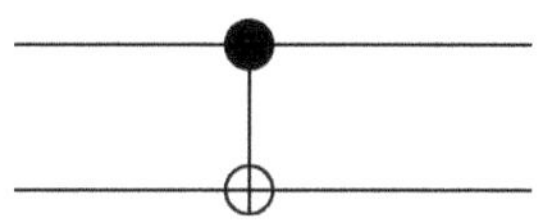

with "control bit" on the first qubit and with "target bit" on the second qubit"

Exercise 1 (a) Describe a unitary matrix for the "inverse CNOT gate" ($|x, y\rangle \leftrightarrow |x \oplus y, y\rangle$) in the standard basis. (b) Determine the unitary matrix for the CNOT gate expressed in the dual basis?

Transfers from quantum gates as mappings to unitary matrices are quite easy. Indeed, in general, if a quantum gate has n inputs and outputs then for the corresponding unitary matrix the entry in the column $x \in \{0,1\}^n$ and in the row $y \in \{0,1\}^n$ is the amplitude for transition from the basis state $|x\rangle$ to the basis state $|y\rangle$.

Quantum (unitary) circuits are defined in a similar way as classical circuits are, only gates of any quantum circuit have to be unitary.

In Figure 4.4 we see a quantum circuit consisting of four Hadamard gates H and one CNOT gate that actually implements an "inverse CNOT gate", at which the control and target qubits are changed.

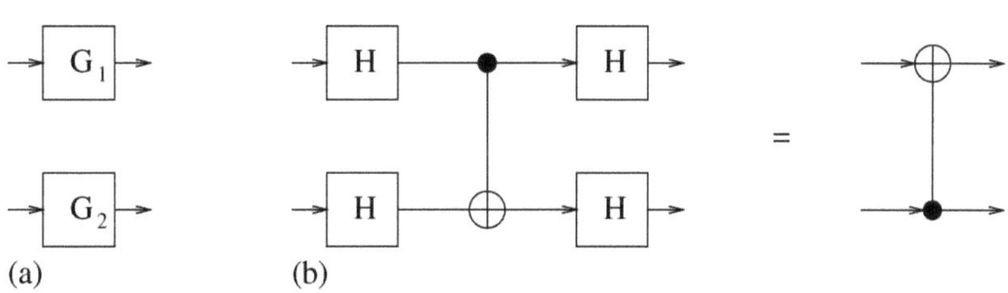

Fig. 4.4. An implementation of the inverse of the XOR gate.

Information processing in the network on the left side of the identity in Figure 4.4b, for the input $|0\rangle|1\rangle$, can be depicted as follows:

$$\begin{array}{rcl}
|0\rangle|1\rangle & \stackrel{H\text{-gates}}{\longrightarrow} & |0'\rangle|1'\rangle \\
& = & \frac{1}{\sqrt{2}}(|0\rangle + |1\rangle)\frac{1}{\sqrt{2}}(|0\rangle - |1\rangle) \\
& = & \frac{1}{2}(|0\rangle|0\rangle + |1\rangle|0\rangle - |0\rangle|1\rangle - |1\rangle|1\rangle) \\
& \stackrel{\text{XOR gate}}{\longrightarrow} & \frac{1}{2}(|0\rangle|0\rangle + |1\rangle|1\rangle - |0\rangle|1\rangle - |1\rangle|0\rangle) \\
& = & \frac{1}{\sqrt{2}}(|0\rangle - |1\rangle)\frac{1}{\sqrt{2}}(|0\rangle - |1\rangle) \\
& = & |1'\rangle|1'\rangle \\
& \stackrel{H \text{ gates}}{\longrightarrow} & |1\rangle|1\rangle.
\end{array}$$

Special Gates and Universal Sets of Gates

The following gates play especially important role in quantum information processing:

$$\sigma_x = X, \sigma_y = Y, \sigma_z = Z, K = \sigma_z^{\frac{1}{2}}, T = \sigma_z^{\frac{1}{4}}.$$

where σ_x, σ_y and σ_z are Pauli operators;

$$CNOT, \text{TOF, HADAMARD} = H = \frac{1}{2}(\sigma_x + \sigma_z), \text{ SWAP and } \sqrt{\text{SW AP}}.$$

where TOF-gate (Toffoli-gate) is a three-qubit gate performing the mapping $|x, y, z\rangle \rightarrow |x, y, x \bigwedge y) \oplus z\rangle$ and the SWAP gate is a two qubit gate that exchanges two input qubits and can be implemented using the following circuit with three CNOT gates:

Of importance are the following finite, interesting and important universal sets of gates:

- SHOR=$\{\text{TOF}, H, \sigma_z^{\frac{1}{2}}\}$, see Shor (1996).
- KLZ1 = $\{CNOT, \Lambda_1(\sigma_z^{\frac{1}{2}}), \sigma_z^{\frac{1}{2}}\}$, see Knill et al. (1998?).
- KITAEV = $\{\Lambda_1(\sigma_z^{\frac{1}{2}}), H\}$, see Kitaev (1997).
- BMPRV=$\{CNOT, H, \sigma_z^{\frac{1}{4}}\}$, see Boykin et al. (1999).

4.2.8 Quantum Teleportation

The so called *No-teleportation theorem*, one of the fundamental laws of quantum mechanics, says that *(classical) teleportation* is impossible. This means

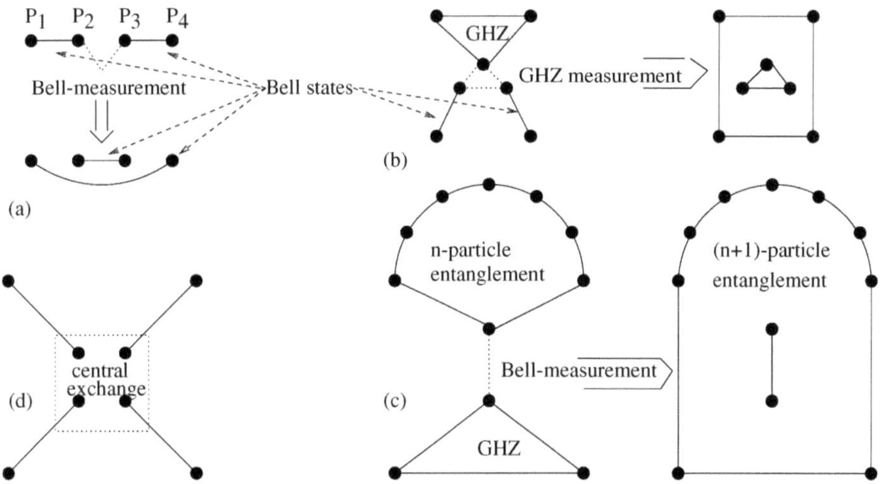

Fig. 4.5. A circuit with two CNOT gates and one reverse CNOT gate to realize the swap of two qubits.

that there is no way to use classical channels to transmit faithfully quantum information.[4]

In more technical terms, there is no possibility to measure, in general, a single copy of a quantum state in such a way that the classical outcomes of the measurement would be sufficient to reconstruct faithfully the state. Indeed, the classical teleportation would imply that *quantum cloning is possible* and this would imply that *super-luminal communication* is possible.

On the other hand, quantum teleportation, one of the most peculiar and most surprising and useful outcomes of quantum information processing, is possible, what will now be illustrated.

Quantum teleportation allows to transmit unknown quantum information to a very distant place in spite of impossibility to measure or to broadcast information to be transmitted. We show how one party, usually called Alice, can teleport unknown state of her particle to another participant, called usually Bob, under the assumption that both Alice and Bob share two particles in so called "$|\text{EPR-state}\rangle = \frac{1}{\sqrt{2}}(|00\rangle + |11\rangle)$. The total state of these three particles can be expressed as follows:

$$|\psi\rangle|EPR - state\rangle = |\Phi^+\rangle\frac{1}{\sqrt{2}}(\alpha|0\rangle + \beta|1\rangle) + |\Psi^+\rangle\frac{1}{\sqrt{2}}(\beta|0\rangle + \alpha|1\rangle)$$
$$+|\Phi^-\rangle\frac{1}{\sqrt{2}}(\alpha|0\rangle - \beta|1\rangle) + |\Psi^-\rangle\frac{1}{\sqrt{2}}(-\beta|0\rangle + \alpha|1\rangle))$$

[4] I looked to several dictionaries for term teleportation. Webster 's New World Dictionary of the American Language says *Teleportation is theoretical transportation of matter through space by converting it first into energy and then converting it back at the terminal point*;

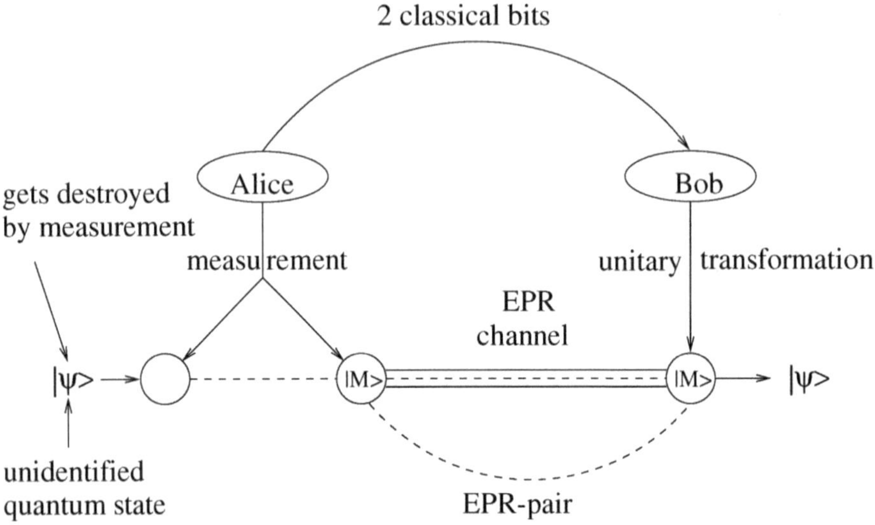

and therefore measurement of the first two particles, with respect to the Bell basis, $\{|\Phi^+\rangle, |\Phi^-\rangle, |\Psi^+\rangle, |\Psi^+\rangle\}$ projects the state of the Bob's particle into a "small Pauli-modification " $|\psi_1\rangle$ of the unknown state $|\psi\rangle = \frac{1}{\sqrt{2}}(\alpha|0\rangle + \beta|1\rangle)$.

The unknown state $|\psi\rangle$ can therefore be obtained from the state $|\psi_1\rangle$ by applying one of the four operations

$$\sigma_x, \sigma_y, \sigma_z, I$$

and the result of Alice's Bell measurement provides two bits specifying which of the above four operations should be applied.

These four bits Alice needs to send to Bob using a classical channel (by email, for example) so that Bob knows which of the Pauli operator to apply to his particle in order to get it into the teleported state $\alpha|0\rangle + |\beta|1\rangle$.

Observe that if the first two particles are measured with respect to the Bell basis, then Bob's particle gets into the mixed state

$$(\frac{1}{4}, \alpha|0\rangle + \beta|1\rangle), (\frac{1}{4}, \alpha|0\rangle - \beta|1\rangle), (\frac{1}{4}, \beta|0\rangle + \alpha|1\rangle), (\frac{1}{4}, \beta|0\rangle - \alpha|1\rangle)$$

to which corresponds the density matrix

$$\frac{1}{4}\begin{pmatrix}\alpha\\ \beta*\end{pmatrix}(\alpha^*, \beta^*) + \frac{1}{4}\begin{pmatrix}\alpha\\ -\beta\end{pmatrix}(\alpha^*, -\beta^*) + \frac{1}{4}\begin{pmatrix}\beta\\ \alpha\end{pmatrix}(\beta^*, \alpha^*) +$$

$$\frac{1}{4}\begin{pmatrix}\beta\\ -\alpha\end{pmatrix}(\beta^*, -\alpha^*) = \frac{1}{2}I$$

The resulting density matrix is therefore identical to the density matrix for the mixed state corresponding to the random bit

$$(\frac{1}{2}, |0\rangle) \oplus (\frac{1}{2}, |1\rangle).$$

It means that before Bob's receives two classical bits specifying classical result of Alice's measurement in Bell basis, Bob has received "nothing", represented by the density matrix of the completely mixed state.

4.3 Quantum Automata

For most of the main classical models of automata, see for example Gruska (1997), there have been introduced and explored also their quantum versions. For example, for finite automata, pushdown automata, stack automata, Turing machines and cellular automata.

Models of quantum automata are used:

- To get an insight into the power of different quantum computing models and modes, using formal language/automata theoretic methods.
- To discover the simplest models of computation at which one can demonstrate large (or huge) difference in the power of quantum versus classical models.
- To explore mutual relations between different quantum and classical computation models and modes.
- To discover, in a transparent and elegant form, limitations of quantum computations and communications.

The basic formal way to develop a quantum version of a classical automata model is to replace in its probabilistic version probabilities of transitions by probability amplitudes. The main problem is to make this replacement in such a way that a to-be-quantum automaton is really quantum, that is that its evolution is unitary.

This section gives only a very brief introduction to quantum automata. For more see book Gruska (1999) and its web extension as well as paper Gruska (2000).

4.3.1 Quantum Finite Automata

One way to "imagine" a quantum finite automaton is as a multi-head finite automaton where each head has its own control unit and at each step each head can be replaced by a set of new heads located at some of the "neighbouring" cells.

Observe that to simulate classically such a quantum model of automata it is necessary to store information about positions of particular heads and for that space proportional to the length of the input is needed. (With the exception of one important case of so called **one-way quantum automata**, to be discussed next mainly, the term "finite" for such models of quantum automata is therefore quite questionable.)

Formally, a model of quantum finite automaton is defined, for a given input w, as follows.

Let Σ be an input alphabet, $\Gamma = \Sigma \cup \{\#, \$\}$, $w = w_1 \dots w_n$, $|w| = n$, be an input word over Σ, $Q = Q_a \cup Q_r \cup Q_n$ be the set of states decomposed into the accepting, rejecting and nonterminating states; q_0 be the initial state. Let (q, i), with a state q and position on tape $0 \leq i \leq n+1$, be a configuration and let $C(Q, w)$ be the set of all configurations. (All that is classical.)

For a quantum automaton $\mathcal{A}$ with an input w and a set Q of classical states, the underlying Hilbert space will be $\mathcal{H}_{C(Q,w)}$ with the set of basis states $\{|(q,i)\rangle \mid q \in Q, 0 \leq i \leq n+1$ labeled by classical configurations (q, i). Evolution is given by a unitary mapping δ such that

$$\delta(|(q,i)\rangle) = \sum_{q' \in Q, 0 \leq j \leq n+1} \alpha_{q',j} |q', j\rangle,$$

where $\alpha(q', j) = 0$ for all j that do not represent next possible position of "a new head created by a head on the cell i". Computation is assumed to start at the configuration $|q_0, 0\rangle$ with "one head on the left end-marker" and proceeds from one quantum state to another. Measurement is then done, as discussed below, either after each evolution step or "at the end of computation/evolution".

Measurement results in a projection to one of the following subspaces:
The subspace S_a generated by basis states $|(q,i)\rangle$, where $q \in Q_a$;
The subspace S_r generated by basis states $|(q,i)\rangle$, where $q \in Q_r$;
The subspace S_n generated by basis states $|(q,i)\rangle$, where $q \in Q_n$;
that is either to the subspace generated by basis states corresponding to configurations with accepting states only, or to the subspace generated by basis states corresponding to configurations with rejecting states only or, finally, to the subspace generated by basis states corresponding to configurations with nonterminating states only.

Two basic modes of computation are called MM-mode (*many measurements* mode) and MO-mode (*measurement once* mode). In the many measurements mode, a measurement is done after each evolution step and computation continues only in the case the current state is projected into the subspace generated by the basis states (configurations) with nonterminating states only. In the measurement once mode, measurement is done after the last step of the evolution.

There are four basic types of quantum finite automata:

2QFA – Two-way quantum finite automata (at which heads "can move and create new heads" in both directions).

1.5QFA – One-and-half-way quantum automata (at which heads "can move and create new heads" only in one direction, but they "do not have always to move").

1QFA — One-way (real time) quantum automata (at which all heads " move and create new heads" always and in the same direction.)

RFA – reversible finite deterministic automata – automata at which for any state q and any input symbol a there is exactly one state q' from which the automaton can get into the state q under the input a.

There are several natural ways how language acceptance by quantum finite automata can be defined.

A language L is said to be accepted by a QFA $\mathcal{A}$ with probability $p > \frac{1}{2}$, if every word $x \in L$ ($x \notin L$) is accepted (rejected) with probability at least p.

A language L is said to be accepted by a QFA $\mathcal{A}$ with the cut point $\lambda > \frac{1}{2}$, if for all $x \in L$ ($x \notin L$) the probability that $\mathcal{A}$ accepts x is $> \lambda$ ($\leq \lambda$).

A language L is said to be accepted by a QFA $\mathcal{A}$ with the **cut point** $\lambda > \frac{1}{2}$ and **bounded error** if there is an $\varepsilon > 0$ such that any $x \in L$ ($x \notin L$) is accepted by $\mathcal{A}$ with probability $\geq \lambda + \varepsilon$ ($\leq \lambda - \varepsilon$). If $\mathcal{A}$ accepts a language L with a cut point λ, but not with a bounded error, then $\mathcal{A}$ is said to accept L with **unbounded error.**

A language L is said to be Monte Carlo accepted by a QFA $\mathcal{A}$ with bounded error ε if every $x \in L$ ($x \notin L$) is accepted (rejected) with probability 1 ($1-\varepsilon$).

4.3.2 One-way (Real-time) Quantum Automata

This is the most basic and really finite model of quantum finite automata.

Definition 2 One-way (real-time) quantum finite automaton *(1QFA) $\mathcal{A}$ is given by: Σ — the input alphabet; Q — the set of states; q_0 – the initial state; $Q_a \subseteq Q$, $Q_r \subseteq Q$ — sets of accepting and rejecting states and the transition function*

$$\delta : Q \times \Gamma \times Q \rightarrow C_{[0,1]},$$

where $\Gamma = \Sigma \cup \{\#, \$\}$ and $\#$, $\$$ are end-markers.

The evolution (computation) of $\mathcal{A}$ is performed in the Hilbert space $\mathcal{H}_Q$ with the basis states $\{|q\rangle \,|\, q \in Q\}$ using the operators $V_\sigma, \sigma \in \Gamma$, defined by

$$V_\sigma|q\rangle = \sum_{q' \in Q} \delta(q, \sigma, q')|q'\rangle$$

and the transition function δ has to be such that all operators V_σ are unitary.

Measurement of states in $\mathcal{H}_Q$ is defined by the following orthogonal decomposition of the Hilbert space $\mathcal{H}_Q$ to the subspaces:

$$E_a = \text{span}\{|q\rangle \,|\, q \in Q_a\}$$
$$E_r = \text{span}\{|q\rangle \,|\, q \in Q_r\}$$

E_n is the orthogonal complement of $E_a \oplus E_r$.

Example 4 We show a 1QFA $\mathcal{A}$, see Ambainis and Freivalds (1998), accepting the $L = \{0^i 1^j \mid i \geq 0, j > 0\}$ with probability $p = 1 - p^3$, that is with probability $p = 0.68$, in the sense that each binary word that is (is not) in the language L is accepted (rejected) with probability at least p.

$\mathcal{A}$ has the set of states $Q = \{q_0, q_1, q_2, q_a, q_r\}$, with the initial state q_0 and $Q_a = \{q_a\}$, $Q_r = \{q_r\}$. For $s \in \{\#, 0, 1, \$\}$, the unitary transformations V_s are defined as follows (actually, we define results of application of transformations V_s only to some of the basis states; for remaining ones these transformations can be defined in an arbitrary way provided the resulting transformation matrices are unitary):

$$V_{\#}|q_0\rangle = \sqrt{1-p}|q_1\rangle + \sqrt{p}|q_2\rangle,$$
$$V_0|q_1\rangle = (1-p)|q_1\rangle + \sqrt{p(1-p)}|q_2\rangle + \sqrt{p}|q_r\rangle,$$
$$V_0|q_2\rangle = \sqrt{p(1-p)}|q_1\rangle + p|q_2\rangle - \sqrt{1-p}|q_r\rangle,$$
$$V_1|q_1\rangle = |q_r\rangle, V_1|q_2\rangle = |q_2\rangle, \quad V_{\$}|q_1\rangle = |q_r\rangle, V_{\$}|q_2\rangle = |q_a\rangle.$$

We show first that each word $0^i 1^j$, where $0 \leq i, 1 < j$, is accepted by $\mathcal{A}$. Indeed, the first computation step has the form

$$V_{\#}|q_0\rangle = \sqrt{1-p}|q_1\rangle + \sqrt{p}|q_2\rangle$$

and the resulting state is not changed by the measurement because both basis states in the superposition are non-terminating. In the next steps, transformation V_0^i is applied and it is easy to verify that no application of the operation V_0 changes the state $\sqrt{1-p}|q_1\rangle + \sqrt{p}|q_2\rangle)$, which is a superposition of nonterminating basis states. After that a sequence of j transformations V_1 should be applied. The first one yields

$$V_1(\sqrt{1-p}|q_1\rangle + \sqrt{p}|q_2\rangle) = \sqrt{1-p}|q_r\rangle + \sqrt{p}|q_2\rangle.$$

If now a measurement is applied, the resulting state is rejected with the probability $1-p$ and projected into the nonterminating state $|q_2\rangle$ with probability p. Next sequence of $j-1$ V_1 operations will not change that state and only when the operation $V_{\$}$, corresponding to the situation that the head is on the right end-marker, is applied the state $|q_a\rangle$ is produced which is then accepted at the subsequent measurements with the probability 1.

On the other hand, each input $0^i 1^j 0x$, where $i \geq 0, j > 0, x \in \{0,1\}^*$, is rejected with the probability at least p. Indeed, as analysed above, the state after the input $0^i 1$ is $\sqrt{1-p}|q_r\rangle + \sqrt{p}|q_2\rangle$ and it is rejected with probability $1-p$; with probability p computation proceeds in the state $|q_2\rangle$ until the first 0 is read. The resulting state is then $\sqrt{p(1-p)}|q_1\rangle + p|q_2\rangle - \sqrt{1-p}|q_r\rangle$, which is rejected with the total probability $p(1-p)$ and with total probability p^2 the computation proceeds to process x. If $x \in L$, then it is accepted with probability p and rejected with probability $1-p$ and therefore the total probability of rejection in this case is

$$1 - p + p(1-p) + p^2(1-p) = 1 - p^3 = p;$$

if $x \notin L$, then x is rejected with total probability at least p^3 and therefore the total probability of rejection is

$$(1-p) + p(1-p) + p^3 > p.$$

Power of 1QFA

Let us denote by BMO (BMM) the family of languages accepted by 1QFA working in the measurement once (measurements many) mode, so called MO-1QFA (so called MM-1QFA), with a bounded error and by UMO (UMM) the family of languages accepted by OM-1QFA (MM-1QFA) with an unbounded error.

The class of languages accepted by MO-1QFA is already quite well understood, especially due to Moore and Crutchfiled (1997) and Brodsky and Pippenger, (1999).

Theorem 1 *The class BMO is exactly the class of group languages and therefore a proper subclass of the class of regular languages.*[5] *It is closed under Boolean operations, inverse homomorphism and word quotients, but not under homomorphism.*

Theorem 2 *It is decidable whether a regular language is accepted by a MO-1QFA, and it is decidable whether two MO-1QFA are equivalent.*

Concerning the class MM-1QFA situation is quite different. The following results are due to Watrous (1997); Kondacs and Watrous (1997); Ambainis and Freivalds (1998).

Theorem 3 *MM-1QFA can accept only regular languages, but not all of them. For example, not the language $L = \{0,1\}^*0$.*

The family of languages accepted by MM-1QFA is closed under complement, inverse homomorphism and word quotients, but not under homomorphism.

Open problem 1 Is the equivalence problem for MM-1QFA decidable?

It has been shown by Valdats (2000) that the class of languages accepted by 1QFA is not closed under union and, consequently, not under any binary Boolean operation. Namely, he has showed that the languages

$$L_1 = (aa)^*bb^*a(b^*ab^*a)^*b^* \cup (aa)^*$$

and $L_2 = aL_1$ can be accepted by 1QFA with probability $\frac{2}{3}$, but their union is not acceptable by a 1QFA.

[5] It is the class of languages accepted by group finite automata, or, equivalently, the class of regular languages syntactical monoids of which are groups.

In addition, it has been shown that the above example represents a border case in the following sense. If two languages L_1 and L_2 can be accepted by 1QFA with probabilities p_1 and p_2 such that $\frac{1}{p_1} + \frac{1}{p_2} < 3$, then their union is accepted by a 1QFA with probability

$$\frac{2p_1p_2}{p_1 + p_2 + p_1p_2}.$$

Interesting relations have been showed between MM-1QFA and classical reversible finite automata (RFA).

Theorem 4 *(Ambainis, Kiktus, 2001) (a) If a language L can be accepted by a 1QFA which gives the correct answer with probability greater than* $\frac{52+4\sqrt{7}}{81}$*, then L is accepted by a 1RFA; (b) There is a language that is accepted by a 1QFA with the probability* $\frac{52+4\sqrt{7}}{81}$*, but that cannot be accepted by any 1RFA.*

For more about classical reversible finite automata and their power see Pin (1987).

Succinctness of 1QFA

Since 1QFA do not accept all regular languages, and therefore do not beat classical finite automata concerning recognition power, it is natural to explore whether they do not beat classical finite automata at least concerning size. The result are not conclusive. Indeed, it holds that

- In some cases (sequential) 1QFA can be, likely due to the parallelism in their evolution, exponentially more succinct than classical DFA;
- In some cases quantum one-way finite automata can be, likely due to the requirement on unitarity of their evolution, exponentially larger, with respect to the number of states, as the corresponding DFA.

Specific results along these lines are the following ones:

Theorem 5 *(Ambainis and Freivalds, 1998) The languages* $L_n = \{0^n\}$*, each DFA of which has to have clearly n states, can be recognized by a MM-1QFA with* $\mathcal{O}(\lg n)$ *states.*

Proof (Sketch) The proof uses methods that have been developed for probabilistic automata. At first a special reversible probabilistic finite automaton is constructed using $\mathcal{O}(\frac{\lg n}{\lg \lg n})$ primes of the size $\mathcal{O}(\lg n)$. The automaton first randomly chooses one of the primes p and computes, using only $\mathcal{O}(\lg p)$ states, as discussed above, for a given input word the remainder modulo p of the length of the input word. This remainder is then compared with the correct one.

The resulting reversible PFA is then converted into a 1QFA in a way that increases the number of states at most by the factor of 2. The resulting number of states is therefore

$$\mathcal{O}\left(\frac{\lg n}{\lg\lg n}\lg p\right) = \mathcal{O}\left(\frac{\lg n}{\lg\lg n}\lg\lg n\right) = \mathcal{O}(\lg n).$$

□

Theorem 6 *(Ambainis et al., 1998, Nayak, 1999) For any integer n let $L_n = \{wa \mid w \in \{0,1\}^*, |w| \leq n\}$. It holds:*

1. *The language L_n can be recognized by a DFA of size $\mathcal{O}(n)$ and also by an MM-1QFA.*
2. *Any MM-1QFA recognizing L_n with a constant probability greater than $\frac{1}{2}$ has to have $2^{\Omega(n)}$ states.*

It is also natural to compare expressive power of RFA, as perhaps the simplest version of 1QFA, and deterministic finite automata. Here is one result:

Theorem 7 *(Ambainis, Freivalds, 1998) For any integer n let $L_n = \{xy|zy\}^n \cup \{\{xy|zy\}^i xx \mid 0 \leq i \leq n-1\}$. Then, it holds*

1. *L_n can be recognized by a DFA with $3n+2$ states.*
2. *L_n can be recognized by a RFA but it has to have at least $3(2^n - 1)$ states.*

Open problem 2 Is the increase in the size from n to $2^{\Omega(n)}$, when going from a DFA to a MM-1QFA, the worst possible?

4.3.3 1.5QFA

They are QFA heads of which can move "only in one direction", but "a head does not have move at each step, sometimes it can stay idle for some time".

Theorem 8 *(1) 1.5QFA can accept non-regular languages with respect to the unbounded error acceptance*
(2) (Amano, Iwama, 1999) The emptiness problem is undecidable for generalized 1QFA.

The last result is quite surprising because emptiness problem is decidable even for push-down automata. Surprisingly, recognition power of 1.5QFA is still not well determined. To do that is an interesting research challenge.

Open problem 3 Is every regular language accepted by a 1.5QFA?

4.3.4 Two-way Quantum Finite Automata

In the case of classical finite automata, two-way automata are not more powerful than classical ones. This is not the case for quantum finite automata (see Watrous (1997), Kondacs and Watrous (1998)).

A two-way quantum finite automaton $\mathcal{A}$ can be specified, in a standard way, by a finite (input) alphabet Σ, a finite set of states Q, an initial state q_0, sets $Q_a \subset Q$ and $Q_r \subset Q$ of accepting and rejecting states, respectively, with $Q_a \cap Q_r = \emptyset$, and a transition function

$$\delta : Q \times \Gamma \times Q \times \{\leftarrow, \downarrow, \rightarrow\} \longrightarrow \mathbf{C}_{[0,1]},$$

where $\Gamma = \Sigma \cup \{\#, \$\}$ is a tape alphabet of $\mathcal{A}$ and $\#$ and $\$$ are end-markers not in Σ, which satisfies the condition that the corresponding evolution is unitary. It can be shown that this is equivalent to the following conditions (of well-formedness) for any $q_1, q_2 \in Q$, $\sigma, \sigma_1, \sigma_2 \in \Gamma$, $d \in \{\leftarrow, \downarrow, \rightarrow\}$:

1. **Local probability and orthogonality condition.**
   $\sum_{q',d} \delta^*(q_1, \sigma, q', d)\delta(q_2, \sigma, q', d) = \begin{cases} 1, & \text{if } q_1 = q_2; \\ 0, & \text{otherwise.} \end{cases}$
2. **Separability condition I.**
   $\sum_{q'} \delta^*(q_1, \sigma_1, q', \rightarrow)\delta(q_2, \sigma_2, q', \downarrow) + \sum_{q'} \delta^*(q_1, \sigma_1, q', \downarrow)\delta(q_2, \sigma_2, q', \leftarrow) = 0.$
3. **Separability condition II.**
   $\sum_{q'} \delta^*(q_1, \sigma_1, q', \rightarrow)\delta(q_2, \sigma_2, q', \leftarrow) = 0.$

It is not easy to verify the above well-formedness conditions. However, fortunately, in order to explore recognition power of 2QFA it is sufficient to consider a simplified model of 2QFA because it holds that to each two-way quantum finite automaton there is an equivalent one (so-called *unidirectional* or *simple*) 2QFA in which

1. for each pair of states q and q' a probability amplitude is assigned that the automaton moves from the state q to the state q'.
2. To each state q a head movement $D(q)$ — to right, to left or no movement — is defined with the interpretation that if the automaton comes to a state q, then the head always moves in the direction $D(q)$.

2QFA accept any regular language and also some non-regular (even non-context free) languages.

Example 5 2QFA accepting the language $\{0^i1^i \mid i \geq 0\}$.

$$Q = \{q_0, q_1, q_2, q_3\} \cup \{s_j \mid 1 \leq j \leq n\} \cup \{r_{j,k} \mid 1 \leq j \leq n, 1 \leq k \leq n - j + 1\},$$

$Q_a = \{s_n\}$; the initial state is $|q_0\rangle$. Transitions and movements of the heads are shown in the following table:

$V_{\#}\|q_0\rangle = \|q_0\rangle$,	$V_{\$}\|q_0\rangle = \|q_3\rangle$,
$V_{\#}\|q_1\rangle = \|q_3\rangle$,	$V_{\$}\|q_2\rangle = \frac{1}{\sqrt{n}} \sum_{j=1}^{n} \|r_{j,0}\rangle$,
$V_{\#}\|r_{j,0}\rangle = \frac{1}{\sqrt{n}} \sum_{l=1}^{n} e^{\frac{2\pi i}{n} jl} \|s_l\rangle, 1 \leq j \leq n$,	
$V_0\|q_0\rangle = \|q_0\rangle$,	$D(q_0) = \rightarrow$,
$V_0\|q_1\rangle = \|q_2\rangle$,	$D(q_1) = \leftarrow$,
$V_0\|q_2\rangle = \|q_3\rangle$,	$D(q_2) = \rightarrow$,
$V_0\|r_{j,0}\rangle = \|r_{j,j}\rangle, 1 \leq j \leq n$,	$D(q_3) = \downarrow$,
$V_0\|r_{j,k}\rangle = \|r_{j,k-1}\rangle, 1 \leq k \leq j, 1 \leq j \leq n$,	
$V_1\|q_0\rangle = \|q_1\rangle$,	$D(r_{j,0}) = \leftarrow, 1 \leq j \leq n$,
$V_1\|q_2\rangle = \|q_2\rangle$,	$D(r_{j,k}) = \downarrow, 1 \leq j \leq n, k \neq 0$,
$V_1\|r_{j,0}\rangle = \|r_{j,n-j+1}\rangle, 1 \leq j \leq n$,	$D(s_j) = \downarrow, 1 \leq j \leq n$,
$V_1\|r_{j,k}\rangle = \|r_{j,k-1}\rangle, 1 \leq k \leq j \leq n$.	

Informally, the automaton, which has an acceptance parameter n, works in the following way that is demonstrated in Figure 4.3.4. At first, the automaton works using one head, always in one of the states $|q_0\rangle, |q_1\rangle, |q_2\rangle, |q_3\rangle$ to check whether the input has the form 0^i1^j for some i, j. If not, the input is rejected. If yes, the checking stage ends in the state $|q_2\rangle$, with the head on the right end-marker. After that the head is "destroyed" and n new heads are created, with the k-th head in the state $|k, 0\rangle$, and all these heads move one cell left. In the next stage, each of the heads move to the left with a special speed. The k-th head stays on symbol 0 for k steps and then moves left and stays on symbol 1 for $n-k+1$ steps. Due to these arrangements, in case $i = j$ all heads arrive to the left end-marker at the same time, with the k-th head in the state $|r_{k.o}\rangle$. The k-th head then goes to a very special state $\frac{1}{\sqrt{n}} \sum_{l=1}^{n} e^{\frac{2\pi i}{n} jl} |s_l\rangle, 1 \leq j \leq n$, by applying a quantum version of the Fourier transform. The overall state is then the sum (superposition) of all such states,

$$\frac{1}{\sqrt{n}} \sum_{k=1}^{n} \frac{1}{\sqrt{n}} \sum_{l=1}^{n} e^{\frac{2\pi i}{n} kl} |s_l\rangle = |s_n\rangle$$

what, surprisingly, sums up to a very simple state, $|s_n\rangle$, what is the single accepting state and therefore the input is then accepted with probability 1. In case $i \neq j$, one of the heads comes first and the state is accepted only with probability $\frac{1}{n}$.

4.3.5 Finite Automata With Classical and Quantum States

Models of QFA considered so far have all been natural quantum versions of the classical models of automata.

Of a different type is the model introduced by Ambainis and Watrous (1999), and called *two-way finite automata with quantum and classical states* (2QCFA).

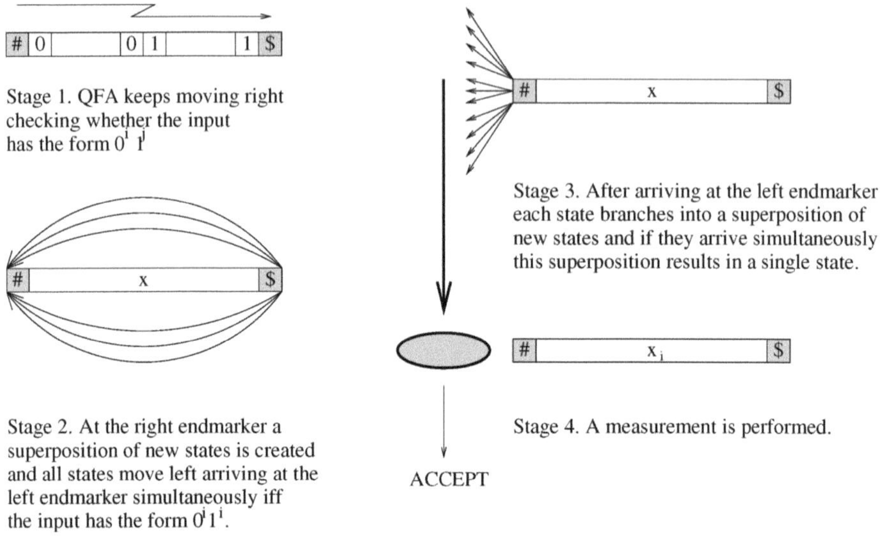

This model is also more powerful than classical (probabilistic) 2FA and at the same time it seems to be more realistic, and really more "finite" than 2QFA because 2QFA need quantum memory of size $\mathcal{O}(\lg n)$ to process an input of the size n. 2QCFA can be seen as an intermediate model between 1QFA and 2QFA. A 2QCFA is defined similarly as a classical 2FA, but, in addition, it has a fixed size quantum register (which can be in a mixed state) upon which the automaton can perform either a unitary operation or a measurement.

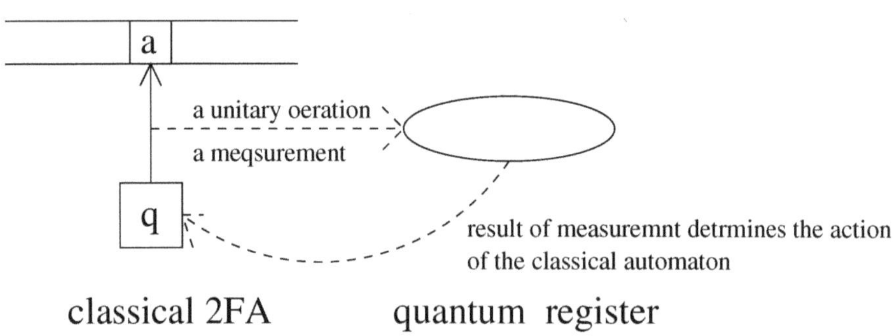

A 2QCFA has a classical initial state q_0 and an initial quantum state $|\phi_0\rangle$. The evolution of a 2QCFA is specified by a mapping Θ that assigns to each classical state q and a tape symbol σ an action $\Theta(q, \sigma)$.

One possibility is that $\Theta(q, \sigma) = (q', d, U)$, where q' is a new state, d is next movement of the head (to left, no movement or to right), and U is a unitary operator to be performed on the current state of the quantum register.

The second possibility is that

$$\Theta(q,\sigma) = (M, m_1, q_1, d_1, m_2, q_2, d_2, \ldots, m_k, q_k, d_k)$$

where M is a measurement, $m_1, \ldots, m_k$ are its possible classical outcomes, and for each measurement outcome a new state of the control unit and a new movement of the head is determined. In such a case both the state transmission and the head movement are probabilistic.

Ambainis and Watrous (1999) have shown that 2QCFA with only 1 qubit of quantum memory are already very powerful. Such 2QCFA can accept with bounded error the language of palindromes over the alphabet $\{0,1\}$, which cannot be accepted by probabilistic 2FA at all, and also the language $\{0^i 1^i \mid i \geq 0\}$, in polynomial time — this language can be accepted by probabilistic 2FA, but only in exponential time.

4.3.6 Quantum Turing Machines

We will discuss two main ways quantum versions of Turing machines have been defined. The first one is a "usual" quantum version of the classical concept of the probabilistic one tape Turing machine. Multitapes case have been studied in details by Nishimura and Ozawa (1999).

We present the "standard" definition of the quantum Turing machines for the case of one-tape Turing machines only.

Definition 3 *A (one-tape)* **quantum Turing machine** $\mathcal{M} = \langle \Sigma, Q, q_0, q_f, \delta\rangle$, *QTM in short, is defined by sets of states and tape symbols, an initial state* q_0 *and an final state* q_f, *and the transition amplitude mapping*

$$\delta : Q \times \Sigma \times \Sigma \times Q \times \{\leftarrow, \downarrow, \rightarrow\} \longrightarrow \mathbf{C}_{[0,1]}$$

which is required to be such that quantum evolution of $\mathcal{M}$ *is unitary.*

A **configuration** of $\mathcal{M}$ is determined by the content τ of the tape, $\tau \in \Sigma^{\mathbf{Z}}$, by an $i \in \mathbf{Z}$ which specifies the position of the head, and by a $q \in Q$, the current state of the tape.

Let $C_{\mathcal{M}}$ denote the set of all configurations of $\mathcal{M}$. Computation (evolution) of $\mathcal{M}$ is performed in the Hilbert space space $H_{C_{\mathcal{M}}}$ with the basis $\{|c\rangle \mid c \in C_{\mathcal{M}}\}$. The transition function δ uniquely determines a mapping $a : C_{\mathcal{M}} \times C_{\mathcal{M}} \to \mathbf{C}$ such that for $c_1, c_2 \in C_{\mathcal{M}}$, $a(c_1, c_2)$ is the amplitude of the transition of $\mathcal{M}$ from the basis state $|c_1\rangle$ to $|c_2\rangle$.

Quantum evolution mapping $U_{\mathcal{M}} : H_{\mathcal{M}} \to H_{\mathcal{M}}$ is defined for basis states by

$$U_{\mathcal{M}}|c\rangle = \sum_{c' \in C_{\mathcal{M}}} a(c, c')|c'\rangle.$$

It is in general very difficult to verify that the transition mapping of a to-be-quantum Turing machine is really unitary and therefore that it is really quantum Turing machine. It can be shown that this is indeed the case when the transition function specifies the following well-formedness conditions:

Definition 4 *A QTM* $\mathcal{M} = \langle \Sigma, Q, q_0, q_f, \delta \rangle$ *with the transition mapping*

$$\delta : Q \times \Sigma \times \Sigma \times Q \times \{\leftarrow, \downarrow, \rightarrow\} \longrightarrow \mathbf{C}$$

is said to be strongly well-formed if the following conditions are satisfied.

1. **Local probability condition.** *For any* $(q_1, \sigma_1) \in Q \times \Sigma$;

$$\sum_{(\sigma,q,d)\in\Sigma\times Q\times\{\leftarrow,\downarrow,\rightarrow\}} |\delta(q_1, \sigma_1, \sigma, q, d)|^2 = 1.$$

2. **Separability condition I.** *For any two different pairs* $(q_1, \sigma_1), (q_2, \sigma_2)$ *from the set* $Q \times \Sigma$:

$$\sum_{(q,\sigma,d)\in Q\times\Sigma\times\{\leftarrow,\downarrow,\rightarrow\}} \delta^*(q_1, \sigma_1, \sigma, q, d)\delta(q_2, \sigma_2, \sigma, q, d) = 0.$$

3. **Separability condition II.** *For any* $(q, \sigma, d), (q', \sigma', d')$ *from the set* $Q \times \Sigma \times \{\leftarrow, \downarrow, \rightarrow\}$ *such that* $(q, \sigma, d) \neq (q', \sigma', d')$:

$$\sum_{(q_1,\sigma_1)\in Q\times\Sigma} \delta^*(q_1, \sigma_1, \sigma, q, d)\delta(q_1, \sigma_1, \sigma', q', d') = 0.$$

4. **Separability condition III.** *For any* $(q_1, \sigma_1, \sigma_1'), (q_2, \sigma_2, \sigma_2') \in Q\times\Sigma\times\Sigma$ *and* $d_1 \neq d_2 \in \{\leftarrow, \downarrow, \rightarrow\}$:

$$\sum_{q\in Q} \delta^*(q_1, \sigma_1, \sigma_1', q, d_1)\delta(q_2, \sigma_2, \sigma_2', q, d_2) = 0.$$

It has been shown (see Bernstein and Vazirani, 1993) that there exist universal quantum Turing machines that can efficiently simulate any other quantum Turing machine. Quantum Turing machines and (uniform families of) quantum circuits are polynomially equivalent models of quantum computers. Well-formedness conditions have been formulated also for multitape quantum Turing machines (Nishimura, Ozawa, 2000). A variety of normal forms for one-tape QTM have been established. For example, the so-called unidirectional QTM at which the movement of the head is uniquely determined by the state the QTM comes into.

Measurement Based Turing Machines

Universal is also so called measurement based quantum Turing machine, Perdrix and Jorrand (2004), composed of (a) control unit; (b) one-qubit memory; (c) bi-infinite tape with qubits in cells and with a transition function

$$\text{States} \times \text{Measurement outcomes} \rightarrow \text{States} \times \text{Observables} \times \text{Head moves}$$

and with the set of observables

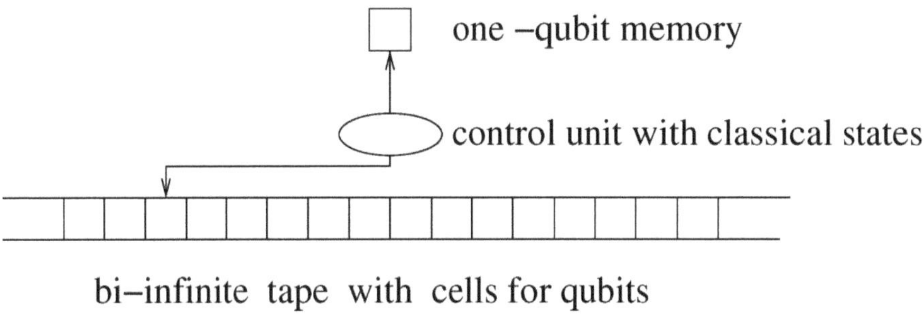

Fig. 4.6. Universal QTM (Perdrix and Jorrand (2003)

$$\{X\otimes X, Z\otimes Z, X\otimes Z, X\otimes Z, X\otimes I, Z\otimes I, I\otimes X, I\otimes Z, \frac{1}{\sqrt{2}}(X\otimes X + X\otimes Y)\}.$$

This new type of Turing machines has not been yet explored sufficiently well.

Quantum Cellular Automata

Classical cellular automata are a very exciting and intriguing model of computation, evolution and also of complex behaviour. Even more intriguing seems to be the model of reversible classical cellular automata, especially of two-dimensional and three-dimensional cellular automata. Such cellular automata are an important model of reversible computation performed by interactions of neighbouring elements and are of significant importance in several areas of physics. Quantum cellular automata are a natural generalization of reversible cellular automata.

The problem to define properly quantum cellular automata in full generality seems to be intriguing. The difficulty is in defining locally behaviour of an automaton global evolution of which is unitary. For initial approaches to quantum cellular automata see Gruska (1999).

Other Types of Models of Quantum Automata

A more abstract approach to quantum finite automata, more along the lines with algebraic theory of classical finite automata, has been developed by Gudder (2000).

Quantum one-counter automata have been introduced by Kravtcev (1999) and by Yamasaki et al. (2000). Bonner et al. (2000a) explored their power in comparison with probabilistic one-counter automata. Moreover, Bonner et al. (2000) have shown that the emptiness problem is undecidable for quantum one-counter automata. Moore and Crutchefeld (1997) and Golovkins (1999) introduced and explored the concept of quantum pushdown automata. Moore and Crutchefeld introduced, in addition, the concept of quantum context-free grammars. These models are natural, from mathematical and informatics point of view. However, a "deeper quantum" importance of these models is not yet clear, or not yet discovered?

The concept of 2-tape quantum finite automata introduced by Bonner et al. (2000) has been re-casted and studied by Freivalds and Winter (2000) as a model of quantum finite state transducers ( ) that can be used to study the acceptance of the class of binary relations on strings. Quantum finite state transducers are defined in such a way that if the output tape is thrown away one gets quantum finite (one-way) automata. Namely, it holds

Theorem 9 *A language L is accepted by a 1-way quantum finite automaton with probability bounded away from $1/2$ if and only if the relation $L \times \{0\} \cup \bar{L} \times \{1\}$ is computed with an isolated cut-point.*

Ambainis et al. (1999) have studied quantum finite multitape automata.

References

1. M. Amano and K. Iwama. Undecidability on quantum finite automata, *Proceedings of the 31st ACM STOC*, ACM, 1999, 368–375.
2. A. Ambainis, A. Nayak, A. Ta-Shma and U. Vazirani. Dense quantum coding and a lower bound for 1-way quantum finite automata, Technical report, quant-ph/9804043, 1998.
3. A. Ambainis, R. Freivalds. 1-way quantum finite automata: strengths, weaknesses and generalizations, *Proceedings of* 39*th IEEE FOCS*, 1998, 332–341.
4. A. Ambainis, R. Bonner, R. Freivalds, M. Golovkins and M. Karpinski. Quantum finite multitape automata, Technical report, quant-ph/9905026, 1999.
5. A. Ambainis and J. Watrous. Two-way finite automata with quantum and classical states, Technical report, quant-ph/9911009, 1999.
6. A. Ambainis, R. Bonner, R. Freivalds and A. Ķikusts. Probabilities to accept languages by quantum finite automata, Technical report, quant-ph/9904066, 1999.
7. A. Ambainis and A. Kikusts. Exact results for acceepting probabilities of quantum automata, *Lecture Notes in Computer Science 2136,* Springer-Verlag, Berlin, 2001, 135–147.
8. E. Bernstein and U. Vazirani. Quantum complexity theory, *SIAM Journal of Computing*, 26, 5 (1997), 1411–1473.
9. R. Bonner, R. Freivalds and M. Rikards. Undecidability on quantum finite-one-counter automata, *Proceedings of the International Workshop on Quantum Computing and Learning, Sundbyholms Slott, Sweden, May 2000* (R. Bonner and R. Freivalds, eds.), Mälardalen University, 2000, 65–71.
10. R. Bonner, R. Freivalds and M. Kravtsev. Quantum versus probabilistic one-way finite automata with counter, *Proceedings of the International Workshop on Quantum Computing and Learning, Sundbyholms Slott, Sweden, May 2000* (R. Bonner and R. Freivalds, eds.), Mälardalen University, 2000, 80–88.

11. P. O. Boykin, T. Mor, M. Pulver, V. Roychowdhury and F. Vatan. On universal and fault-tolerant quantum computing, Technical report, quant-ph/9906054, 1999.
12. A. Brodsky and N. Pippenger. Characterizations of 1-way quantum finite automata, Technical report, quant-ph/9903014, 1999.
13. R. Freivalds and A. Winter. Quantum finite state transducers, Technical report, quant-ph/0011052, 2000.
14. M. Golovkins. An introduction to quantum pushdown automata, *Proceedings of the International Workshop "Quantum computation and learning", Riga, September 11-13*, 1999, 44–52.
15. J. Gruska. *Foundations of computing*, Thomson International Computer Press, 1997.
16. J. Gruska. *Quantum computing*, McGraw-Hill, 1999.
17. J. Gruska. Quantum challenges, *Proceedings of SOFSEM'99*, LNCS 1523, Springer Verlag, Berlin, 1999, 2–29.
18. J. Gruska. Descriptional complexity issues in quantum computing, *Journal of Automata, Languages and Combinatorics*, 5, 3 (2000), 198–218.
19. J. Gruska. Quantum computing challenges, *Mathematics Unlimited, 2001 and beyond*, Springer Verlag, Berlin, 2001, 529–564.
20. S. Gudder. Basic properties of quantum automata, Technical report, University of Denver, 2000.
21. A. Kitaev. Quantum computations: algorithms and error correction, *Russian Mathematical Survey*, 52 (1997), 1191–1249.
22. E. Knill, R. Laflamme and W. H. Zurek. Resilent quantum computation: error models and thresholds, *Proceedings of the Royal Society of London, Series A*, 454 (1998), 375–384.
23. A. Kondacs and J. Watrous. On the power of finite state automata, *Proceedings of 36th IEEE FOCS*, 1997, 66–75.
24. M. Kravtsev. Quantum finite one-counter automata, Technical report, quant-ph/9905092, 1999.
25. C. Moore and J. P. Crutchfield. Quantum automata and quantum grammars, Technical report, Santa Fe University, 1997.
26. A. Nayak. Optimal lower bounds for quantum automata and random access codes, *Proceedings of 40th IEEE FOCS*, 1999, 369–376.
27. H. Nishimura and M. Ozawa. Computational complexity of uniform circuit families and quantum Turing machines, Technical report, University of Osaka, 1999.
28. S. Perdrix and P. Jorrand. Measurement-based quantum Turing machines and questions of universalities, Technical report, quant-ph/0402156, 2004.
29. J.E. Pin. On the languages accepted by finite reversible automata, *Proceedings of 14th ICALP*, LNCS 267, Springer-Verlag, Berlin, 1987, 237–249.
30. P. W. Shor. Fault-tolerant quantum computation, *Proceedings of 37th IEEE FOCS*, 1996, 56–65.
31. M. Valdats. The class of languages recognizable by 1-way quantum finite automata is not closd under union, Technical report, quant-ph/0001105, 2000.
32. J. Watrous. On the power of 2-way quantum finite automata, Technical report, University of Wisconsin, 1997.
33. T. Yaamasaki, H. Kobayashi, Y. Tokunaga and H. Imai. One-way probabilistic reversible and quantum finite one-counter automata, Technical report, Department of Information Science, University of Tokyo, 2000.

5

Splicing and Regularity

Tom Head and Dennis Pixton

Department of Mathematical Sciences
Binghamton University
Binghamton, New York 13902-6000
E-mail: {tom,dennis}@math.binghamton.edu

5.1 Introduction

In 1987 a new formalism for the generation of languages was introduced and has since received extensive theoretical development. This new formalism, *splicing*, was inspired by a study of the recombination of DNA molecules carried out in laboratories of molecular biology. For several decades it has been realized that molecules of DNA, RNA, and proteins may be idealized as strings of symbols over finite alphabets with the symbols chosen to denote deoxyribonucleotides, ribonucleotides, and amino acids, respectively. Let us adopt this view and consider how we might represent the making of one cut in each of two molecules m and m' and the construction of a new molecule w by attaching a left portion of m to a right portion of m': We let $m = pq$ and $m' = uv$ where we plan to cut m between its subsegments p and q and m' between its subsegments u and v. This is easy; the new molecule is represented by the string $w = pv$. It is natural to say that we have cut the molecules represented by m and m' and *spliced* the left segment of m with the right segment of m', producing the recombined molecule w. There are marvelous tools used in molecular biology called *restriction enzymes*. These tools allow us to cut (double stranded) DNA molecules at precisely specifiable positions. The resulting segments can be attached using an enzyme called a *ligase*. Consequently, in the presence of an appropriate restriction enzyme and a ligase, from such molecules $m = pq$ and $m' = uv$, a new molecule $w = pv$ may be produced. This is a greatly simplified indication of a technology used in *gene splicing*, a fundamental feature of *genetic engineering*. A detailed explanation of the origin of the splicing concept from a study of biomolecular science is the content of Section 5.2. The present Chapter has been organized so that a reader who does *not* wish to be concerned with the biomolecular roots of splicing can skip Section 5.2 and hasten to the formal theory of splicing systems and the languages they generate which is resumed in Section 5.3. Other readers may find Section 5.2 to be of great interest in motivating the formal concepts introduced in this Chapter. In the remainder of this section

T. Head and D. Pixton: *Splicing and Regularity*, Studies in Computational Intelligence (SCI) **25**, 119–147 (2006)
www.springerlink.com

definitions that are required in *all* further sections are given. These definitions constitute the beginning of a logical theory that can stand alone independent of the context of its origin in the study of the biomolecular sciences. See [7] for extensive developments and references concerning the theory of splicing systems and languages.

Definition 1. *Let A be a finite set for use as an alphabet. Let A^* be the free monoid consisting of all strings of symbols of A, including the empty string λ. By a* splicing rule *we mean an ordered quadruple $r = (u, u'; v', v)$ with u, u', v', v in A^*. From any* ordered pair *of strings that admit factorizations $puu'q$ and $xv'vy$, with p, q, x, y in A^* the string $puvy$ is said to be* generated by *r.*

For each of the nine Examples below, the alphabet is the set $A = \{a, c, g, t\}$.

Example 1. Let $r = (g, gatcc; a, gatct)$. From the ordered pair of strings $u = aaaggatccgg$, $v = ttagatctccc$ the string $w = aaaggatctccc$ is generated by r. Note that r generates no string at all from the ordered pair v, u, not even the empty string λ.

If we had chosen the rule $r' = (a, gatct; g, gatcc)$ then from the ordered pair v, u the string $w' = ttagattccgg$ would have been generated by r' and no string would have been generated from the ordered pair u, v by this rule r'.

Example 2. Let $r = (cg, cg; cg, cg)$. From the ordered pair of strings $u = aacgcgaacgcgaa$, $v = ttcgcgtt$ there are two strings that are generated by r, one for each of the two occurrences of the substring $cgcg$ in u. Thus $aacgcgtt$ is generated from this ordered pair, but so is $aacgcgaacgcgtt$. Note that from the ordered pair v, u the strings $ttcgcgaacgcgaa$ and $ttcgcgaa$ are generated by r.

Example 3. Let A, r, and u be as in the Example 2. Note that for the ordered pair u, u there are two choices of the substring $cgcg$ in each member of the pair. Thus four pairs of choices are possible. From two of these choices, r generates a repeat of the original string u. From the remaining two choices r generates the new strings $x = aacgcgaa$ and $y = aacgcgaacgcgaacgcgaa = (aacgcg)^3aa$. From the ordered pair u, y there are several choices of segments $cgcg$, but only one new string, $z = (aacgcg)^4aa$, is generated by r. In fact, from each ordered pair of strings u, $(aacgcg)^iaa$ with $i \geq 2$ one new string $(aacgcg)^{i+1}aa$ is generated by r. Apparently from the ordered pair u, u, each string in the regular language $\{\,(aacgcg)^iaa\colon i \geq 1\,\}$ can be generated by *iterating* the application of the rule r.

Definition 2. *Let A be an alphabet, let L be a language in A^*, and let R be a* set *of splicing rules. We define $R(L) = \{\,w \in A^*\colon$ There exist strings $u, v \in L$ and a rule $r \in R$ for which w is generated by r from the ordered pair $u, v\,\}$. Thus $R(L)$ consists of all the strings that can be generated from ordered* *pairs of strings in* L *by a* single *application of* one *of the rules belonging to* R. *Let* $R^0(L) = L$ *and, for each non-negative integer* i, *let* $R^{i+1}(L) = R^i(L) \cup R(R^i(L))$. *The language* $R^*(L) = \bigcup \{ R^i(L) : i \geq 0 \}$ *is said to be* the language generated from L through iterated application of the rule set R.

Example 4. For $L = \{ aa, aacgcgaacgcgaa \}$ and the rule set $R = \{ r \}$, where $r = (cg, cg; cg, cg)$, $R^*(L) = (aacgcg)^*aa$. Note that $(aacgcg)^0aa$ and $(aacgcg)^2aa$ are in $R^*(L)$ since each is in L and $aacgcgaa$ and each $(aacgcg)^iaa$, with $i \geq 2$, are in $R^*(L)$ by Example 3.

Example 5. For $L = \{ c, ac, ca \}$ and $R = \{ (a, c; \lambda, ac), (ca, \lambda; c, a) \}$, $R^*(L) = a^*c + ca^*$. Note that if we adjoin aca to L then $R^*(L) = a^*ca^*$.

Example 6. For $L = \{ c, caa \}$ and $R = \{ r \}$, where $r = (caa, \lambda; c, \lambda)$, $R^*(L) = c(aa)^*$.

Example 7. For $L = \{ a, c, g, t \}$ and $r = (\lambda, \lambda; \lambda, \lambda)$ we have: $R(L) = \{ \lambda \} \cup L \cup L^2$; $R^0(L) = L$, $R^1(L) = \{ \lambda \} \cup L \cup L^2$, $R^2(L) = \{ \lambda \} \cup L \cup L^2 \cup L^3 \cup L^4$, and $R^*(L) = \bigcup \{ L^i : i \geq 0 \} = A^*$. We sketch a computation of $R(L)$: Since $a = \lambda\lambda a = a\lambda\lambda$, applying r to the ordered pair consisting of the latter two factorizations, we see that $\lambda\lambda = \lambda$ must be in $R(L)$. Applying r to the ordered pair $a\lambda\lambda$ and $a\lambda\lambda$ we see that $a\lambda\lambda = a$ must be in $R(L)$. Applying r to the ordered pair $a\lambda\lambda$, and $\lambda\lambda a$, we see that $a\lambda\lambda a = aa$ must be in $R(L)$. Applying r to the ordered pair $a\lambda\lambda$ and $\lambda\lambda c$, we see that $a\lambda\lambda c = ac$ must be in $R(L)$. That $R(L) = \{ \lambda \} \cup L \cup L^2$ follows by replacing the a and c in all possible ways from the alphabet set $A = \{ a, c, g, t \}$.

It seems natural to view the splicing of strings as an operation related to the (iterated) concatenation of strings. In the following example the symbol t is reserved to function as an end marker.

Example 8. For $A = \{ a, c, g \}$, $L \subseteq A^*$ and $R = \{ (s, t; t, s') : s, s' \in A \}$, $R(tLt) = tLLt$ and $R^*(tt \cup tLt) = tL^*t$.

Definition 3. *A* splicing system *is an ordered pair* $S = (R, I)$, *where* R *is a set of splicing rules,* $R \subseteq A^* \times A^* \times A^* \times A^*$, *and* I *is an* initializing language, $I \subseteq A^*$. *The language* $R^*(I)$ *is said to be* the language generated by the splicing system $S = (R, I)$. *A language* L *is a* splicing language *if* $L = R^*(I)$ *for some splicing system* $S = (R, I)$. *A splicing system* $S = (R, I)$ *is said to be a* finite splicing system *if both* R *and* I *are finite sets. By a* finite splicing language *we mean a language that is generated by a finite splicing system.*

We say that a language L *is preserved by a splicing rule* r if, for every ordered pair of strings x, y in L and for every string z that can be obtained by splicing x and y using rule r, we have z in L. This allows us to describe the language $L = R^*(I)$ generated by the splicing system $S = (R, I)$ as *the* *smallest language that contains I and is preserved by each of the splicing rules in the set R.*

Each finite language L is a finite splicing language since it is generated by the finite splicing system $S = (R, I)$, where R is empty and $I = L$. Examples 4, 5, 6 and 7 demonstrate that the five *regular* languages $(aacgcg)^*aa$, $a^*c{+}ca^*$, a^*ca^*, $c(aa)^*$, and A^* itself are all finite splicing languages. That *all finite splicing languages are regular* is a major result of splicing theory. This substantial result is proved in Section 5.3. However, not all regular languages are finite splicing languages: It is not difficult to prove that the regular language $L = (aa)^*$ is *not* a finite splicing language even though it is a very close relative, even a homomorphic image, of the finite splicing language $c(aa)^*$ of Example 6. In fact, for every regular language L and any symbol that does not occur in any string in L, say c, cL is a splicing language and L is an image of cL under the homomorphism that erases the symbol c and leaves all other symbols unchanged. See Corollary 2 in Section 5.5.

Example 9. That the regular $L = a^*ca^*ca^*$ is *not* a finite splicing language can be seen as follows: Suppose that $r = (u, u'; v', v)$ is a rule that preserves L. If either uu' or $v'v$ is not a subsegment of any string in L, then r trivially preserves L, but it generates no string at all from any pair of strings in L. Thus if r generates any string at all from a pair of strings in L, then the number of occurrences of the symbol c in each of uu' and $v'v$ is either 0, 1 or 2. If the number of occurrences of c in either uu' or $v'v$ is either 0 or 1, then one can easily specify two strings in L from which r generates a string containing more than two occurrences of the symbol c. The same is true if uu' and $v'v$ each contain 2 occurrences of the symbol c, unless each of the four substrings u, u', v', v contains precisely one occurrence of c. In this last case there are at most a finite number of non-negative integers n for which new strings of the form $a^*ca^nca^*$ can be generated by r. Consequently, for every finite set R of rules that preserve L and any finite subset I of L, there is a positive integer N for which the language generated by $S = (R, I)$ is contained in $\{ a^*ca^kca^* : 0 \le k \le N \}$. Consequently L is not generated by any finite splicing system.

Definition 4. *A splicing system $S = (R, I)$, is said to be* reflexive *if, for each rule $r = (u, u'; v', v)$ in R, the language that S generates is preserved by the rules $\dot{r} = (u, u'; u, u')$ and $\ddot{r} = (v', v; v', v)$.*

In Section 5.2 it is explained how biomolecular considerations suggest that the reflexive systems be given special attention. Note that a splicing system (R, I) is surely reflexive if, for any rule r in R, both $\dot{r}$ and $\ddot{r}$ are also in R, although they do not need to be listed in R if they do not directly contribute to the generation of the splicing language. The splicing systems from which the five previously listed splicing languages $(aacgcg)^*aa$, $a^*c + ca^*$, a^*ca^*, $c(aa)^*$, and A^* were generated can be verified to be reflexive.

Exercises

1. Let $A = \{a, c, g, t\}$ serve as the alphabet. For $r = (a, a;\ c, c)$ and $R = \{r\}$, find the three languages $R^*(\{aacc\})$, $R^*(\{aaa, ccc\})$, and $R^*(\{aaccaacc\})$.
2. Let $A = \{a, c, g, t\}$. For $r = (a, a;\ c, c)$, $r' = (a, c;\ a, c)$ and $R = \{r, r'\}$, find the languages $R^*(\{aacc\})$, $R^*(\{aaa, ccc\})$, and $R^*(\{aaccaacc\})$.
3. Let $A = \{0\}$. For $R = \{(00, \lambda;\ \lambda, 00)\}$, find $R^*(\{\lambda, 00\})$. [Be careful here.]
4. Let $A = \{1\}$. Find three different splicing systems that generate 1^*.
5. Let $A = \{0, 1\}$. Find a splicing system $S = (R, I)$ that generates the language 1^*0^*.
6. Which of the splicing systems in the first five exercises is reflexive?
7. Let $A = \{0\}$. Prove that the language $(00)^*$ is not the language generated by any finite slicing system.

5.2 The Biomolecular Inspiration for the Splicing Concept

This section is provided for those readers who wish to be acquainted with the molecular behaviors from which the splicing concept was developed as an abstraction. This section is organized to allow the reader who has completed Section 5.1, but does *not* wish to be concerned with the relevant biomolecular science, to proceed immediately to Section 5.3, skipping this entire Section 5.2. For brevity we focus attention on linear fully double stranded DNA molecules (ds-DNA). Although ds-DNA molecules tend to occur in the famous double helical form discovered by J. Watson and F. Crick, it is adequate for our purposes here to ignore the helical form and represent ds-DNA molecules in the form exemplified in this display:

$$\begin{array}{ll} 5'\text{-ATCGTGAGGC-}3' & 5'\text{-GCCTCACGAT-}3' \\ 3'\text{-TAGCACTCCG-}5' & 3'\text{-CGGAGTGCTA-}5' \end{array}$$

For this paragraph only, consider the display as representing *four* single stranded DNA (ss-DNA) molecules. Each of the four upper case letters, A, C, G, T, represents one of the four deoxyribonucleotides of which each ss-DNA molecule is viewed as a chain. Each deoxyribonucleotide has the structure of a ribose (pentagonal) sugar that contains five carbon atoms that are numbered $1'$, $2'$, $3'$, $4'$, and $5'$. A phosphate is attached at the $5'$ carbon. At the $1'$ carbon the essential information is attached in the form of one of the four nuclear bases: either adenine, A, cytosine, C, guanine, G, or thymine, T. The individual nucleotides are linked by the phosphate at the $5'$ carbon of one ribose ring being attached to the $3'$ carbon of the next ribose ring. This allows us to speak of an ss-DNA molecule as consisting of a 'sugar-phosphate backbone' having one of the four bases, A, C, G, or T attached to each sugar.

One end of an ss-DNA has a $3'$ carbon to which no phosphate is attached. The other end has a phosphate protruding from a $5'$ carbon but not attached to any other carbon. It is very important to understand the distinction between the $5'$ end and the $3'$ end of an ss-DNA molecule. Consider the upper left ss-DNA in the display above. The left most nucleotide incorporates adenine and has a phosphate protruding from its $5'$ end. The right most nucleotide incorporates cytosine and has no phosphate attached at its $3'$ carbon. Each of the four ss-DNAs in the display can now be understood. All the chemical bonds in the ss-DNA molecules are strong (covalent) chemical bonds. When we wish to break one of these bonds we use special molecular tools (enzymes). Mild changes of conditions, such as moderate increases in temperature, will not break these bonds.

In this paragraph consider the display above as representing *two* ds-DNAs. The first principle concerning the formation of ds-DNA molecules from two ss-DNAs is that the $5' \to 3'$ orientations of the two strands must be opposites. Note that this holds in the display above. The bonds that hold the two ss-DNAs in association are weak (hydrogen) bonds. This allows the two strands of a ds-DNA to be separated into its component ss-DNAs by a moderate increase in temperature. In (perfectly formed) ds-DNA, a nucleotide A in one strand bonds only to a T in the other strand and vice versa. Likewise a C in one strand bonds only to a G in the other strand and vice versa. These conditions are the so-called *Watson-Crick pairing* conditions. In the display above each of the two ds-DNAs consists of a pair of adjacent ss-DNAs that are perfectly matched as Watson-Crick hydrogen bonded pairs.

For the reader who progresses to more subtle considerations than are required here, it will be worth noting that A and T are bonded through two hydrogen bonds, but C and G are bonded through three. As a consequence, less energy is required to pull A-T pairs apart than to pull C-G pairs apart.

In modeling there sometimes arise subtle points that must be taken into account that need not arise in the formal theory of splicing. Here is the *first such point*: A formal string of symbols 'lives' in a one-dimensional space (the ordered line). A molecule 'lives' in three-dimensional (not linearly ordered) space. However, consider again the display above. As *formal symbol strings* over the alphabet, D, consisting of the *four* compound *bi-level* symbols:

$$\boxed{\begin{matrix}A\\T\end{matrix}}\quad\boxed{\begin{matrix}C\\G\end{matrix}}\quad\boxed{\begin{matrix}G\\C\end{matrix}}\quad\boxed{\begin{matrix}T\\A\end{matrix}}$$

they are *quite different*. However, as representations of *molecules* they are *the same*: If either of these two molecules

$$\begin{matrix}5'\text{-ATCGTGAGGC-}3' & \qquad & 5'\text{-TAGCACTCCG-}3'\\ 3'\text{-TAGCACTCCG-}5' & \qquad & 3'\text{-CGGAGTGCTA-}5'\end{matrix}$$

is rotated $180°$ about its mid-point while remaining in the plane of this page, the 'other' molecule is obtained. This means that ds-DNA *molecules* are actually represented by *equivalence classes of strings*, where each equivalence class consists of either exactly *two strings* or, sometimes only *one string*: The string

5′-CCCCCCGGGGGG-3′
3′-GGGGGGCCCCCC-5′

when rotated 180° about its center, remains the same string. Chemists say that such a ds-DNA molecule has *dyadic symmetry*. The equivalence class of such a molecule is a *singleton*.

As tools to cut ds-DNA we use enzymes that are called *restriction endonucleases*, or, more casually, *restriction enzymes*. Such enzymes are naturally occurring products produced by bacteria. There are currently over 200 different restriction enzymes available from laboratory supply houses. One such enzyme is *Bam*H I. When this enzyme encounters a ds-DNA molecule m in an aqueous (water) solution it can cut m into two pieces if m has a six base pair segment with the sequence:

5′-GGATCC-3′
3′-CCTAGG-5′

Suppose that m has the form

5′-XXXXXXXXXXGGATCCYYYYYYYYYY-3′
3′-XXXXXXXXXXCCTAGGYYYYYYYYYY-5′

where each of the symbols X and Y denotes an unspecified symbol (a 'don't care symbol') from $\{A, C, G, T\}$ and where we require only that each vertically displayed pair is Watson-Crick compatible. A cut by *Bam*H I at the indicated site in the molecule m results in two molecules neither of which is fully double stranded:

5′-XXXXXXXXXXG|GATCCYYYYYYYYYY-3′
3′-XXXXXXXXXXCCTAG|GYYYYYYYYYY-5′

Notice that the (strong) covalent bonds GG in both the upper strand and the lower strand have been cut by *Bam*H I. With these two covalent bonds cut, there is insufficient strength provided by the (weak) hydrogen (vertically displayed) bonds to hold the left and right portions of the molecule together. Thermal agitation in the solution is adequate to disrupt such a four term sequence of hydrogen bonds. It is still true that these two halves are attracted by their potential for forming again these hydrogen bonds. However, if they do form again, without a means to have the covalent GG bonds restored, they will separate once again. The two projecting four term single stranded segments are called *overhangs* or *sticky ends*. If we remove *Bam*H I from the solution and add in its place an enzyme called a *ligase* then, when two mutually compatible sticky ends and a ligase enzyme come sufficiently close together the ligase will re-establish the two previously cut strong GG bonds, restoring the original fully double stranded DNA molecule.

Here is the *second point* that must be taken into account in modeling the behavior of molecules in aqueous solution: Suppose that (at least) two molecules are present that have the sequence given above for m. Then after each is cut with *Bam*H I, there are two distinct left segments (and two distinct right segments) having sticky ends. Since these segments are in water (in three dimensional space, not on a line on a page), when *Bam*H I is replaced by a ligase, two left segments can relate as illustrated:

5′-XXXXXXXXXXG|GATCCXXXXXXXXXXX-3′
3′-XXXXXXXXXXCCTAG|GXXXXXXXXXXX-5′

which allows them to form hydrogen bonds and be ligated to form a ds-DNA molecule. Two right segments can likewise pair together to form a ds-DNA. Consequently the cut operation with *Bam*H I, followed by the paste operation provided by a ligase, can potentially yield any one of the three ds-DNA molecules:

1. the original molecule m,
2. the molecule:

   5′-XXXXXXXXXXGGATCCXXXXXXXXXXX-3′
   3′-XXXXXXXXXXCCTAGGXXXXXXXXXXX-5′

3. the molecule:

   5′-YYYYYYYYYYGGATCCYYYYYYYYYYY-3′
   3′-YYYYYYYYYYCCTAGGYYYYYYYYYYY-5′

In wet lab experiments the number of ds-DNA molecules of a given sequence used is vast (perhaps trillions). Consequently molecules of each sequence that are *enabled to appear* are *expected to appear*. In the present case the three molecules indicated above would surely arise.

When the restriction enzyme *Bgl* II encounters a ds-DNA molecule m' in solution it can cut m' into two pieces if m' has a six base pair segment with the sequence:

5′-AGATCT-3′
3′-TCTAGA-5′

Suppose that m' has the form

5′-UUUUUUUUUUAGATCTVVVVVVVVVV-3′
3′-UUUUUUUUUUTCTAGAVVVVVVVVVV-5′

where each of the symbols U and V denotes an unspecified symbol from { A, C, G, T } and each vertically displayed pair is Watson-Crick compatible. A cut by *Bgl* II at the indicated site in m' results in the two molecules

5′-UUUUUUUUA|GATCTVVVVVVV-3′
3′-UUUUUUUUTCTAG|AVVVVVVV-5′

The covalent bonds between A and G (5′-AG-3′ in the upper strand and 3′-GA-5′ in the lower strand) have been cut by *Bgl* II.

Suppose now that we have a solution containing the molecules m and m' for which the sequences have been suggested in the two preceding paragraphs. Suppose that both *Bam*H I and *Bgl* II are also present in the solution. Then the four segments, two each from m and m' will result. We have chosen this pair of restriction enzymes because they produce *the same sticky ends.* Consequently when these enzymes are replaced by a ligase the following (recombinant) molecule may arise:

5′-XXXXXXXXXXGGATCTVVVVVVV-3′
3′-XXXXXXXXXXCCTAGAVVVVVVV-5′

There are several other molecules that may arise, but this one immediately above will motivate the precise definition of *the splicing operation* applied to the *ordered pair* m, m'. As a valuable exercise the reader may wish to determine the seven further ds-DNA molecules that may arise, in addition to the two original molecules and the one exhibited above. Note that, when a segment of m is combined with a segment of m', the result is a molecule which no longer has a site that can be cut by either *Bam*H I or *Bgl* II.

We make one more simplification of the notation for ds-DNA molecules before tying our molecular considerations to formal splicing theory as introduced in Section 5.1. In a ds-DNA molecule, with fully and perfectly matched single strands, each single strand determines the other. Consequently we need to record only one strand of a ds-DNA when we know that the strands are fully and perfectly matched (with no sticky ends). We therefore make the convention of representing a ds-DNA by listing only one strand. This would require labeling at least one of the two ends as 3' or 5'. We make the further convention that, when no label is given for either end of a string displayed on a line, the 5' end is the *left* end. Finally, when these conventions are used we employ the lower case. These denotations apply also to enzyme sites.

Summarizing these molecular examples:

1. *aaaggatccgg* denotes the molecule

   5′-AAAGGATCCGG-3′
   3′-TTTCCTAGGCC-5′

2. *ccggatccttt* denotes *the same molecule* as in part 1. [Be careful here.]
3. *ttagatctccc* denotes the molecule

   5′-TTAGATCTCCC-3′
   3′-AATCTAGAGGG-5′

4. *Bam*H I and *Bgl* II cut at segments of the form *ggatcc* and *agatct*, respectively.

We have seen above that *Bam*H I and *Bgl* II cut in such a way that they produce identical sticky ends. We observed that molecules m, m' when cut by *Bam*H I and *Bgl* II and provided with a ligase may produce a fully and perfectly matched ds-DNA *recombinant* molecule consisting of the left portion of m and the right portion of m'. The splicing concept was developed to provide a context in which strings are spliced (recombined) in simulation of the recombination procedures carried out on ds-DNA molecules in the presence of sets of restriction enzymes and a ligase. The original formulation of splicing [5] was kept very close to the biochemistry. There is still merit in thinking in terms of the original model when considering actual molecular processes. However the precision of the original formulation is maintained and made more elementary to handle in mathematical proofs by employing the improved formulation given by George Păun. For developing formal language theoretic splicing, Păun's notation used here (in Sec. 1 and below) is much more appropriate. The result of cutting with enzymes and recombining with a ligase can be expressed by an ordered quadruple of strings. The example appropriate for representing the action of *Bam*H I, *Bgl* II, and a ligase will make clear the general procedure. The formal representation of the generative capacity of any set of restriction enzymes and a ligase acting on a set of ds-DNA molecules can be represented in the fashion illustrated here:

The compatibility of the sticky ends produced by cuts with *Bam*H I and *Bgl* II is conveyed by listing the sites at which these enzymes act, *ggatcc* and *agatct*, respectively, in this order in an ordered tuple. The extra information that specifies which symbols in the sites yield the sticky ends can be included by further segmenting each of the sites to produce the ordered quadruple $r = (g, gatcc; a, gatct)$ that constitutes the splicing rule r that we associate with the *ordered pair* of enzymes *Bam*H I, *Bgl* II. The reader will now profit from rereading Section 5.1, Example 1, which is a purely formal version of the present discussion. The two sites relevant for rule r in the ordered pair of strings (molecules) *aaaggatccgg* and *ttagatctccc* [listed as molecular examples 1 and 3 above] are illustrated by segmenting these two strings appropriately: *aaa-g,gatcc-gg* and *tt-a,gatct-ccc*. From the rule r and the segmentations we see that we splice at the commas to produce the recombinant string (molecule) *aaa-g,gatct-ccc* = *aaaggatctccc*.

We give one more molecular example. Example 2 of Section 5.1 was constructed as a purely formal version of this next molecular example. The restriction enzyme *Bst*U I cuts a ds-DNA molecule m in aqueous solution into two pieces if m has a four base pair segment with the sequence:

5′-CGCG-3′
3′-GCGC-5′

Suppose that m has the form

5′-X...XCGCGY...Y-3′
3′-X...XGCGCY...Y-5′

where each of the symbols X and Y denotes an unspecified symbol (a 'don't care symbol') from $\{A, C, G, T\}$ and where each vertically displayed pair is Watson-Crick compatible. A cut by *Bst*U I at the indicated site in the molecule m results in two molecules both of which are fully double stranded:

$$\begin{array}{l|l} 5'\text{-X...XCG} & \text{CGY...Y-}3' \\ 3'\text{-X...XGC} & \text{GCY...Y-}5' \end{array}$$

Notice that the (strong) covalent bonds 5′-GC-3′ (= 3′-CG-5′) in both the upper strand and the lower strand have been cut by *Bst*U I. Note that this enzyme does not provide single stranded overhangs as do the enzymes *Bam*H I and *Bgl* II. Enzymes such as *Bst*U I are said to leave *blunt ends*. Even though cutting leaves blunt ends, re-attachment of the resulting blunt ends can be done using a ligase. Thus even the blunt ends should be regarded as being sticky. Consequently, when molecules

$$\begin{array}{lcl} 5'\text{-X...XCGCGY...Y-}3' & & 5'\text{-U...UCGCGV...V-}3' \\ 3'\text{-X...XGCGCY...Y-}5' & \text{and} & 3'\text{-U...UGCGCV...V-}5' \end{array}$$

are cut by *Bst*U I, and a ligase is then applied, the following molecules may arise:

$$\begin{array}{lcl} 5'\text{-X...XCGCGV...V-}3' & & 5'\text{-U...UCGCGY...Y-}3' \\ 3'\text{-X...XGCGCV...V-}5' & \text{and} & 3'\text{-U...UGCGCY...Y-}5' \end{array}$$

Don't overlook the fact that the two original molecules may be reconstructed and that six additional new molecules may arise, exemplified by the two molecules:

$$\begin{array}{lcl} 5'\text{-X...XCGCGX...X-}3' & & 5'\text{-X...XCGCGU...U-}3' \\ 3'\text{-X...XGCGCX...X-}5' & \text{and} & 3'\text{-X...XGCGCU...U-}5' \end{array}$$

The astute reader may wish to ask whether when a ligase is added to a solution that is *initialized* to contain molecules such as

$$\begin{array}{l} 5'\text{-X...XCGCGY...Y-}3' \\ 3'\text{-X...XGCGCY...Y-}5' \end{array}$$

concatenations of these molecules, such as

$$\begin{array}{l} 5'\text{-X...XCGCGY...YX...XCGCGY...Y-}3' \\ 3'\text{-X...XGCGCY...YX...XGCGCY...Y-}5' \end{array}$$

and

$$\begin{array}{l} 5'\text{-(X...XCGCGY...Y)}^n\text{X...XCGCGY...Y-}3' \\ 3'\text{-(X...XGCGCY...Y)}^n\text{X...XGCGCY...Y-}3' \end{array}$$

may arise. Would blunt end ligation allow these additional molecules to be produced? *This point is subtle and may be glossed over by readers interested only in the formal theory.* The answer is 'no', but only by the convention of assuming that the *initially given* DNA molecules are not provided with a phosphate attached at the 5′ end. The ligase requires the presence of a phosphate at the 5′ end in order to complete the necessary covalent bond. When a restriction enzyme cuts a ds-DNA molecule it leaves the phosphate attached at the freshly produced 5′ end. This allows re-ligation at ends created by such cuts. Researchers working with DNA have the biochemical means of either *having* or *not having* phosphates attached at the 5′ ends of the *initial* DNA molecules. Thus our 'default' assumption is that our *initial* DNA molecules have no phosphates at their 5′ ends. (In short, a blunt end is 'sticky' if and only if a phosphate is present on its 5′ strand.) If one wishes to have all possible concatenations to be potentially present, then simply begin with DNA for which phosphates are present at the 5′ ends. (Note: blunt end ligation is quite effective in circularizing linear molecules having blunt ends.)

We have noted that the site at which the restriction enzyme *Bst*U I cuts is

$$\begin{array}{c}5'\text{-CGCG-}3'\\3'\text{-GCGC-}5'\end{array}$$

Using our convention for compressing ds-DNA sequences, we specify this site in the short form: *cgcg*. Attending to the fact that *Bst*U I cuts exactly in the middle of its site, it is natural for us to model its recombining capacity (when accompanied by a ligase) with the rule: $r = (cg, cg; cg, cg)$. This is the rule used in Example 2 of Section 5.1. As Example 2 is re-read now one sees that it is an abstract representation of the fact that when *Bst*U I and a ligase are added to an aqueous solution containing the molecules

$$u = \begin{array}{c}5'\text{-AACGCGAACGCGAA-}3'\\3'\text{-TTGCGCTTGCGCTT-}5'\end{array} \text{ and } v = \begin{array}{c}5'\text{-TTCGCGTT-}3'\\3'\text{-AAGCGCAA-}5'\end{array}$$

the following two recombinant ds-DNA molecules may arise (from this ordered pair u,v):

$$\begin{array}{c}5'\text{-AACGCGTT-}3'\\3'\text{-TTGCGCAA-}5'\end{array} \text{ and } \begin{array}{c}5'\text{-AACGCGAACGCGTT-}3'\\3'\text{-TTGCGCTTGCGCAA-}5'\end{array}.$$

Moreover (from the ordered pair v,u) these two recombinant molecules may also arise:

$$\begin{array}{c}5'\text{-TTCGCGAACGCGAA-}3'\\3'\text{-AAGCGCTTGCGCTT-}5'\end{array} \text{ and } \begin{array}{c}5'\text{-TTCGCGAA-}3'\\3'\text{-AAGCGCTT-}5'\end{array}.$$

One may note that the ordered pairs u,u and v,v provide yet more recombinant molecules.

Examples 3 and 4 of Section 5.1 should now be thought through again, this time as providing abstract descriptions of recombinant behaviors of ds-DNA molecules made possible by the presence of *Bst*U I and a ligase.

An astute reader may observe that we could just as well use for the rule r any one of several rules that will provide exactly the same generating power.

Two such alternates are $(cgcg, \lambda;\ cgcg, \lambda)$ and $(\lambda, cgcg;\ \lambda, cgcg)$. These rules do not suggest the same cutting behavior as $(cg, cg;\ cg, cg)$. However, if one c is not considered different from another and likewise for g, then each of these rules is exactly equivalent in generative power to another. The rule $(cgcg, \lambda;\ cgcg, \lambda)$ is said to have only *left context* and the rule $(\lambda, cgcg;\ \lambda, cgcg)$ is said to have only *right context*. Splicing systems for which all rules have one sided context generate especially simple types of languages [6]. If some bases are labeled (by radioactivity, fluorescence, or otherwise) then a difference in effect of such rules can be detected, but not otherwise.

When one wishes to model with a splicing system the full generative power of a set of ds-DNA molecules under the action of a set of restriction enzymes and a ligase, then an awareness of all the considerations discussed in Section 5.2 is required. This is illustrated with a single example: Suppose that it is desired to model the set of fully well formed linear ds-DNA molecules (with no sticky ends) that can potentially arise in an appropriate aqueous solution containing *Bam*H I, *Bgl* II and a ligase as ds-DNA molecules, each having the sequence $m = a^{20}ggatcca^{20}agatcta^{20}ggatcca^{20}$, are added to the solution. Recalling the sites at which *Bam*H I and *Bgl* II act one might too quickly respond by listing the splicing system $S = (R, I)$ where $R = \{\, r \,\}$, $r = (g, gatcc;\ a, gatct)$, and $I = \{\, m \,\}$. This would be woefully inadequate. By recalling that molecules live in three-dimensional space (not on a line), we must include the alternate linear representation of m, namely $m' = t^{20}ggatcct^{20}agatctt^{20}ggatcct^{20}$, in the initial language. When two molecules are cut by the same enzyme, the left segment of one can inevitably be ligated to the right segment of the other. Consequently the presence of *Bam*H I enables the rule $r_1 = (g, gatcc;\ g, gatcc)$ and *Bgl* II enables $r_2 = (a, gatct;\ a, gatct)$. (*This is the consideration that guarantees that the splicing systems that arise in the modeling discussed here are inevitably reflexive.*) Finally, two molecules in solution, when modeled on a line, must be allowed to occur in either order. Thus the rule $r_3 = (a, gatct;\ g, gatcc)$ is enabled. The appropriate splicing model for the representation of the generative activity of *Bam*H I, *Bgl* II and a ligase acting on molecules having the sequence $\{\, m \,\}$ is: $S = (R, I)$ where $R = \{\, r, r_1, r_2, r_3 \,\}$ and $I = \{\, m, m' \,\}$.

Summary: What precisely is claimed when a splicing system is given as a model of the generative activity of a specified set of restriction enzymes and a ligase acting on ds-DNA molecules having sequences belonging to a specified set? *The answer includes some subtleties*: Let R be the set of rules that express the cutting and rejoining actions of the enzymes (understood as independent of the choice of the initial set). Let I be the set of strings that represent the initial ds-DNA molecules (understood as closed under 180° rotation). The language $R^*(I)$ that is generated consists of the strings that represent sequences of the fully double stranded DNA molecules (*without sticky ends*) which can potentially arise through the action of the given enzymes on *sufficiently* many ds-DNA molecules each having a sequence belonging to I. There is no need to make the unpleasant assumption of the existence of an *actual infinity* of initial molecules although this is sometimes done informally. (Recall that the set of positive integers may be described as those numbers that can potentially arise through *sufficiently* many additions of one to an initial set consisting of a single one.) Finally, we remark that we have idealized away conditions that are too detailed to merit inclusion here (temperatures, buffering, etc.). For further details see [5] and [4].

Example Application: Given a finite set of restriction enzymes and a finite set of ds-DNA sequences, can a specified ds-DNA molecule, M, be constructed from sufficiently many ds-DNA molecules having sequences in the given set through the application of the given restriction enzymes and a ligase? General procedure: Construct the splicing system model for the given data. Section 5.3 contains an algorithm that produces a finite automaton that recognizes the language generated by the splicing system. M can be constructed by the specified means if and only if the automaton recognizes the string that represents M. In many special cases the general algorithm can be by-passed and simpler procedures can be used; see Section 5.5 for more discussion and further references. Procedures for contracting long sub-sequences to single auxiliary alphabet symbols may allow the longer ds-DNA molecules to be treated.

Exercises

1. List all the sequences of the ds-DNA molecules (without sticky ends) that can arise as ds-DNA molecules having the sequence

$$\begin{array}{c}5'\text{-}\mathrm{C}^{19}\mathrm{GGATCCC}^{17}\text{-}3'\\ 3'\text{-}\mathrm{G}^{19}\mathrm{CCTAGGG}^{17}\text{-}5'\end{array}$$

are added to an aqueous solution containing *Bam*H I and a ligase.

2. Find all the sequences of the ds-DNA molecules that can arise as ds-DNA molecules having the sequence

$$\begin{array}{c}5'\text{-}\mathrm{G}^{19}\mathrm{GGATCCC}^{15}\mathrm{GGATCCG}^{17}\text{-}3'\\ 3'\text{-}\mathrm{C}^{19}\mathrm{CCTAGGG}^{15}\mathrm{CCTAGGC}^{17}\text{-}5'\end{array}$$

are added to an aqueous solution containing *Bam*H I and a ligase.

3. Find all the sequences of the ds-DNA molecules that can arise as ds-DNA molecules having the sequence

$$\begin{array}{c}5'\text{-}\mathrm{G}^{19}\mathrm{CGCGC}^{15}\mathrm{GGATCCG}^{17}\text{-}3'\\ 3'\text{-}\mathrm{C}^{19}\mathrm{GCGCG}^{15}\mathrm{CCTAGGC}^{17}\text{-}5'\end{array}$$

are added to an aqueous solution containing *Bam*H I, *Bst*U I, and a ligase.

4. Give splicing models, with alphabet $A = \{a, c, g, t\}$, appropriate for Exercises 2 and 3.

5.3 Regularity of the Splicing Languages

The goal of this section is to prove that finite splicing languages are regular. The proof of regularity requires that the set of rules is finite, but only requires that the initial language is regular. Here is the precise statement:

Theorem 1. *If R is a finite set of splicing rules and I is a regular language then the splicing language $R^*(I)$ is regular.*

This was first proved by Culik and Harju [1]; we follow the proof given in [9].

We shall prove the theorem by constructing an automaton which recognizes the splicing language $R^*(I)$. The automaton will be non-deterministic, with null transitions, and will be constructed in a number of stages. We start with an automaton $M = (Q, A, s_0, F, \delta)$ which accepts the initial language I, and we augment this automaton as follows. Suppose $r = (u, u';\ v', v)$ is a rule in R, and let B_r be an automaton with initial state i_r and terminal set $\{ t_r \}$ which accepts exactly the string uv. The details of B_r are not important; the obvious choice is just a linear graph leading from i_r to t_r with edges labeled by the symbols in uv. We arrange that the states of the automata B_r for different rules are disjoint from each other and from Q. Then we define a new automaton $M_0 = (Q_0, A, s_0, F, \delta_0)$ so that Q_0 is the union of Q and the state sets of the various B_r and δ_0 is the "union" of δ and the transition relations for the various B_r.

Notice that the initial state is not changed, nor is the set of terminal states. These are in Q, and there are no transitions between Q and the new states, so the new states cannot be used in accepting any string. Hence M_0 accepts the language I. We shall refer to the automaton B_r as the r-bridge, and to i_r and t_r as the entry and exit states for this bridge.

To illustrate the construction we use a very simple example. The alphabet is $\{ a, b, c \}$, the initial language contains only *aacbacbbc*, and there are only two rules, $r_1 = (ab, c;\ a, a)$ and $r_2 = (ba, c;\ b, b)$. Then we can represent M_0 graphically as follows:

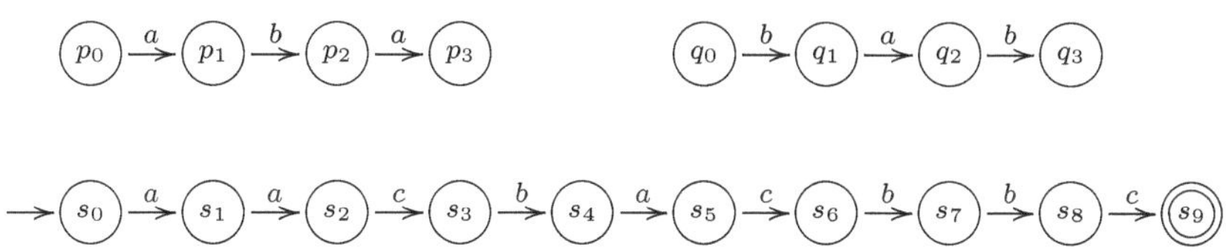

Here the original state sets in Q are s_j, $0 \leq j \leq 9$, with s_9 being the only accepting state. The states $p_0, \ldots, p_3$ are the bridge states for r_1, so $i_{r_1} = p_0$ and $t_{r_1} = p_3$. Similarly $q_0, \ldots, q_3$ are the bridge states for r_2, so $i_{r_2} = q_0$ and $t_{r_2} = q_3$.

Now we construct a sequence of automata $M_k = (Q_0, A, s_0, F, \delta_k)$ by recursion, starting with M_0 as above. So we suppose that M_{k-1} has been constructed and we explain the construction of M_k. The only difference is that

M_k may contain certain null transitions that are not present in M_{k-1}. Suppose $r = (u, u'; v', v)$ is a rule in R. If i is a state in M_{k-1} satisfying

1. i is not an exit state of any bridge, and
2. some accepting path in M_{k-1} is in state i immediately before reading uu' as a substring,

then we add a null transition from i to i_r; and if t is a state in M_{k-1} satisfying

1. t is not an entry state of any bridge, and
2. some accepting path in M_{k-1} is in state t immediately after reading $v'v$ as a substring,

then we add a null transition from t_r to t. The automaton M_k is the result of adding all such null transitions to the automaton M_{k-1}.

The first type of transition, leading to some entry state i_r, is called an *entry transition*; and the second type, leading from some exit state t_r, is called an *exit transition*. Note that entry transitions cannot start at exit states and exit transitions cannot end at entry states, so this classification is unambiguous. An entry or exit transition is said to be of *level k* if it is not already present in the automaton M_{k-1}. This notion will be very important in the final phase of the proof.

Continuing our example above, the only string accepted by the automaton M_0 is *aacbacbbc*. This contains the string *bac*, corresponding to the site uu' of rule r_2. There is only one accepting path for this string, and that path is in the state s_3 just before reading the substring *bac*, so we add an entry transition from s_3 to q_0, the r_2 entry state. We also find the string *bb* in *aacbacbbc*, which is the site $v'v$ of r_2, and the only accepting path is in the state s_8 just after reading the substring *bb*. Thus we add an exit transition from q_3, the r_2 exit state, to s_8. Examining the other rule r_1, we see that *abc*, the uu' site, does not occur in *aacbacbbc*, but that *aa*, the $v'v$ site, does occur. As above, we add an exit transition from p_3, the r_1 exit state, to s_2. Here is the resulting automaton M_1; we have labeled the new transitions with 1, but it should be remembered that these are null transitions.

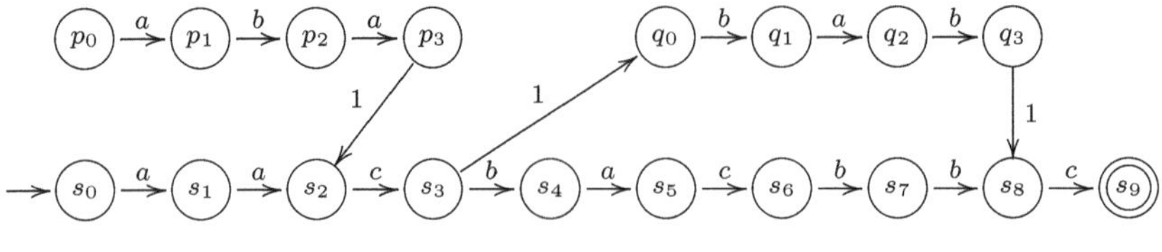

Now consider the language accepted by the automaton M_1. Of course the initial string *aacbacbbc* is still accepted. However, there is now another accepting path: From s_0 to s_3 reading *aac*, along the entry transition to the bridge for r_2, from q_0 to q_3 along the bridge reading *bab*, along the exit transition to s_8, and then from s_8 to s_9 reading *c*. Hence the string *aacbabc* is accepted by M_1. This is the only new string accepted by M_1, since the states of the r_1 bridge do not participate in an accepting path. Also notice that *aacbabc* is the result of splicing two copies of *aacbacbbc*, using r_2 at the sites (ba, c) and (b, b).

We now repeat the construction: We must examine the new string *aacbabc* for any sites which require adding more entry or exit transitions. This string contains *abc*, the uu' site of r_1. The only accepting path for this string is in the state q_1 just before reading *abc*, so we must add an entry transition from q_1 to p_0, the r_1 entry state. There are no other new transitions, so the next automaton M_2 is as follows, where we have labeled the new transition with 2:

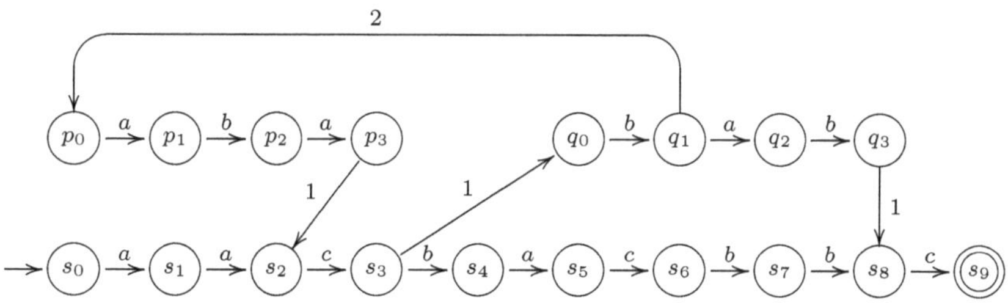

Notice that M_2 now has an accepting path which contains a loop (labeled by the non-empty string *babac*. Hence the language accepted by M_2 is infinite; it corresponds to the regular expression $aac(babac)^*(bacbbc + babc)$. In particular the string *aacbabacbacbbc* is accepted by M_2 and contains two copies of the uu' site *bac* corresponding to r_2. The only accepting path for this string is in the state p_1 just before reading the first *bac*, so we need to add an entry transition from p_1 to the r_2 entry state q_0. This accepting path is in the state s_3 just before the second copy of *bac*, but we do not need to add an entry transition from s_3 to q_0 since one already exists. Further examination of M_2 does not yield any other sites requiring entry or exit transitions, so our next automaton M_3 appears as follows, where we have labeled the new entry transition with 3:

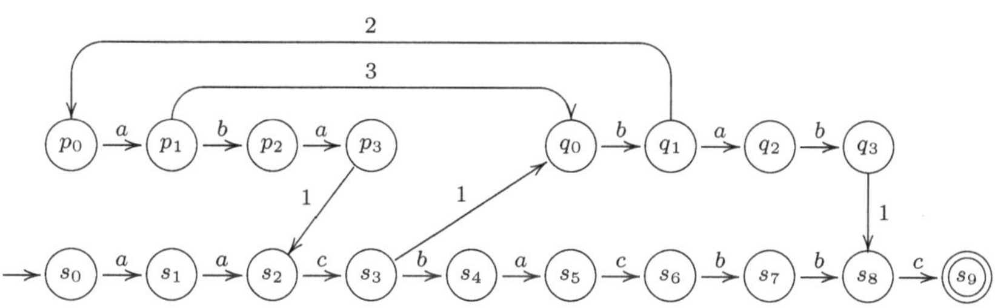

There is another loop in the graph, following the path $q_0 \to q_1 \to p_0 \to p_1 \to q_0$ while reading the string *ba*, and the language accepted by M_3 is larger than the language accepted by M_2. However, the reader should verify that there are no other opportunities for building entry or exit transitions, so our process stops here. We claim that the language accepted by M_3 is the splicing language defined by the splicing system (R, I).

Rather than verifying that M_3 recognizes the splicing language in the example, we now explain why the process works in general. There are three steps in the proof:

1. The successive construction of automata terminates after a finite number of steps with an automaton M_n which does not allow further construction of entry or exit transitions.
2. The language L_n accepted by M_n contains I and is closed under splicing by rules in R, so L_n contains the splicing language $R^*(I)$.
3. Every string accepted by M_n is in the splicing language $R^*(I)$.

For the first step, consider that M_k differs from M_{k-1} only in the transition relation. Specifically, M_k allows the same transitions as M_{k-1} plus a number of null transitions that were not allowed by M_{k-1}. The total number of such null transitions is limited: Any such null transition represents a connection between a state in Q_0 and either i_r or t_r for some r in R; so the total number of such null transitions is bounded by $2N_Q N_R$ where N_Q is the number of states in Q_0 and N_R is the number of rules in R. Hence it is impossible to continue forever finding such null transitions to add to the automaton so, for some n, the automata M_n and M_{n+1} are the same.

For the second step, first notice that M_n still contains all the states and transitions of the original automaton M, so M_n accepts every string that is accepted by M. That is, L_n, the language accepted by M_n, contains I, the language accepted by M. To show that L_n is closed under splicing by rules in R suppose that $r = (u, u'; v', v)$ is a rule in R and suppose that $w = xuu'y'$ and $w' = x'v'vy$ are both accepted by M_n. We must show that the result of splicing w and w' using r at the indicated sites is accepted by M_n. To do this we first analyze w. Choose an accepting path for $w = xuu'y'$ in M_n and follow the *longest* prefix of this path which reads the string x; this leaves the automaton in the state i. If i is the exit state of any bridge then i is not a terminal state of M_n so our accepting path continues beyond i. But the only transitions leading from exit states are null transitions, so we can continue the path at least one more step while still reading just x, contradicting the choice of i. Hence i is not the exit state of any bridge, so, according to the construction of M_{n+1} from M_n, there is an entry transition in M_{n+1} from i to the r entry state i_r. Of course, since $M_{n+1} = M_n$, this entry transition is already available in M_n. Similarly, choose an accepting path for $w' = x'v'vy$ and follow the *shortest* prefix of this path which reads the string $x'v'v$. Arguing as above (but in the reverse direction) we see that t cannot be the initial state of any bridge, and so there is an exit transition from t_r, the r exit state, to t. Now we prescribe a new path in M_n: Follow the accepting path for $w = xuu'v$ until x has been read and the automaton is in state i. Next follow the entry transition from i to i_r, follow the r bridge from i_r to t_r while reading uv, and then follow the exit transition from t_r to t. Finally, follow the rest of the accepting path for $w' = x'v'vy$ from t to an accepting state while reading y. In following this composite path the automaton reads $z = xuvy$, so z is in

L_n. This finishes the argument since z is the result of splicing $w = xuu'y'$ and $w' = x'v'vy$ using r at the indicated sites.

The third step is more difficult. We need to show that every accepting path in M_n accepts a string in the splicing language. The proof works by induction, but it is not clear at first glance what property of the path can be used as a basis for the induction. Obvious things to try are the length of the path, or the number of entry and exit transitions in the path, or the smallest subscript k so that the path lies in M_k. None of these seem to work. We shall, in fact, base the proof on the *complexity* of an accepting path, so we first define this notion and explain how to base an induction proof on it.

Recall that the level of an entry or exit transition is the subscript k so that the transition is in M_k but not in M_{k-1}. If π is a path in the automaton M_n and $1 \le k \le n$ we let $c_k(\pi)$ be the number of entry or exit transitions of level k in π. Then define the *complexity* of π to be the n-tuple $c(\pi) = \langle c_1(\pi), c_2(\pi), \ldots, c_n(\pi) \rangle$. For an example, refer to the diagram for M_3 on page 135, where $n = 3$ and the entry and exit transitions are labeled with their levels. The only accepting path for the string *aacbabc* has complexity $\langle 2, 0, 0 \rangle$ and the only accepting path for *aacbababacbabc* has complexity $\langle 4, 2, 1 \rangle$.

In order to base an induction on complexity we need an order relation, and we use a variant of *lexicographic order*. If c and d are n-tuples of non-negative integers we say c is smaller than d, and write $c \prec d$, if, for some k, we have $c_j = d_j$ for all $j > k$ and, moreover, $c_k < d_k$. In other words, $c \prec d$ means that, when reading the n-tuples from the right, c is less than d in the first index in which they differ. For example, $\langle 9, 4, 3 \rangle$ is smaller than $\langle 1, 5, 3 \rangle$. It is easy to check that this relation is transitive and satisfies the trichotomy law: For any two n-tuples c and d, exactly one of $c \prec d$, $c = d$ or $c \prec c$ holds. In fact, this relation is a *well ordering*, which is the key idea in the proof of the following principle of induction.

Lemma 1. *Suppose P_c is a set of statements indexed by the set of n-tuples of non-negative integers c. Suppose*

If P_c is true for all $c \prec d$ then P_d is true.

Then P_d is true for all n-tuples d.

Proof. Let S be the set of all n-tuples d for which the statement P_d is *false*. The statement of the lemma is equivalent to the assertion that S is empty, so assume that it is *not* empty. The set of all n^{th} components d_n of members d of S is a non-empty set of non-negative integers; let s_n be the minimum of this set of non-negative integers, and let S_n be the set of members d of S for which $d_n = s_n$. Then S_n is not empty, and all the elements of S_n have the same n^{th} component. However, as above, the set of all $(n-1)^{\text{th}}$ components d_{n-1} of elements of S_n has a minimum element s_{n-1}.

After n steps of this process we are left with non-negative integers s_k for $1 \le k \le n$, so that the corresponding n-tuple $s = \langle s_1, s_2, \ldots, s_n \rangle$ is in S and satisfies $s \prec d$ for all other n-tuples d in S. Hence for all c with $c \prec s$ we must have $c \notin S$, and so P_c is true. But from the assumption of the lemma we must then have P_s true, contradicting $s \in S$. This contradiction finishes the proof of Lemma 1. □

So now we are ready to establish the third step of the proof. Suppose π is an accepting path in M_n, and assume that any string accepted by an accepting path of smaller complexity is in the splicing language. We must show that the string accepted by π is in the splicing language.

First consider the possibility that π never traverses an entry transition. Since π starts at s_0 in Q it can never reach a bridge, and so it can never reach an exit transition. Thus π has complexity $\langle 0, 0, \ldots, 0 \rangle$ and it never leaves the automaton M. So the string accepted by π is in the language I accepted by M. Since I is contained in the splicing language, there is nothing more to say.

Alternatively suppose π contains an entry transition. Let the *last* entry transition in the path π be from the state i to the entry state i_r of the r bridge corresponding to some rule $r = (u, u'; v', v)$. Now the only way to leave this bridge is by an entry transition or by an exit transition from the exit state t_r of this bridge to some state t; but we have assumed there are no following entry transitions on the path. Hence we can divide our path π into segments: From s_0 to i while reading some string p; from i to i_r via an entry transition; from i_r to t_r along the r bridge while reading uv; from t_r to t via an exit transition; and finally from t to some state in F while reading some string q. Thus the string accepted by the path is $z = puvq$. We need to identify two strings which are accepted by paths in M_n with smaller complexity and which, when spliced together, produce z.

Suppose the entry transition from i to i_r has level k. Remember the justification for creating this entry transition: It was constructed because there is an accepting path τ in M_{k-1} which accepts a string of the form $xuu'y'$ and is in state i after reading x. We define another path σ in M_n by following the path π from s_0 to i while reading p, and then following the path τ from i to an accepting state while reading $uu'y'$. That is, the string $w = puu'y'$ is read by the accepting path σ in M_n. We can visualize the three paths π, τ and σ (labeled with the strings they accept) as follows:

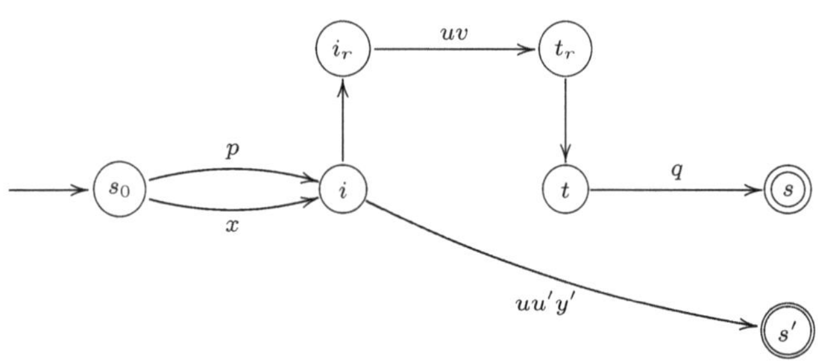

(This is just a schematic diagram; the path labeled p, for example, may contain loops and entry or exit transitions, and may intersect the other paths.)

We denote the subpath of π from s_0 to i as π_1 and the remaining subpath from i to an accepting state as π_2. Similarly we denote the subpaths of τ before and after i as τ_1 and τ_2. Thus σ consists of π_1 followed by τ_2. Now we compare $c(\sigma)$ and $c(\pi)$. We have $c_j(\pi) = c_j(\pi_1) + c_j(\pi_2)$ and $c_j(\sigma) = c_j(\pi_1) + c_j(\tau_2)$. However, if $j \geq k$ we have $c_j(\tau_2) = 0$ since τ is an accepting path in M_{k-1}, so it does not contain any entry or exit transitions of level j. Hence $c_j(\sigma) \leq c_j(\pi)$. Moreover, π_2 contains the entry transition from i to i_r which is of level k, so $c_k(\pi_2) > 0$ and we conclude $c_k(\sigma) < c_k(\pi)$. So the first index j (reading from the right) at which $c(\sigma)$ and $c(\pi)$ differ must satisfy $j \geq k$ and $c_j(\sigma) < c_j(\pi)$; that is, $c(\sigma) \prec c(\pi)$.

Consider also the exit transition from t_r to t. Reasoning as above we first find an accepting path τ' in some $M_{k'}$ which accepts a string $x'v'vy$ and is in state t after reading $x'v'v$; then we connect the first part of this path with the last segment of the path π to form a path σ' in M_n. This path accepts the string $w' = x'v'vq$ and $c(\sigma') \prec c(\pi)$.

Since σ and σ' have smaller complexity than π the inductive hypothesis implies that w and w' are in the splicing language. If we splice $w = puu'y'$ and $w' = x'v'vq$ using the rule r at the indicated sites we get z, so z is in the splicing language.

Now the induction principle of Lemma 1 finishes step 3 of the proof, and so finishes the proof of Theorem 1.

Exercises

1. Following the proof of Theorem 1, construct an automaton which accepts the splicing language generated from the initial string $bbaaac$ using the rules $(b, a;\ ba, a)$, $(b, ac;\ ba, c)$ and $(bb, c;\ b, bc)$. [You will need to construct M_5.] As a check on your construction you should see that the splicing language is infinite and consists of words of the form $b^m a^n c$. What are the restrictions on m and n?
2. Find the complexity of each accepting path in the automaton of Exercise 1.
3. Suppose every rule in R has the form $(a, \lambda;\ a, \lambda)$, where a is in the alphabet. Show that in this case the splicing language is the language accepted by M_1. In fact it may not be true that $M_2 = M_1$, since there may be "null loops" introduced in the second stage.
4. Find an example to show that M_1 does not necessarily define the splicing language if all rules have the form $(a, \lambda;\ b, \lambda)$, where a and b are in the alphabet.
5. Suppose the alphabet has cardinality N. Is there an integer n (in terms of N, but independent of I and R) so that $M_n = M_{n+1}$ for all splicing systems of the form described in Exercise 4? [This is a fairly hard problem.] Note that Exercise 1 can be adapted, by replacing the initial string with bba^pc, to show that no such bound exists in general, even with a fixed choice of R.

5.4 Sets of Splicing Rules

The regularity theorem leads to two natural questions: What happens if the initial language I is not finite? What happens when the set of splicing rules R is not finite?

The first question has a simple answer. We already remarked that the proof of the regularity theorem does not require that the initial language be finite, but just that it be regular. A generalization is that if the initial language I belongs to a *full AFL* $\mathcal{A}$ and the rule set is finite then the splicing language $R^*(I)$ also belongs to $\mathcal{A}$. We do not pursue this further but refer the reader to [8]. Note that this implies, for example, that if I is context-free then so is $R^*(I)$.

The second question is more interesting. To formulate it properly we need to consider a classification of infinite rule sets. For this we introduce an alternative notation for splicing rules: We identify a rule $(u, u'; v', v)$ with the string $u\#u'\$v'\#v$, where $\#$ and $\$$ are two new symbols not in our original alphabet A. This correspondence between rules and strings in $A^*\#A^*\$A^*\#A^*$ is obviously a bijection, and we can use it to define, for example, what it means for a set of rules to be regular.

One reason we might want to look at infinite rule sets is in a type of recognition problem, in which we ask whether a given language is a splicing language. Thus, for a language L, we might look at $\mathcal{R}(L)$, the set of all rules r that preserve L. Clearly, if $L = R^*(I)$ is the splicing language generated by the rule set R and the initial language I then $R \subseteq \mathcal{R}(L)$. If we are working on the recognition problem for regular languages then the following is very helpful.

Theorem 2. *If L is a regular language then so is $\mathcal{R}(L)$.*

We shall prove this theorem as an introduction to the use of the *syntactic monoid* in splicing. The definitions and basic properties of the syntactic monoid are given in the appendix, Section 5.6. We shall use the notation $\operatorname{Syn} L$ for the syntactic monoid of the language L.

Since we are proving the regularity of a set of rules we shall use the string representation in the proof.

Suppose $r = u\#u'\$v'\#v$ is in $\mathcal{R}(L)$. We first notice the following: If u is syntactically congruent to $\tilde{u}$ then $\tilde{r} = \tilde{u}\#u'\$v'\#v$ is in $\mathcal{R}(L)$. To prove this suppose $\tilde{z}$ and z' are two words in L which can be spliced using the rule $\tilde{r}$; we must show that any result $\tilde{w}$ of such splicing must be in L. Thus we have factorizations $\tilde{z} = x\tilde{u}u'y$ and $z' = x'v'vy'$, so $\tilde{w} = x\tilde{u}vy'$. Since $\tilde{z}$ is in L and $\tilde{u}$ is syntactically congruent to u we can replace $\tilde{u}$ with u in $\tilde{z}$ and the result, $z = xuu'y$, is in L. Now z and z' are in L and r is in $\mathcal{R}(L)$, so the result $w = xuvy'$ of splicing z and z' using r is also in L. Now we can repeat the argument above, in the opposite direction: Since w is in L and u is syntactically congruent to $\tilde{u}$ we can replace $\tilde{u}$ with u in w and conclude that $x\tilde{u}vy' = \tilde{w}$ is in L, which is what we wanted.

The same observation, with essentially the same proof, applies to the other components u', v', and v of r. Thus we have shown that, whenever r is in $\mathcal{R}(L)$, we must have $C_r = U\#U'\$V'\#V \subseteq \mathcal{R}(L)$, where U is the syntactic class containing u, U' is the syntactic class containing u', and so on. That is, $\mathcal{R}(L)$ is the union of all such concatenations C_r, for $r \in \mathcal{R}(L)$. By Theorem 7 and closure under concatenation, each C_r is a regular language; and by Theorem 6 there are only finitely many such concatenations, so $\mathcal{R}(L)$ is regular.

In light of Theorem 2 it is natural to conjecture that the splicing language generated by a regular set of rules and a regular (or finite) initial language should be regular. This conjecture fails spectacularly:

Theorem 3. *If L is any RE language then there are a finite set I, a regular set of rules R, and a regular language L_0 so that $L = L_0 \cap R^*(I)$.*

We do not prove this here (see [4]), but we give a simple example to illustrate what can happen:

The alphabet is $A = \{a, b, X, X', Y, Y', Z\}$ and the initial set is $I = \{XY, ZbY', X'aZ\}$. The rule set R consists of $(\lambda, Y;\ Y, \lambda)$, together with all rules of the form $(\lambda, Z;\ X, wY)$, $(X'w, Y;\ Z, bY')$, $(X, \lambda;\ X', wY')$, or $(Xw, Y';\ \lambda, Y)$, where w varies over all words in $\{a, b\}^*$. Then $\{a, b\}^* \cap R^*(I)$ is the non-regular language $\{a^n b^n : n \geq 0\}$.

Exercises

1. Show that $\mathcal{R}(L) = \emptyset$ where $L = (a^2)^*$ and $A = \{a\}$.
2. A rule r is *useful* for a language L if there are words w_1 and w_2 in L which may be spliced using r. Let $\mathcal{R}_0(L)$ denote the set of rules in $\mathcal{R}(L)$ which are useful for L. Show that $\mathcal{R}_0(L)$ is a regular set of rules if L is regular.
3. Find $\mathcal{R}_0(L)$ where $L = b(a^2)^*$ and $A = \{a, b\}$.
4. Provide the details for the example at the end of the section.

5.5 Constants and Reflexive Splicing Systems

Recall that a splicing system (R, I) is *reflexive* if the splicing language $R^*(I)$ is preserved by both $\dot{r} = (u, u';\ u, u')$ and $\ddot{r} = (v', v;\ v', v)$ whenever $r = (u, u';\ v', v)$ is in R. A language L is called a *reflexive* splicing language if it is generated by a finite reflexive splicing system. Our understanding of such languages is based on a notion introduced (in a very different context) by Schützenberger: We say a word z is a *constant* of the language L if, for any strings x, y, x' and y', if xzy and $x'zy'$ are in L then xzy' is in L.

The following is the basic connection between constants and splicing:

Proposition 1. *A rule of the form $(u, v;\ u, v)$ preserves the language L if and only if the string uv is a constant of L.*

Using this and the regularity of $\mathcal{R}(L)$ from Section 5.4 it is easy to deduce the (well-known) result that the set of constants of a regular language is regular. We shall need a simple fact about constants:

Lemma 2. *If c is a constant of L and c is a factor of a string c' then c' is a constant of L.*

The proofs of Proposition 1 and Lemma 2 are exercises for the reader.

Suppose (R, I) is a reflexive splicing system which generates the reflexive splicing language L. Write S for the set of all sites uv of rules in R. This is a finite set, and, by Proposition 1, every word of S is a constant of S.

The words of L can be divided into two classes. The words which contain a site of some rule of R are called "live": they can participate in further splicing operations. The ones which do not contain such a site are called "dead": they are inert as far as further splicing goes. The set of live words L_0 is just $L \cap A^*SA^*$, so it is regular; and so the set of dead words L_1, being the complement of L_0 in L, is also regular. If a word in L is not in the initial set I then it is the result of a splicing operation, and the inputs to such a splicing operation contain sites so they are in L_0. Hence each word of L_1 is either in I or is the result of splicing two elements of L_0.

In fact this picture of L can be converted into a characterization of reflexive splicing languages. The key is to characterize languages with only finitely many "dead" words, in the following sense:

Theorem 4. *Suppose L is a regular language and S is a finite set of constants of L. If every element of L, with finitely many exceptions, has a factor in S then L is a reflexive splicing language.*

To prove this we must construct a reflexive splicing system (R, I) from L and S, and then we must show that $L = R^*(I)$. We write $L_0 = L \cap A^*SA^*$ and $L_1 = L - L_0$; by hypothesis L_1 is finite. We first define (R, I) as follows.

Let K be the cardinality of the syntactic monoid $\operatorname{Syn} L$ and let N be the maximum length of a word in S. Our definitions are:

1. R is the set of all rules of the form $(u, \lambda;\ v, \lambda)$ or $(\lambda, u;\ \lambda, v)$, where u and v are syntactically congruent constants of L of length less than $2K + N$.
2. I is the set of words of L of length less than $4K + N$, together with the words of L_1.

It is clear that R is reflexive, since the condition that u and v be syntactically congruent is satisfied if $u = v$.

Now we need to show that $L = R^*(I)$; we start with showing that any z in $R^*(I)$ is in L. This is immediate if $z \in I$, since $I \subseteq L$. We proceed by induction, so we may assume z is in $R^k(I)$ for some $k > 0$ but not in $R^{k-1}(I)$, and that every word in $R^{k-1}(I)$ is in L. But then z is the result of splicing two elements z_1 and z_2 of $R^{k-1}(I)$ using a rule r of R. Hence z_1 and z_2 are in L. Consider the case that r has the form $(u, \lambda;\ v, \lambda)$ (the alternate form is handled similarly). We can factor $z_1 = x_1uy_1$ and $z_2 = x_2vy_2$, so $z = x_1uy_2$. Since u and v are syntactically congruent and $z_2 = x_2vy_2$ is in L we conclude that $z_2' = x_2uy_2$ is in L. Since both $z_1 = x_1uy_1$ and $z_2' = x_2uy_2$ are in L and u is a constant we conclude that $x_1uy_2 = z$ is in L.

The rest of the proof relies on the following:

Lemma 3. *Suppose L is a regular language and K is the cardinality of the syntactic monoid* Syn L. *If z is any string of length at least $2K$ then z has three distinct syntactically congruent prefixes and three distinct syntactically congruent suffixes, each of length at most $2K$.*

Proof. For $0 \leq k \leq 2K$ let p_k be the prefix of z of length k; this is well defined since $2K \leq |z|$. Let η be the quotient map from A^* to $\mathrm{Syn}\, L$. Since there are $2K + 1$ prefixes and only K elements of $\mathrm{Syn}\, L$ we can apply the pigeonhole principle to find three indices $i < j < k$ so that $\eta(p_i) = \eta(p_j) = \eta(p_k)$. That is, the three prefixes p_i, p_j and p_k are syntactically congruent, and they have length at most $2K$.

A similar argument works for suffixes. □

Now to conclude the proof of Theorem 4 we start with $z \in L$ and prove that z is in $R^*(I)$. We proceed by induction on the length of z. If $|z| < 4K+N$ or $z \in L_1$ then $z \in I \subseteq R^*(I)$, so we only need to consider the case that z is not in L_1 and has length at least $4K + N$. Since $z \notin L_1$ it must be in L_0, so it has an element s of S as a factor: $z = xsy$. Since $|z| \geq 4K + N$ and $|s| \leq N$ we must have either $|x| \geq 2K$ or $|y| \geq 2K$. We give the rest of the proof assuming that $|y| \geq 2K$; the alternative is handled similarly.

We now apply Lemma 3 to y, so $y = tuvw$ where the three prefixes t, tu and tuv are distinct and syntactically congruent, with lengths at most $2K$. In particular, u and v cannot be the empty word. Now, using the facts that tuv is syntactically congruent to tu and that $z = xstuvw$ is in L we conclude that $z_1 = xstuw$ is in L, and it is shorter than z since $|v| > 0$.

We need another string to splice with z_1. Let η be the projection of A^* onto $\mathrm{Syn}\, L$. Since t and tu are syntactically congruent we have $\eta(t) = \eta(tu)$. Then, since η is a homomorphism, $\eta(tv) = \eta(t)\eta(v) = \eta(tu)\eta(v) = \eta(tuv)$, and so tv and tuv are syntactically congruent. Hence, by the same argument as above, $z_2 = xstvw$ is in L and is shorter than z.

By the induction hypothesis, z_1 and z_2 are in $R^*(I)$. Moreover, both stu and st are constants (by Lemma 2), they have length less than $N + 2K$ since $|s| \leq N$ and $|tu| < |tuv| \leq 2K$, and they are syntactically congruent since $\eta(stu) = \eta(s)\eta(tu) = \eta(s)\eta(t) = \eta(st)$. Therefore $r = (stu, \lambda;\, st, \lambda)$ is in R, and z is the result of splicing z_1 and z_2 using r, so z is in $R^*(I)$.

This completes the induction argument, and hence the proof of Theorem 4.

The splicing system constructed in the proof is, generally, enormous. Following the details of the proof shows that we only need to consider u and v which have a prefix or suffix in S. As a practical matter, much smaller sets of rules and initial strings are usually sufficient. In any case it is worthwhile stating the form of the rules explicitly:

Corollary 1. *The language L of Theorem 4 may be written in the form $R^*(I)$ where each rule of R has one of the forms $(sp, \lambda;\ sq, \lambda)$ or $(\lambda, ps;\ \lambda, qs)$ with s in S.*

As an example of the power of Theorem 4 we mention the following:

Corollary 2. *If L_1 and L_2 are regular languages in A^* and c is a symbol which is not in A then L_1cL_2 is a reflexive splicing language.*

We have now done all the work to prove the following characterization theorem for reflexive splicing languages.

Theorem 5. *Suppose L is a regular language. Then L is a reflexive splicing language if and only if there are finite sets S and I of strings and a finite set of rules R so that:*

1. *Each element of S is a constant of L.*
2. *Each site of each rule of R is in S.*
3. *$L = L_0 \cup R(L_0) \cup I$ where $L_0 = L \cap A^*SA^*$.*

Proof. Suppose (R, I) is a reflexive splicing system and $L = R^*(I)$. We define S to be the set of sites of rules in R. Then, following the discussion before the statement of Theorem 4, S is a finite set of constants of L and every "dead" word in $L_1 = L - L_0$ is in I or in $R(L_0)$.

For the converse, assume we have such sets S, I and R. Clearly the words in S are constants of L_0, so we can apply Theorem 4 to produce a reflexive splicing system (R_0, I_0) so that $L_0 = R_0^*(I_0)$. Now define the splicing system (R_1, I_1) by $R_1 = R_0 \cup R$ and $I_1 = I_0 \cup I$. According to Corollary 1, we may arrange that each site of a rule of R_0 has a factor in S and, by condition 2, the same is true for R. Hence the rules of R_1 can only operate on words of L_0. So we have $R_0(L) = R_0(L_0) \subseteq L_0 \subseteq L$ and, by condition 3, $R(L) = R(L_0) \subseteq L$. Hence the rules of R_1 preserve L and $I_1 \subseteq L$, and it follows that $R_1^*(I_1) \subseteq L$. But $L_0 = R_0^*(I_0) \subseteq R_1^*(I_1)$, so, again using condition 3, $L \subseteq R_1^*(I_1) \cup R(R_1^*(I_1)) \cup I \subseteq R_1^*(I_1)$. So we have shown $L = R_1^*(I_1)$. □

This characterization theorem points the way to an algorithm for determining whether a given regular language L is a reflexive splicing language. We fix an integer N and let S be the set of all constants of L of length at most N. Next, for each pair s and t of constants in S and for each choice of factorizations $s = uu'$, $t = v'v$ we decide whether the rule $(u, u';\ v', v)$ is in $R(L)$. This is algorithmically feasible since the proof of Theorem 2 determines the regular set $R(L)$ explicitly in terms of the syntactic monoid (and a similar argument applies to determine the set of constants). We define R to be the set of all rules selected in this way. We also define I to be the set of all "short" words in L; for example, the set of words of length at most N. It is easy to construct an automaton to accept $R(L_0)$ for regular L_0 and finite R, so we can calculate the regular set $L_0 \cup R(L_0) \cup I$ and check whether it is equal to L. If the answer to this is "Yes" then we have determined that L is a reflexive splicing language. The problem, of course, is that if the answer is "No" then we cannot conclude that L is *not* a reflexive splicing language, since we might not have chosen N "large enough". To make this procedure into a decidability algorithm we must determine a value of N which is guaranteed to be large enough. This determination is carried out in [3], which gives a sufficiently large value of N in terms of the size of the syntactic monoid of L.

The more general question of deciding whether a given regular language is a splicing language remains open.

Exercises

1. Prove Proposition 1.
2. Prove Lemma 2.
3. Exhibit a reflexive splicing system (R, I) so that $a(b^2)^* = R^*(I)$. (Note that a is a constant.)
4. Same question for $a(b^2)^*(c^2)^*$.
5. Let L be the language of words in $\{a, b\}^*$ which contain at most 2 b's. Show that L is a splicing language but not a reflexive splicing language.

5.6 Appendix: The Intrinsic Automaton and the Syntactic Monoid of an Arbitrary Language.

As usual: A is the alphabet; A^* is the set of all strings over A and L is a language over A, i.e., a subset of A^*. Relative to the language L we have two constructions, of the automaton recognizing L and the syntactic monoid of L, defined as follows.

The (minimal) automaton M that recognizes L: The *states* of M are the equivalence classes defined, for each x in A^*, by $[x] = \{ y \in A^* \colon R(y) = R(x) \}$ where $R(x)$ is the set of *right contexts* accepted by x relative to L. Specifically: $R(x) = \{ v \in A^* \colon xv \in L \}$. Each symbol a in A *acts on* the state $[x]$ by $[x]a = [xa]$. M has a specified start state, $[1]$, and a specified set of terminal states, $\{ [x] \colon x \in L \}$. M may be viewed as a directed graph having the states of L as its vertices, having directed edges $([x], a, [xa])$, each considered to be labeled by the alphabet symbol a. The automaton M is considered to *recognize* each string in A^* which is the label of a path from a start state to a terminal state. This M recognizes precisely those strings that are in L. A language L is regular if its automaton has only finitely many states.

The syntactic monoid S of L: The elements of S are the equivalence classes defined, for each x in A^*, by $[\![x]\!] = \{ y \in A^* \colon LR(y) = LR(x) \}$, where $LR(x)$

is the set of *two-sided contexts* accepted by x relative to the language L. Specifically: $LR(x) = \{ \langle u, v \rangle \in A^* \times A^* : uxv \in L \}$. S has an associative binary operation that is *well defined* by setting $[\![x]\!][\![y]\!] = [\![xy]\!]$. Since $[\![1]\!]$ serves as a two-sided identity for this operation, S has the structure of a *monoid*, i.e., a *semigroup* with an identity element. The partition of A^* into the classes $[\![x]\!]$ refines the partition of A^* into the classes $[x]$ since $LR(x) = LR(y)$ implies $R(x) = R(y)$. Consequently when S is finite L is regular.

The action of A on the states of M extends, inductively, to an action of A^* on the states of M. Consequently each string y in A^* determines a function from the state set of L into itself defined by $[u]x = [ux]$. Two strings x and y determine the same function precisely if, for every u in A^*, $[ux] = [uy]$. But this holds precisely if, for all v in A^*, uxv is in L if and only if uyv is in L. Thus x and y determine the same function precisely if $[\![x]\!] = [\![y]\!]$, and this construction defines a faithful representation of S as self-maps of the states of M. When L is regular there can be only a finite number of functions from the state set of L into itself. Consequently when L is regular S is finite.

Summary:

Theorem 6. *For every language L we have an intrinsically associated automaton that recognizes the language and we have an associated syntactic monoid. The following are equivalent:*

1. *L is a regular language;*
2. *The intrinsic automaton for L has only finitely many states; and*
3. *The syntactic monoid of L is finite.*

In the same spirit, construct a directed graph with states given by the elements of S and directed edges $([\![x]\!], a, [\![xa]\!])$, each considered to be labeled by the alphabet symbol a. Then $[\![x]\!]$ is recognized by the automaton based on this graph with start state $[\![1]\!]$ and a single terminal state, $[\![x]\!]$. Hence:

Theorem 7. *If L is regular then each syntactic class $[\![x]\!]$ is regular.*

References

1. K. Culik II and T. Harju. Splicing Semigroups of Dominoes and DNA, *Discrete Applied Mathematics*, 31 (1991), 261–277.
2. E. Goode and D. Pixton. Semi-simple Splicing Systems, *Where Mathematics, Computer Science, Linguistics and Biology Meet* (C. Martin-Vide, V. Mitrana, eds.), Kluwer Academic Publishers, Dordrecht, 2001, 343–352.
3. E. Goode and D. Pixton. Recognizing Splicing Languages: Syntactic Monoids and Simultaneous Pumping, to appear.
4. T. Head, G. Păun and D. Pixton. Generative Mechanisms Suggested by DNA Recombination, *Handbook of Formal Languages vol. 2* (G. Rozenberg and A. Salomaa, eds.), Springer-Verlag, Berlin, 1997.

5. T. Head. Formal Language Theory and DNA: an Analysis of the Generative Capacity of Specific Recombinant Behaviors, *Bulletin of Mathematical Biology*, 49, 6 (1987), 737–759.
6. T. Head. Splicing Languages Generated With one Sided Context, *Computing With Bio-molecules–Theory and Experiments* (Gh. Păun ed.), Springer-Verlag, Singapore, 1998, 269–282.
7. Gh Păun, G. Rozenberg and A. Salomaa. *DNA Computing - New Computing Paradigms*, Springer–Verlag, Berlin, 1998.
8. D. Pixton. Splicing in Abstract Families of Languages, *Theoretical Computer Science*, 234 (2000), 135–166.
9. D. Pixton. Regularity of Splicing Languages, *Discrete Appl. Math.*, 69, 1-2 (1996), 99–122.
10. M.P. Schützenberger. Sur Certaines Operations de Fermeture dans les Langages Rationnels, *Symposia Math.*, 15 (1975), 245–253.

6

Combinatorial Complexity Measures for Strings*

Lucian Ilie**

Department of Computer Science
University of Western Ontario
London, ON, N6A 5B7, CANADA
E-mail: ilie@csd.uwo.ca

Summary. We present a number of results and open problems about some important combinatorial complexity measures for strings. The list includes most basic ones, such as factor complexity and repetition threshold, old but not enough investigated ones, like Lempel-Ziv complexity, and newer ones, such as the number of squares or runs, and repetition complexity. The issues addressed concern mainly bounds, relations with periodicity of infinite strings and fixed points of morphisms.

6.1 Introduction

This brief survey includes a number of results and open problems concerning several important combinatorial complexity measures for strings. We are interested only in computable measures and therefore do not discuss complexity measures such as the ones of Kolmogorov [44] and Chaitin [12].

We focus on new results and problems. Factor complexity, the most investigated combinatorial complexity measure for strings is presented essentially for comparison. The de Bruijn strings are an intrinsic part of it and serve as good examples of complex strings in many places.

An old, but not enough investigated, measure is the one of Lempel and Ziv [49], connected with their famous compression methods. Various repetitions, such as squares, runs, total amount or order of repetitions give corresponding complexity measures.

Such complexities are considered for both finite and infinite strings. Lower and/or upper bounds are presented and, in the case of infinite strings, the relation with their ultimate periodicity is considered. As one of the most

* The material included in this paper was presented by the author at the 5th International Ph.D. School in Formal Languages and Applications, Tarragona, Spain, October 2005.

** Research supported in part by NSERC.

L. Ilie: *Combinatorial Complexity Measures for Strings*, Studies in Computational Intelligence (SCI) **25**, 149–170 (2006)
www.springerlink.com

important classes of infinite strings, fixed points of morphisms are given special attention.

This survey does not intend to be complete in any way, nor it attempts to include the most important results and problems. It merely tries, within the limited space, to bring to attention a number of interesting results and, especially, problems which should be addressed. Algorithms for computing the considered measures, a whole topic on its own, are almost completely left out of discussion.

We point out a number of open problems, others are presented implicitly. For instance, very little is known about Lempel–Ziv complexity or the relations among various complexity measures.

6.2 Basic Definitions

An *alphabet* A is a finite non-empty set. The elements of A are called *letters*. The free monoid generated by A is denoted A^*; its elements are all finite *strings*, or *words*[3], over A. Its unit element is the empty string, denoted ε. The set A^* is a monoid with respect to (con)catenation, that is, juxtaposition of strings, denoted by a dot $\cdot$ which we usually omit. The length of a string w is denoted $|w|$ and represents the number of letters, where each letter is counted as many time as it occurs; thus, $|\mathsf{aababb}| = 6$ and $|\varepsilon| = 0$. For $n \geq 0$, the set of all strings of length n over A is denoted A^n.

Given three strings $w, x, y, z \in A^*$ such that $w = xyz$ we say that x, y, and z is a *prefix*, *factor*[4] and *suffix*, resp.; when they are different from w, they are called *proper* prefix, factor, and suffix, resp. A *subsequence* of w is a string $v_1 \ldots v_\ell$, such that $w = u_1 v_1 \ldots u_\ell v_\ell u_{\ell+1}$, where $u_i, v_j \in A^*$.

An *infinite string* over A is a function $w : \mathbb{N} \to A$; notice that we consider right-infinite strings only. The set of all infinite strings over A is denoted by A^ω.

An infinite string w is called *ultimately periodic* if there exist two finite strings u, v, such that v is non-empty and w can be written as $w = uv^\omega$, where $v^\omega = vvv\ldots$.

The prefix of length $n \geq 0$ of $w \in A^* \cup A^\omega$ is denoted $\mathsf{pref}_n(w)$. The prefix, factor, and suffix relations can be used between finite and infinite strings whenever they make sense; for instance, a finite string can be a prefix or factor of an infinite one but not a suffix.

A morphism is a function $h : A^* \to B^*$ such that $h(\varepsilon) = \varepsilon$ and $h(uv) = h(u)h(v)$, for all $u, v \in A^*$. Clearly, a morphism is completely defined by the images of the letters in the domain. For all our morphisms, $A = B$.

---

[3] Both terms are used in literature. For uniformity, we shall use only the term *string* throughout this paper.

[4] A factor is sometimes called *subword* or *substring* which we shall avoid as it is used in the literature to denote also a scattered factor (or subsequence, as called here) which may create confusion.

A morphism $h : A^* \to A^*$ is called *non-erasing* if $h(a) \neq \varepsilon$, for all $a \in A$, *uniform* if $|h(a)| = |h(b)|$, for all $a, b \in A$, and *prolongeable* on $a \in A$ if a is a proper prefix of $h(a)$.

If h is prolongeable on a, then $h^n(a)$ is a proper prefix of $h^{n+1}(a)$, for all $n \in \mathbb{N}$. We say that the morphism h has a *fixed point* $w \in A^\omega$ given by

$$w = \lim_{n \to \infty} h^n(a) = h^\infty(a) .$$

6.3 Factor Complexity

The factor complexity is probably the most investigated combinatorial complexity measure on strings. We refer to [52, 15, 4, 3] and the references therein for more information.

We consider here the factor complexity of both finite and infinite strings. Also, the factor complexity can be defined in two different, related, ways, depending on whether we count all factors or only those of a given length.

Therefore, for a string w and an integer $n \geq 0$, we define

$$\begin{aligned}
\mathsf{fact}_w(n) &= \mathrm{card}\{u \in A^* \mid u \text{ a factor of } w, |u| = n\},\\
\mathsf{all_fact}(w) &= \mathrm{card}\{u \in A^* \mid u \text{ a factor of } w\},\\
\mathsf{all_fact}_w(n) &= \mathsf{all_fact}(\mathsf{pref}_n(w)).
\end{aligned}$$

Notice that the all definitions make sense for finite strings but only the first and the third are well defined for infinite strings; the third will be used for infinite strings only.

The highest possible factor complexities are reached for the well known de Bruijn strings which we describe in the next subsection.

6.3.1 De Bruijn Strings

Assume we are given a $n \geq 0$. We want to construct a finite string which has all strings of length n over A as factors. Obviously, such a string would have to have the length at least $|A|^n + n - 1$ and the next result says that it is always possible to construct such strings; see [10] and the notes of [4] for the history of this strings and further references.

Theorem 1. *For all $n \geq 0$, there exists a string $\mathsf{dB}_n \in A^*$ such that*
(i) $|\,\mathsf{dB}_n\,| = |A|^n + n - 1$ *and*
(ii) dB_n *contains as factors all strings of length n over A.*

We explain next very briefly the idea of the proof for the binary alphabet as it will help us discuss infinite de Bruijn strings in the next section.

For a fixed $n \geq 1$, consider the *graph* G_n whose set of vertices is A^n and egdes are $\{(aw, wb) \mid a, b \in A, w \in A^{n-1}\}$. The graph G_3 over the alphabet $\{0, 1\}$ is shown in Fig. 6.1.

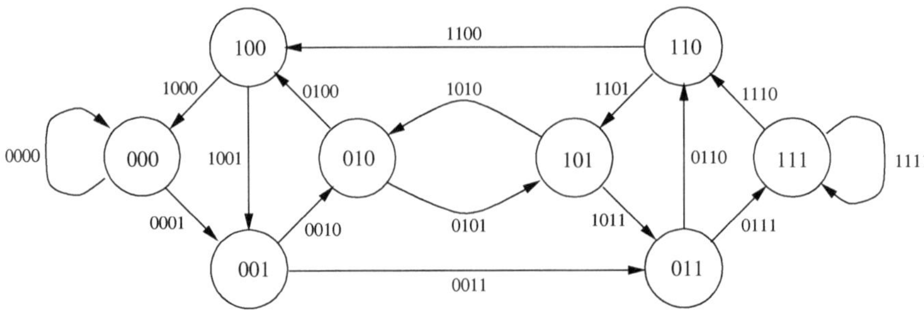

Fig. 6.1. The de Bruijn graph G_3

It can be shown that G_n is Eulerian and then any Eulerian cycle gives a de Bruijn string dB_{n+1}; for instance, for $n = 3$, we have such a string

$$\mathsf{dB}_4 = \mathsf{0001111011001010000}.$$

6.3.2 Infinite de Bruijn Strings

The following generalization of de Bruijn strings appears naturally: Are there infinite strings such that all their prefixes contain the maximum possible number of factors? The problems has been solved in [43] and then again, independently, in [30]. The answer, rather surprising, is given in the next theorem.

Theorem 2. *If A contains at least three letters, then there exits an infinite string w over A such that, for all $n \geq 0$, $\mathsf{pref}_n(w)$ has the highest possible number of factors among all strings of the same length over A. Such a string does not exist if A is binary.*

For its proof, de Bruijn graphs are used again. One Eulerian cycle of G_n becomes a Hamiltonian path in G_{n+1} and, for three letters or more, it can be extended to another Eulerian cycle. For two letters however, this is not possible.

Remark 1. Notice that [30] solves also the same problem for subsequences; their proof for subsequences is interesting but the answer is almost trivial: the number of subsequences is maximized, over $|A| = \mathsf{k}$ letters, by the infinite string (and its prefixes)

$$\mathsf{123} \cdots \mathsf{k123} \cdots \mathsf{k} \cdots .$$

We shall use de Bruijn strings in many of the sections below as a good example of strings with high complexity.

6.3.3 Bounds on Factor Complexity

Some simple bounds for these complexities in the case of finite strings are given in the next result.

Theorem 3. *For any string w and any $n \geq 0$ we have*

(i) $|w| + 1 \leq \mathsf{all\_fact}(w) \leq 1 + \frac{1}{2}|w|(|w| + 1)$,
(ii) $1 \leq \mathsf{fact}_w(n) \leq \min(|w| - n + 1, |A|^n)$.

The above bounds are easy to prove. The upper bounds are reached for strings whose letters are pairwise distinct. Interesting is that the two upper bounds (inside the min) at (ii) are simultaneously reached for de Bruijn strings.

For the number of all factors, we have the following result of Shallit [68]. It says that the number of all factors is the sum of the upper bound at (ii) above and it can be reached.

Theorem 4. *For any string $w \in \{0, 1\}^*$, we have*

$$\mathsf{all\_fact}(w) \leq \binom{|w| - k + 1}{2} + 2^{k+1} - 1,$$

where k is the unique integer such that $2^k + k - 1 \leq |w| \leq 2^{k+1} + k$. The bound is optimal.

In the case of infinite strings we present first a number of conditions which are equivalent with ultimate periodicity.

Theorem 5. *For any infinite string w, the following assertions are equivalent:*

(i) w is ultimately periodic,
(ii) $\mathsf{fact}_w(n) = \mathcal{O}(1)$,
(iii) there exists $n \geq 0$ such that $\mathsf{fact}_w(n) \leq n$,
(iv) there exists $n \geq 0$ such that $\mathsf{fact}_w(n) = \mathsf{fact}_w(n + 1)$,
(v) $\mathsf{all\_fact}_w(n) = \mathcal{O}(n)$.

The equivalence between conditions (i)-(iv) is due to Morse and Hedlund [55]. The last one is from [41].

Notice that conditions (ii) and (iii) imply that as soon as $\mathsf{fact}_w(n)$ is less than or equal to n for some n, it is bounded and the infinite string is ultimately periodic. There are strings which are non-ultimately periodic but have exactly the minimum factor complexity; they are called *Sturmian.* An infinite string w is called Sturmian if $\mathsf{fact}_w(n) = n + 1$, for all $n \geq 0$. Sturmian strings have been very much investigated in the literature; see [8] for a good survey on this topic.

Example 1. We give here one example, one of the most famous infinite strings: Fibonacci. Consider the Fibonacci morphism

$$\phi : \{0,1\}^* \to \{0,1\}^*,$$
$$\phi(0) = 01,$$
$$\phi(1) = 0.$$

We see that ϕ is prolongeable on 0 and so we can define the Fibonacci infinite string

$$\mathbf{f} = \lim_{n\to\infty} \phi^n(0) = 0100101001001010010100\cdots .$$

The string has not been introduced by Fibonacci; it is called this way because the numbers $f_n = |\phi^n(1)|$ are precisely the Fibonacci numbers – if we start from 0 then we obtain the Fibonacci numbers starting with the second. (The strings $\phi^n(1)$ are the finite Fibonacci strings.) The Fibonacci infinite string is Sturmian. The Fibonacci strings, finite and the infinite, are used as extremal objects for many properties and algorithms; see again [8].

6.3.4 Fixed Points of Morphisms

The last result which we present for factor complexity is the characterization of the factor complexity for fixed points of morphisms. Such strings are very important in string combinatorics; many of the important infinite strings are obtained this way. (See above for Fibonacci.)

The investigation of factor complexity for the fixed points of morphisms has been initiated by Ehrenfeucht, Lee, and Rozenberg in [22] (they actually considered the closely related D0L-systems) and continued by Ehrenfeucht and Rozenberg in a series of papers, see [23, 24, 25, 26, 27, 64]. The classification was completed by Pansiot [61, 62] who found also the missing complexity class $\Theta(n \log\log n)$.

We need several definitions.

The *growth function* of the letter $x \in A$ in h is the function $h_x : \mathbb{N} \to \mathbb{N}$ defined by

$$h_x(n) = |h^n(x)| \ .$$

Recall also the following result from [67, 65].

Lemma 1. *There exist an integer* $e_a \geq 0$ *and an algebraic real number* $\rho_a \geq 1$ *such that*

$$h_a(n) = \Theta(n^{e_a}\rho_a^n) \ .$$

The pair (e_a, ρ_a) is called the *growth index* of a in h. We say that h_a (and a as well) is:

- *bounded* if a's growth index w.r.t. h is $(0,1)$,
- *polynomial* if a's growth index w.r.t. h is $(e_a, 1)$, $e_a > 1$, and
- *exponential* if a's growth index w.r.t. h is (e_a, ρ_a), $e_a \geq 0$, $\rho_a > 1$.

The following definitions appear, with different names, in [15]. The morphism h is called:[5]

- *non-growing* if there exists a bounded letter in A,
- *u-exponential* if $\rho_a = \rho_b > 1$, $e_a = e_b = 1$, for all $a, b \in A$,
- *p-exponential* if $\rho_a = \rho_b > 1$, for all $a, b \in A$ and $e_a > 1$, for some $a \in A$, and
- *e-exponential* if $\rho_a > 1$, for all $a \in A$ and $\rho_a > \rho_b$, for some $a, b \in A$.

Here is Pansiot's characterization:

Theorem 6. *Let $w = h^\omega(a)$ be an infinite non-periodic string of factor complexity $f_w(\cdot)$.*

1. *If h is growing, then $\mathsf{fact}_w(n)$ is either $\Theta(n)$, $\Theta(n \log n \log n)$ or $\Theta(n \log n)$, depending on whether h is u-, p- or e-exponential, resp.*
2. *If h is not-growing, then either*
   a) *w has arbitrarily large factors over the set of bounded letters and then $\mathsf{fact}_w(n) = \Theta(n^2)$ or*
   b) *w has finitely many factors over the set of bounded letters and then $\mathsf{fact}_w(n)$ can be any of $\Theta(n)$, $\Theta(n \log n \log n)$ or $\Theta(n \log n)$. (In fact Pansiot proved more, namely that w is the image through a non-erasing morphism of another string w' generated by a growing morphism h' and the factor complexities of w and w' are the same.)*

6.4 Lempel-Ziv Complexity

Before publishing their famous papers introducing the well-known compression schemes LZ77 and LZ78 in [76] and [77], resp., Lempel and Ziv introduced a complexity measure for strings in [49] which attempted to detect "sufficiently random looking" sequences. In contrast with the fundamental measures of Kolmogorov [44] and Chaitin [12], Lempel and Ziv's measure is computable. The definition is purely combinatorial; its basic idea, splitting the string into minimal never seen before factors, proved to be at the core of the well-known compression algorithms, as well as subsequent variations. Another, closely related, variant is to decompose the string into maximal already seen factors.

Lempel–Ziv-type complexity and factorizations have important applications in many areas, such as, data compression [76, 77], string algorithms [19, 54, 45, 66], cryptography [57], molecular biology [35, 29, 14], and neural computing [71, 1, 72].

---

[5] Following [17], we call *u-*, *p-*, and *e-exponential* what are *quasi-uniform*, *polynomially diverging*, and *exponentially diverging* in [61, 62, 15]. The terminology has been changed in [17] so that it does not conflict with the notions for h_a in the next section.

Lempel and Ziv [49] investigate various properties which are expected from a complexity measure which intends to approach randomness. They prove it to be subadditive and also that most sequences are complex but still not too many; see [49] for details.

6.4.1 Lempel–Ziv Factorizations

For a string w, we define the *Lempel–Ziv factorization*[6] of w as the (unique) factorization $w = w_1 w_2 \cdots w_k$ such that, for any $1 \leq i \leq k$ (with the possible exception of $i = k$), w_i is the shortest prefix of $w_i w_{i+1} \cdots w_k$ which does not occur before in w; that is, w_i does not occur as a factor of $\pi(w_1 w_2 \cdots w_i)$, where the application π removes the last letter of its argument.

The complexity measure introduced by Lempel and Ziv represents the *number of factors in the Lempel–Ziv factorization* of w; we denote it by $\text{LZ}(w)$.

Example 2. Consider the string $w = \mathsf{aababbabbabb}$. The Lempel–Ziv factorization of w is

$$w = \mathsf{a.ab.abb.abbabb},$$

where the factors are marked by dots. Therefore, $\text{LZ}(w) = 4$.

Notice that the Lempel–Ziv complexity of finite strings can be computed in linear time by using suffix trees; see [19, 36].

The Lempel–Ziv factorization can be defined in the same way for infinite strings; at each step we take the longest prefix of the remaining infinite suffix which does not appear before; this prefix may be the remaining suffix of the infinite string, in which case it is the last element of the factorization.

Example 3. For the ultimately periodic string $\mathsf{00110101010101010}\ldots$ we have the Lempel–Ziv factorization

$$\mathsf{0.01.10.101010101010}\ldots\ .$$

The *Lempel–Ziv complexity of an infinite string* w is the function $\text{LZ}_w : \mathbb{N} \to \mathbb{N}$ defined by

$$\text{LZ}_w(n) = \text{LZ}(\mathsf{pref}_n(w))$$

as the complexity of finite prefixes of w.

---

6 Lempel and Ziv [49] called this factorization the *exhaustive production history* of w. A closely related variant, which is mostly used in applications, is obtained by decomposing the string into maximal already seen factors; this is called *f-factorization* in [20] and *s-factorization* in [54].

6.4.2 Bounds

The following bounds for finite strings were proved by Lempel and Ziv in the original paper [49]. Again, the de Bruijn strings are most complex with respect to Lempel–Ziv measure.

Theorem 7. *(i)* $\mathrm{LZ}(w) = \mathcal{O}(\frac{|w|}{\log |w|})$,
(ii) $\mathrm{LZ}(\mathsf{dB}_n) \geq \frac{|\,\mathsf{dB}_n\,|}{\log |\,\mathsf{dB}_n\,|}$, *for all* n.

In the case of infinite strings, we have the following characterization from [41] of ultimate periodicity in terms of Lempel–Ziv complexity of infinite strings.

Theorem 8. *For any infinite string* w*, the following assertions are equivalent:*

(i) w *is ultimately periodic,*
(ii) $\mathrm{LZ}_w(n) = \mathcal{O}(1)$.

6.4.3 Fixed Points of Morphisms

A characterization of the Lempel–Ziv complexity classes for fixed points of morphisms, similar with the one of Pansiot for factor complexity, has been very recently discovered by Constantinescu and the author [17].

Theorem 9. *For a fixed point infinite string* $w = h^\omega(a)$*, we have:*

1. *The Lempel–Ziv complexity of* w *is* $\Theta(1)$ *if and only if* w *is ultimately periodic.*
2. *If* w *is not ultimately periodic, then the Lempel–Ziv complexity of* w *is either* $\Theta(\log n)$ *or* $\Theta(n^{1/k})$, $k \in \mathbb{N}$, $k \geq 2$, *depending on whether* h_a *is exponential or polynomial, resp.*

We give next examples from [17] showing that all the above complexities are indeed possible.

Example 4. For the polynomial case, we have the morphism

$$\begin{aligned}
&h_k : \{\mathsf{a}_1, \mathsf{a}_2, \ldots, \mathsf{a}_k\}^* \rightarrow \{\mathsf{a}_1, \mathsf{a}_2, \ldots, \mathsf{a}_k\}^*,\\
&h_k(\mathsf{a}_i) = \mathsf{a}_i\mathsf{a}_{i+1}, \text{ for all } 1 \leq i \leq k-1,\\
&h_k(\mathsf{a}_k) = \mathsf{a}_k,
\end{aligned}$$

for which it can be shown that $(h_k)_{\mathsf{a}_1}(n) = \Theta(n^{k-1})$. This implies that the Lempel–Ziv complexity of the infinite fixed point $h_k^\infty(\mathsf{a}_1)$ is $\Theta(\sqrt[k-1]{n})$.

Example 5. With respect to the exponential case, any uniform morphism with all images of length ℓ has a growth function of exactly ℓ^n. One such famous example is the Thue–Morse morphism

$$t(\mathsf{a}) = \mathsf{ab}, \quad t(\mathsf{b}) = \mathsf{ba}.$$

The Lempel–Ziv complexity of both infinite fixed points $t^\infty(\mathsf{a})$ and $t^\infty(\mathsf{b})$ is $\Theta(\log n)$.

Another exponential case, but non-uniform, is given by the Fibonacci morphism defined above.

6.4.4 Lempel–Ziv vs Factor Complexity for Fixed Points of Morphisms

We presented above the characterizations of complexity classes for fixed points of morphisms with respect to factor and Lempel–Ziv complexity measures. A very natural problem which appears is the connection between the two. This was clarified also in [17].

Theorem 10. *The correspondence between Lempel–Ziv and factor complexities for fixed points of morphisms is shown in Table 6.1, where all intersections are indeed possible.*

<table>
<tr><th></th><th>Lempel–Ziv complexity</th><th>Factor complexity</th></tr>
<tr><td>$h^\infty(a)$ is ultimately periodic</td><td>$\Theta(1)$</td><td>$\Theta(1)$</td></tr>
<tr><td rowspan="5">$h^\infty(a)$ is not ultimately periodic and h_a is polynomial</td><td>$\Theta(n^{\frac{1}{2}})$</td><td rowspan="5">$\Theta(n^2)$</td></tr>
<tr><td>$\Theta(n^{\frac{1}{3}})$</td></tr>
<tr><td>$\vdots$</td></tr>
<tr><td>$\Theta(n^{\frac{1}{k}})$</td></tr>
<tr><td>$\vdots$</td></tr>
<tr><td rowspan="4">$h^\infty(a)$ is not ultimately periodic and h_a is exponential</td><td rowspan="4">$\Theta(\log n)$</td><td>$\Theta(n^2)$</td></tr>
<tr><td>$\Theta(n \log n)$</td></tr>
<tr><td>$\Theta(n \log \log n)$</td></tr>
<tr><td>$\Theta(n)$</td></tr>
</table>

Table 6.1. Lempel–Ziv vs. factor complexity

We see that both measures of complexity recognize ultimately periodic strings as having bounded complexity, the lowest class of complexity.

In the nontrivial case of non-periodic fixed points, the Lempel–Ziv complexity groups together all strings $h^\infty(a)$ with h_a exponential, whereas the factor complexity distinguishes four different complexities. On the other hand, the factor complexity does not make any distinction among strings with h_a polynomial, whereas Lempel–Ziv gives an infinite hierarchy.

For examples showing that all intersections in Table 6.1 are non-empty, see [17].

6.5 Repetitions

By repetition in a string we mean a factor which has several adjacent occurrences. The simplest repetition is a square. Consider for instance the following example due to [4]: the English word hotshots = $(\mathsf{hots})^2$ is a *square*.

We can have naturally higher repetitions. Consider another example, also from [4]: the English word alfalfa contains the square alfalf = $(\mathsf{alf})^2$ but we see that it contains also the beginning of a third occurrence of alf, that is, the letter a; the repetition alfalfa = $(\mathsf{alf})^{\frac{7}{3}}$ is called a $\frac{7}{3}$-repetition.

One remarkable example is the Finish sentence (some spaces were ommitted) otavaltavaltavaltavaltiolta which contains the $\frac{21}{5}$-repetition $(\mathsf{taval})^4\mathsf{t}$.

In general, a *α-repetition* is a string $y = x^n x'$ such that x' is a prefix of x and $|y| = \alpha|x|$; α is the *exponent* and $|x|$ is the *period* of the repetition.

6.5.1 Squares

One of the most natural questions one can ask about a string is how many squares it contains. The question can be asked in several ways. First, if we indiscriminately count all squares, then the solution in trivial as we have a quadratic number in a unary string. Two variations are most important. In the first, we restrict the set to primitively rooted squares, that is, of the form w^2 where w cannot be written as a non-trivial integer power. In this case, Crochemore [18] gives the answer.

Theorem 11. *The number of primitively rooted squares in a string of length n is $\mathcal{O}(n \log n)$. The bound is optimal, being reached for Fibonacci strings.*

The second natural question asks for distinct squares. An answer to this problem was given rather recently by Frankel and Simpson [32]. Somewhat unexpected, the number is strictly lower than the previous one.

Theorem 12. *Any string of length n has at most $2n$ distinct squares.*

Their proof was a bit complicated and a very simple proof was given by the author in [37]. The main idea of the proof is to count each square at the position of its last occurrence in the string – a string of integers, say $s(w)$, of the same length as the initial string w is obtained; the ith component of $s(w)$ is the number of squares whose last occurrence in w is at position i. It is then proved that no three squares can have their last occurrence at the same position, that is, $s(w) \in \{0,1,2\}^*$.

The conjecture in [32, 48], supported by computations, is that the number of distinct squares in a string of length n is at most n.

This bound, if correct, would be asymptotically optimal as Fraenkel and Simpson gave in [32] a family of strings

$$w_m = \bigodot_{i=1}^{m} 0^{i+1}10^i10^{i+1}1$$

with length $\frac{3}{2}m^2 + \frac{13}{2}m$ and number of distinct squares $\frac{3}{2}m^2 + 4m - 3 + \frac{odd(m)}{2}$. The string w_4 is shown in Fig. 6.2 together with the last occurrences of all distinct squares; the bottom array contains the string $s(w_4)$.

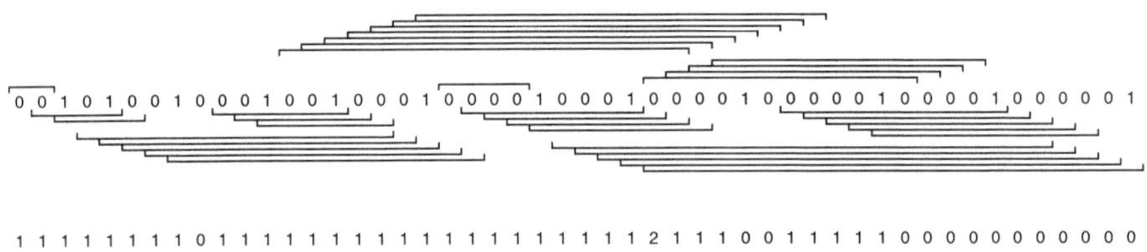

Fig. 6.2. The squares of the string w_4

The strings $s(w)$ were investigated further by the author in [38] where the best bound to date was obtained. Essentially, it says that the number we are looking for is bounded away from $2n$.

Theorem 13. *The number of distinct squares in a string of length n is at most $2n - \Theta(\log n)$.*

A noticed by [38], it is interesting that the strings $s(w_m)$ do not contain many 2s as one might expect but in fact very few; see, e.g., the bottom array of Fig. 6.2.

6.5.2 Runs

In the search for linear time algorithms which find all repetitions, see [19, 54], a major problem was that there are strings which have superlinear number of repetitions (see the above result on primitively rooted squares). This brought the idea of *runs*.[7] A run is a repetition of order at least two which cannot be extended to the left or right without breaking its period. For example, in the string 00010100010 we have the run 01010. None of the squares 0101 and 1010

[7] The term *run*, which we use here, comes from [42]; it is called *maximal repetition* by Main [54] and *m-repetitions* by [45].

is a run as they can be extended so that the period 2 is preserved. However, if we extend 01010 by one letter to the left or to the right, the period becomes 5.

Clearly, all repetitions in a string are encoded in the runs. Kolpakov and Kucherov [45, 47] proved that the number of runs is linear, which enabled them to modify Main's algorithm [54] to find all repetitions in linear time.

Theorem 14. *The number of runs in a string of length n is $\mathcal{O}(n)$.*

More precisely, they proved that the number of runs in a string of length n is at most $c_1 n - c_2\sqrt{n}\log n$. However, nothing is known about the constants c_1 and c_2.

Later, in [46], Kolpakov and Kucherov proved the following stronger result.

Theorem 15. *The sum of exponents of all runs in a string of length n is $\mathcal{O}(n)$.*

6.5.3 Repetition Complexity

A complexity measure attempting to capture the amount of repetition in a string was introduced in [41]. The idea is very simple: any integer repetition $uu\ldots u = u^n$ is replaced, somehow in the spirit of Lempel-Ziv algorithms, by the pair (u, n). The process is repeated until no longer possible. The *repetition complexity* of w, denoted $\text{RC}(w)$, is given by the shortest string that can be obtained in this way. Here is an example.

Example 6. Consider the string $w = \mathsf{ababaabababbbabb}$. Several possible reductions for w are shown below (the first is optimal and so $\text{RC}(w) = 10$):

$$w \Rightarrow (\mathsf{ababa}, 2)\mathsf{bbbabb} \Rightarrow (\mathsf{ababa}, 2)(\mathsf{b}, 3)\mathsf{abb} \Rightarrow ((\mathsf{ab}, 2)\mathsf{a}, 2)(\mathsf{b}, 3)\mathsf{abb}$$
$$w \Rightarrow^* (\mathsf{ab}, 2)\mathsf{aaba}(\mathsf{babb}, 2)$$
$$w \Rightarrow^* \mathsf{a}(\mathsf{ba}, 2)(\mathsf{ab}, 2)\mathsf{a}(\mathsf{b}, 3)\mathsf{a}(\mathsf{b}, 2)$$

The first result compares the repetition complexity with Lempel–Ziv complexity.

Theorem 16. *(i) For any string w, $\text{RC}(w) \geq \text{LZ}(w)$.*
(ii) There are arbitrarily long strings w for which $\text{RC}(w) = |w|$ and $\text{LZ}(w) = \mathcal{O}(\log |w|)$.

It is not known whether the result at (i) can be improved.

One more characterization of ultimate periodicity for infinite strings is given using repetition complexity; as before, we define the repetition complexity for infinite strings in terms of prefixes: $\text{RC}_w(n) = \text{RC}(\mathsf{pref}_n(w))$.

Theorem 17. *For any infinite string w, the following assertions are equivalent:*
(i) w is ultimately periodic,
(ii) $\text{RC}_w(n) = \log n + \mathcal{O}(1)$.

Also, as expected, the repetition complexity of de Bruijn strings is high.

Theorem 18. $\mathrm{RC}(\mathsf{dB}_n) = \Omega\left(\frac{|\,\mathsf{dB}_n\,|\ \log\log(|\,\mathsf{dB}_n\,|)}{\log(|\,\mathsf{dB}_n\,|)}\right)$.

It is also open whether the above result can be improved.

6.5.4 Repetition Threshold

Our last topic concerns the repetition orders which can appear in infinite strings, one of the oldest topics in combinatorics on words; such questions have been already investigated by Thue in [73, 74] (that is, one hundred years ago).

It is easy to see that any string of length at least 4 over a binary alphabet must contain a square. That is, we cannot construct infinite strings without squares over two letters. We say that squares are *unavoidable* over two letters. However, as shown already by Thue's work, [73, 74], squares can be avoided over three letters. This idea can be made more precise by considering fractional powers, as defined above.

For real $\alpha > 1$, we say a string *avoids* α*-powers* if it contains no factor that is an α'-power for any rational $\alpha' \geq \alpha$. Brandenburg [9] and Dejean [21] considered the problem of determining the *repetition threshold*; that is, the least exponent $\alpha = \alpha(k)$ such that there exist infinite strings over $A_k = \{a_1, \ldots, a_k\}$ that avoid $(\alpha + \epsilon)$-powers for all $\epsilon > 0$.

Dejean proved that $\alpha(3) = \frac{7}{4}$. She also conjectured that $\alpha(4) = \frac{7}{5}$ and $\alpha(k) = \frac{k}{k-1}$ for $k \geq 5$.

The next theorem summarizes what is known; we add the corresponding references in parentheses.

Theorem 19. *(i)* $\alpha(2) = 2$ *(Thue [73, 74]);*
(ii) $\alpha(3) = \frac{7}{4}$ *(Dejean [21]);*
(iii) $\alpha(4) = \frac{7}{5}$ *(Pansiot [63])*
(iv) $\alpha(k) = \frac{k}{k-1}$, *for* $5 \leq k \leq 11$ *(Moulin-Ollagnier [56]).*

In spite of the results of Pansiot and Moulin-Ollagnier confirming some particular cases, Dejean's conjecture, in its full generality, is still open. For more information, see [15].

The complete status of Dejean's conjecture is shown in Table 6.2. The values in bold have been proved, the others are still open.

6.5.5 Generalized Repetition Threshold

Instead of avoiding *all* squares, one interesting variation is to avoid *all sufficiently large* squares. Entringer, Jackson, and Schatz [28] showed that there exist infinite binary strings avoiding all squares xx with $|x| \geq 3$. Furthermore,

$\lvert A\rvert = k$	2	3	4	5	6	7	8	9	10	11	$k \geq 12$
$\alpha(k)$	**2**	$\mathbf{\frac{7}{4}}$	$\mathbf{\frac{7}{5}}$	$\mathbf{\frac{5}{4}}$	$\mathbf{\frac{6}{5}}$	$\mathbf{\frac{7}{6}}$	$\mathbf{\frac{8}{7}}$	$\mathbf{\frac{9}{8}}$	$\mathbf{\frac{10}{9}}$	$\mathbf{\frac{11}{10}}$	$\frac{k}{k-1}$?

Table 6.2. Dejean's conjecture

they proved that every binary string of length ≥ 18 contains a factor of the form xx with $|x| \geq 2$, so the bound 3 is best possible.

In [39, 40], these two variations are combined. Let $\alpha > 1$ be a rational number, and let $\ell \geq 1$ be an integer. A string w is a *repetition of order* α *and length* ℓ if we can write it as $w = x^n x'$ where x' is a prefix of x, $|x| = \ell$, and $|w| = \alpha|x|$. For brevity, we also call w an (α, ℓ)-*repetition*. Notice that an α-power is an (α, ℓ)-repetition for some ℓ. We say a string is (α, ℓ)-*free* if it contains no factor that is a (α', ℓ')-repetition for $\alpha' \geq \alpha$ and $\ell' \geq \ell$. We say a string is (α^+, ℓ)-*free* if it is (α', ℓ)-free for all $\alpha' > \alpha$.

For integers $k \geq 2$ and $\ell \geq 1$, we define the *generalized repetition threshold* $R(k, \ell)$ as the real number α such that either

(a) over A_k there exists an (α^+, ℓ)-free infinite string, but all (α, ℓ)-free strings are finite; or
(b) over A_k there exists a (α, ℓ)-free infinite string, but for all $\epsilon > 0$, all $(\alpha - \epsilon, \ell)$-free strings are finite.

Notice that $R(k, 1)$ is essentially the repetition threshold of Dejean and Brandenburg.

Table 6.3 contains a number of values from [40]; the ones in bold are known and the others are conjectured based (weakly) on numerical results. Notice that the conjecture for $R(k, 1)$, for $k \geq 12$, is Dejean's.

There does not exist a conjecture for the general form of the numbers $R(k, \ell)$.

The proved results are as follows:

Theorem 20. *(i)* $R(2, 1) = 2$ *(follows from Thue [73, 74, 7]);*
(ii) $R(2, 2) = 2$ *(follows from Thue [73, 74, 7] and Entringer, Jackson and Schatz [28]);*
(iii) $R(3, 1) = \frac{7}{4}$ *(Dejean [21]);*
(iv) $R(4, 1) = \frac{7}{5}$ *(Pansiot [63]);*
(v) $R(k, 1) = \frac{k}{k-1}$ *for* $5 \leq k \leq 11$ *(Moulin-Ollagnier [56]);*
(vi) The remaining (bold) ones are new and are proved in [40].

In [40], several uniform morphisms wer built to prove the values at (vi) in the above theorem. The one proving $R(2, 3) = \frac{8}{5}$ is 992-uniform.

$R(k,\ell)$	$\ell=1$	2	3	4	5	6	7	8	≥ 9
$k=2$	$\mathbf{2}$	$\mathbf{2}$	$\mathbf{\frac{8}{5}}$	$\mathbf{\frac{3}{2}}$	$\mathbf{\frac{7}{5}}$	$\mathbf{\frac{4}{3}}$	$\frac{31}{24}?$	$\frac{24}{19}?$	
3	$\mathbf{\frac{7}{4}}$	$\mathbf{\frac{3}{2}}$	$\mathbf{\frac{4}{3}}$	$\frac{5}{4}?$	$\frac{6}{5}?$	$\frac{7}{6}?$	$\frac{8}{7}?$	$\frac{9}{8}?$	$\frac{\ell+1}{\ell}?$
4	$\mathbf{\frac{7}{5}}$	$\frac{5}{4}?$	$\frac{6}{5}?$	$\frac{7}{6}?$	$\frac{8}{7}?$	$\frac{9}{8}?$	$\frac{10}{9}?$	$\frac{11}{10}?$	$\frac{\ell+3}{\ell+2}?$
5	$\mathbf{\frac{5}{4}}$	$\frac{6}{5}?$	$\frac{8}{7}?$	$\frac{9}{8}?$	$\frac{10}{9}?$				
6	$\mathbf{\frac{6}{5}}$	$\frac{36}{31}?$							
7	$\mathbf{\frac{7}{6}}$	$\frac{8}{7}?$							
8	$\mathbf{\frac{8}{7}}$								
9	$\mathbf{\frac{9}{8}}$								
10	$\mathbf{\frac{10}{9}}$								
11	$\mathbf{\frac{11}{10}}$								
≥ 12	$\frac{k}{k-1}?$								

Table 6.3. Known and conjectured values of $R(k,\ell)$.

6.6 Other Complexity Measures

We briefly enumerate here other complexity measures for strings, some of which are famous, some very little known or investigated.

Kolmogorov complexity.

(also Chaitin's, [12] and Solomonoff's, [69, 70]) Introduced by Kolmogorov [44], it is the most important complexity measure for strings. It represents the length of a shortest program to compute the given string. We have not discussed it at all since it is not combinatorial and non-computable.

It has been very much investigated and the best reference is the book of Li and Vitani [50].

Logical depth.

First described by Chaitin [13] and investigated at greater length by Bennet [5, 6], its definition takes into account both the length of the program and the time needed to generate the given string. It is also non-computable.

Automaticity.

Automatic infinite strings are defined as outputs of finite state automata. An infinite string w is k-automatic if there is a k-state deterministic automaton which outputs w's nth letter on input the base k representation of n.

These strings have been first studied systematically by Cobham [16] and very much investigated ever since; the best reference is the book of Allouche and Shallit [4].

Linguistic complexity.

The notion is used in [75] for complexity of biological sequences. For a string w, it is defined as the ratio between $\mathsf{all\_fact}(w)$ and the maximum of $\mathsf{all\_fact}(\cdot)$ for all strings of the same length, that is,

$$\mathsf{ling}(w) = \frac{\mathsf{all\_fact}(w)}{\max_{|x|=|w|} \mathsf{all\_fact}(x)} .$$

As an example, consider the string $w = \mathsf{ACGGGAAGCTGATTCCA}$ over the alphabet of nucleotides $\{\mathsf{A}, \mathsf{C}, \mathsf{G}, \mathsf{T}\}$; w has length 17 and contains 4 of 4 possible nucleotides, 15 of 16 possible dinucleotides (for a string of length 17), 15 of 15 possible trinucleotides, etc. Therefore, the linguistic complexity of w is $\mathsf{ling}(w) = \frac{139}{140}$. (This example is considered in [75] but the complexity computed there is incorrect.)

In [75] algorithms are given to compute it using suffix trees.

Spectrum.

It has been introduced and investigated by [33] where it is called *combinatorial complexity*; we changed the name for obvious reasons.

Let A be an alphabet, d a positive integer, and w a string of length n over A. Denote $S_d(w)$ the list of 2^d entries, numbered $0, 1, \ldots, 2^d - 1$ such that $S_d(w)(i)$ is the number of factors u of w such that u is the representation of i in binary. For instance, $S_2(001101101) = (1, 3, 2, 2)$. For a fixed d, the *spectrum complexity* is defined then as

$$\mathsf{spec}_d(w) = \log(\mathrm{card}\{x \mid S_d(x) = S_d(w)\}).$$

Because of results from Kolmogorov complexity, to which the spectrum complexity is compared, the case $d = \lfloor \log n - 2 \log \log n + 1 \rfloor$ is given a special importance. From combinatorial point of view, all such complexities seem interesting.

Another spectrum is mentioned in [15, p. 394]; the *k-spectrum* of a string w is a function which associates with each string no longer than w its number of occurrences in w.

Linear complexity.

Linear complexity of an infinite string is the size of the shortest linear feedback shift register (LFSR) which can produce that string. (It is defined for binary ultimately periodic strings.) There are also several related complexity measures, such as *linear complexity profile* and *k-error linear complexity.* All have applications to cryptography. See the survey [59] and the references therein.

Tree complexity.

Introduced in [60], it is defined as follows. For a string $w = w_1w_2w_3\ldots$ and a positive integer h, $\mathsf{tree}_h(w)$ is the number of subsequences of the form

$$w_k w_{2k} w_{2k+1} w_{4k} w_{4k+1} w_{4k+2} w_{4k+3} \cdots w_{2^h k} \cdots w_{2^h k + 2^h - 1}.$$

It turns out that it is closely related to automaticity (see above). See [58] and the references therein.

Arithmetical complexity.

Introduced in [2], this measure is related with both factor and tree complexity. Subsequences are again counted but now the lengths of the gaps form an arithmetical progression. For a string $w = w_1w_2w_3\ldots$ and $n \geq 1$, the arithmetic complexity is defined as

$$\mathsf{arit}_w(n+1) = \mathrm{card}\{w_k w_{k+d} \cdots w_{k+nd} \mid k \geq 0, d \geq 1\}.$$

A number of interesting results have been discovered recently; see [11, 31, 34] and the references therein.

References

1. J. M. Amigo, J. Szczepanski, E. Wajnryb and M. V. Sanchez-Vives, Estimating the entropy rate of spike trains via Lempel-Ziv complexity, *Neural Computation* **16**(4) (2004) 717 – 736.
2. S.V. Avgustinovich, D.G. Fon-Der-Flaass and A. E. Frid, Arithmetical complexity of infinite words, in: Masami Ito and Teruo Imaoka, eds., *Words, Languages and Combinatorics III (ICWLC 2000)*, World Scientific Publishing, Singapore, 2003, 51 – 62.
3. J.-P. Allouche, Sur la complexité des suites infinies, *Bull. Belg. Math. Soc. Simon Stevin* **1**(2) (1994) 133 – 143.
4. J.-P. Allouche and J. Shallit, *Automatic Sequences. Theory, Applications, Generalizations*, Cambridge Univ. Press, Cambridge, 2003.
5. C.H. Bennett, Information, Dissipation, and the Definition of Organization, *Emerging Syntheses in Science* (David Pines, ed.), Santa Fe Institute, 1985, 297 – 313, Addison-Wesley, Reading, Mass., 1987.
6. C.H. Bennett, Logical Depth and Physical Complexity, *The Universal Turing Machine – a Half-Century Survey* (R. Herken, ed.), Oxford University Press 1988, 227 – 257.
7. J. Berstel, *Axel Thue's Papers on Repetitions in Words: a Translation*, Number 20 in Publications du Laboratoire de Combinatoire et d'Informatique Mathématique, Université du Québec à Montréal, February 1995.
8. J. Berstel and P. Séébold, Sturmian Words, in [52], 45 – 110.
9. F.-J. Brandenburg, Uniformly growing k-th power-free homomorphisms, *Theoret. Comput. Sci.* **23** (1983) 69 – 82.

10. N.G. de Bruijn, A combinatorial problem, *Nederl. Akad. Wetensch. Proc.* **49** (1946) 758 – 764.
11. J. Cassaigne and A.E. Frid, On arithmetical complexity of Sturmian words, *Proc. of the 5th International Conference on Combinatorics on Words (WORDS'05)* (S. Brlek, C. Reutenauer, eds.), LaCIM **36**, Montreal, 2005, 197 – 207.
12. G. Chaitin, On the length of programs for computing finite binary sequences, *J. Assoc. Comput. Mach.* **13** (1966) 547 – 569.
13. G. Chaitin, Algorithmic Information Theory, *IBM J. Res. Develop.* **21** (1977) 350 – 359.
14. X. Chen, S. Kwong and M. Li, A compression algorithm for DNA sequences, *IEEE Engineering in Medicine and Biology Magazine* **20**(4) (2001) 61 – 66.
15. C. Choffrut and J. Karhumäki, Combinatorics on Words, *Handbook of Formal Languages, Vol. I* (G. Rozenberg, A. Salomaa, eds.), Springer-Verlag, Berlin, Heidelberg, 1997, 329 – 438.
16. A. Cobham, Uniform Tag Sequences, *Math. Systems Th.* **6** (1972) 164 – 192.
17. S. Constantinescu and L. Ilie, The Lempel-Ziv complexity of fixed points of morphisms, submitted, Aug. 2005.
18. M. Crochemore, An optimal algorithm for computing the repetitions in a word, *Inform. Process. Lett.* **12**(5) (1981) 244 – 250.
19. M. Crochemore, Linear searching for a square in a word, *NATO Advanced Research Workshop on Combinatorial Algorithms on Words* (A. Apostolico, Z. Galil, eds.), 1984, Springer-Verlag, Berlin, New York, 1985, 66 – 72.
20. M. Crochemore and W. Rytter, *Text Algorithms*, Oxford Univ. Press, 1994.
21. F. Dejean, Sur un théorème de Thue, *J. Combin. Theory. Ser. A* **13** (1972) 90 – 99.
22. A. Ehrenfeucht, K.P. Lee and G. Rozenberg, Subword complexities of various classes of deterministic developmental languages without interaction, *Theoret. Comput. Sci.* **1** (1975) 59 – 75.
23. A. Ehrenfeucht and G. Rozenberg, On the subword complexities of square-free D0L-languages, *Theoret. Comput. Sci.* **16** (1981) 25 – 32.
24. A. Ehrenfeucht and G. Rozenberg, On the subword complexities of D0L-languages with a constant distribution, *Theoret. Comput. Sci.* **13** (1981) 108 – 113.
25. A. Ehrenfeucht and G. Rozenberg, On the subword complexities of homomorphic images of languages, *RAIRO Informatique Théorique* **16** (1982) 303 – 316.
26. A. Ehrenfeucht and G. Rozenberg, On the subword complexities of locally catenative D0L-languages, *Information Processing Letters* **16** (1982) 7 – 9.
27. A. Ehrenfeucht and G. Rozenberg, On the subword complexities of m-free D0L-languages, *Information Processing Letters* **17** (1983) 121 – 124..
28. R.C. Entringer, D.E. Jackson, and J.A. Schatz, On nonrepetitive sequences, *J. Combin. Theory. Ser. A* **16** (1974) 159 – 164.
29. M. Farach, M.O. Noordewier, S.A. Savari, L.A. Shepp, A.D. Wyner and J. Ziv, On the entropy of DNA: algorithms and measurements based on memory and rapid convergence, *Proc. of SODA'95*, 1995, 48 – 57.
30. A. Flaxman, A. Harrow, and G. Sorkin, Strings with maximally many distinct subsequences and substrings, *Electron. J. Combin.* **11**(1) (2004) #R8.
31. D.G. Fon-Der-Flaass and A.E. Frid, On periodicity and low complexity of infinite permutations, *EuroComb'05*, 267 – 272.

32. A.S. Fraenkel and J. Simpson, How many squares can a string contain?, *J. Combin. Theory, Ser. A,* **82** (1998) 112 – 120.
33. F. Frayman, V. Kanevsky, and W. Kirchherr, The combinatorial complexity of a finite string, *Proc. of Mathematical Foundations of Computer Science 1994* (Košice, 1994), Lecture Notes in Comput. Sci. **841**, Springer, Berlin, 1994, 364 – 372,
34. A.E. Frid, On possible growths of arithmetical complexity, *Theoretical Informatics and Applications (RAIRO)*, to appear.
35. V.D. Gusev, V.A. Kulichkov, and O.M. Chupakhina, The Lempel-Ziv complexity and local structure analysis of genomes, *Biosystems* **30**(1-3) (1993) 183 – 200.
36. D. Gusfield, *Algorithms on Strings, Trees, and Sequences. Computer Science and Computational Biology*, Cambridge University Press, Cambridge, 1997.
37. L. Ilie, A simple proof that a word of length n has at most $2n$ distinct squares, *J. Combin. Theory, Ser. A* **112**(1) (2005) 163 – 164.
38. L. Ilie, A note on the number of distinct squares in a word, *Proc. of the 5th International Conference on Combinatorics on Words (WORDS'05)* (S. Brlek, C. Reutenauer, eds.), LaCIM **36**, Montreal, 2005, 289 – 294.
39. L. Ilie, P. Ochem, and J. Shallit, A generalization of repetition threshold, *Proceedings of the 29th International Symposium on Mathematical Foundations of Computer Science (MFCS)* (J. Fiala et al., eds.), Lecture Notes in Comput. Sci. **3153**, Springer, Berlin, 2004, 818 – 826.
40. L. Ilie, P. Ochem, and J. Shallit, A generalization of repetition threshold, *Theoret. Comput. Sci.*, to appear, 2005
41. L. Ilie, S. Yu and K. Zhang, Word complexity and repetitions in words, *Internat. J. Found. Comput. Sci.* **15**(1) (2004) 41 – 55.
42. C. Iliopoulos, D. Moore, and W.F. Smyth, A characterization of the squares in a Fibonacci string, *Theoret. Comput. Sci.* **172**(1-2) (1997) 281 – 291.
43. A. Ivanyi, On the d-complexity of words, *Ann. Univ. Sci. Budapest. Sect. Comput.* **8** (1987) 69 – 90.
44. A.N. Kolmogorov, Three approaches to the quantitative definition of information, *Probl. Inform. Transmission* **1** (1965) 1 – 7.
45. R. Kolpakov and G. Kucherov, Maximal repetitions in words or how to find all squares in linear time, Rapport Interne LORIA 98-R-227, Laboratoire Lorrain de Recherche en Informatique et ses Applications, 1998 (available from URL: http://www.loria.fr/ kucherov/res activ.html).
46. R. Kolpakov and G. Kucherov, On the sum of exponents of maximal repetitions in a word, LORIA, 1999, Rapport Interne, 99-R-034, Laboratoire Lorrain de Recherche en Informatique et ses Applications, 1999 (available from URL: http://www.loria.fr/ kucherov/res activ.html).
47. R. Kolpakov and G. Kucherov, Finding maximal repetitions in a word in linear time, *Proc. of the 40th Annual Symposium on Foundations of Computer Science*, IEEE Computer Soc., Los Alamitos, CA, 1999, 596 – 604.
48. R. Kolpakov and G. Kucherov, Periodic Structures in Words, in [53], 399 – 442.
49. A. Lempel and J. Ziv, On the Complexity of Finite Sequences , *IEEE Trans. Inform. Theory* **92**(1) (1976) 75 – 81.
50. M. Li and P. Vitani, *An Introduction to Kolmogorov Complexity and Its Applications*, 2nd ed., New York: Springer-Verlag, 1997.
51. M. Lothaire, *Combinatorics on Words*, Addison-Wesley, Reading, MA, 1983 (2nd ed., Cambridge Univ. Press, 1997).

52. M. Lothaire, *Algebraic Combinatorics on Words*, Cambridge Univ. Press, 2002.
53. M. Lothaire, *Applied Combinatorics on Words*, Cambridge Univ. Press, 2005.
54. M.G. Main, Detecting leftmost maximal periodicities, *Discrete Appl. Math.* **25**(1-2) (1989) 145 – 153.
55. M. Morse and G.A. Hedlund, Symbolic dynamics II. Sturmian trajectories, *Amer. J. Math.* **61** (1940) 1 – 42.
56. J. Moulin-Ollagnier, Proof of Dejean's conjecture for alphabets with $5, 6, 7, 8, 9, 10$ and 11 letters, *Theoret. Comput. Sci.* **95** (1992) 187 – 205.
57. S. Mund and Ziv-Lempel complexity for periodic sequences and its cryptographic application, *Advances in Cryptology – EUROCRYPT '91*, Lecture Notes in Comput. Sci. **547**, Springer-Verlag, 1991, 114 – 126.
58. H. Niederreiter, Some computable complexity measures for binary sequences, *Sequences and Their Applications* (C. Ding, T. Helleseth, and H. Niederreiter, eds.), Springer-Verlag, London, 1999, 67 - 78.
59. H. Niederreiter, Linear complexity and related complexity measures for sequences, *Progress in Cryptology INDOCRYPT 2003* (T. Johansson and S. Maitra, eds.), Lecture Notes in Comput. Sci. **2904**, Springer-Verlag, Berlin, 2003, 1 - 17.
60. H. Niederreiter and M. Vielhaber, Tree Complexity and a Doubly Exponential Gap between Structured and Random Sequences, *J. Complexity* **12** (1996) 187198.
61. J.-J. Pansiot, Bornes inférieures sur la complexité des facteurs des mots infinis engendrés par morphismes itérés, *Proc. of STACS'84*, Lecture Notes in Comput. Sci. **166**, Springer, Berlin, 1984, 230 – 240.
62. J.-J. Pansiot, Complexité des facteurs des mots infinis engendrés par morphismes itérés, *Proc. of ICALP'84*, Lecture Notes in Comput. Sci. **172**, Springer, Berlin, 1984, 380 – 389.
63. J.-J. Pansiot, A propos d'une conjecture de F. Dejean sur les répétitions dans les mots, *Disc. Appl. Math.* **7** (1984) 297 – 311.
64. G. Rozenberg, On subwords of formal languages, *Proc. of Fundamentals of computation theory*, Lecture Notes in Comput. Sci. **117**, Springer, Berlin-New York, 1981, 328 – 333.
65. G. Rozenberg and A. Salomaa, *The Mathematical Theory of L Systems*, Academic Press, 1980.
66. W. Rytter, Application of Lempel-Ziv factorization to the approximation of grammar-based compression, *Theoret. Comput. Sci.* **302**(1-3) (2003) 211 – 222.
67. A. Salomaa and M. Soittola, *Automata-Theoretic Aspects of Formal Power Series*, Springer, New York, 1978.
68. J. Shallit, On the maximum number of distinct factors of a binary string, *Graphs Combin.* **9**(2) (1993) 197 – 200.
69. R.J. Solomonoff, A preliminary report on a general theory of inductive inference, Technical Report ZTB-138, Zator Company, Cambridge, MA, November 1960.
70. R.J. Solomonoff, A formal theory of inductive inference, Parts I and II, *Information and Control* **7** (1964) 1 - 22 and 224 - 254.
71. J. Szczepanski, M. Amigo, E. Wajnryb, and M.V. Sanchez-Vives, Application of Lempel-Ziv complexity to the analysis of neural discharges, *Network: Computation in Neural Systems* **14**(2) (2003) 335 – 350.
72. J. Szczepanski, J.M. Amigo, E. Wajnryb, and M. V. Sanchez-Vives, Characterizing spike trains with Lempel-Ziv complexity, *Neurocomputing* **58-60** (2004) 79 – 84.

73. A. Thue, Über unendliche Zeichenreihen, *Norske vid. Selsk. Skr. Mat. Nat. Kl.* **7** (1906) 1 – 22. (Reprinted in *Selected Mathematical Papers of Axel Thue*, T. Nagell, editor, Universitetsforlaget, Oslo, 1977, 139 – 158.)
74. A. Thue, Über die gegenseitige Lage gleicher Teile gewisser Zeichenreihen, *Norske vid. Selsk. Skr. Mat. Nat. Kl.* **1** (1912) 1 – 67. (Reprinted in *Selected Mathematical Papers of Axel Thue*, T. Nagell, editor, Universitetsforlaget, Oslo, 1977, 413 – 478.)
75. O. Troyanskaya, O. Arbell, Y. Koren, G. Landau, and A. Bolshoy, Sequence complexity profiles of prokaryotic genomic sequences: A fast algorithm for calculating linguistic complexity, *Bioinformatics* **18** (2002) 679 – 688.
76. J. Ziv and A. Lempel, A universal algorithm for sequential data compression, *IEEE Trans. Inform. Theory* **23**(3) (1977) 337 – 343.
77. J. Ziv and A. Lempel, Compression of individual sequences via variable-rate coding, *IEEE Trans. Inform. Theory* **24**(5) (1978) 530 – 536.
78. W. F. Smyth, *Computing Patterns in Strings*, Pearson Addison Wesley, 2003.

7

Image Processing Using Finite Automata

Jarkko Kari*

Department of Mathematics, University of Turku
FIN-20014 Turku, Finland and
Department of Computer Science, University of Iowa
MLH 14, Iowa City, IA 52242, USA
E-mail: jkari@utu.fi

7.1 Introduction

Finite automata are among the simplest and best understood mathematical objects used in computer science. Their applications cover a wide spectrum of the field. In this article we demonstrate the usefulness of finite state techniques in digital image processing and compression. The basic idea is to use words over some alphabet to address image pixels. Then finite automata can be used to specify subsets of pixels, i.e. bilevel images, while weighted automata specify grey-scale images. Simple inference algorithms exist for constructing automata representations of given images, and standard automata minimization techniques can be used to minimize the number of states. This naturally leads to the idea of viewing such automata as compressed representations of images. This idea was introduced by by the author and K.Culik in the early 1990's [3, 5, 6, 7]. Since then other research groups have studied and further improved and generalized the algorithms, see e.g. [1, 2, 8, 9, 10, 12, 13]. In this article the basic concepts and algorithms are presented from the beginning, without assuming any prior knowledge of the field. The purpose is to provide a tutorial suitable for self-study on this topic.

We start in Section 7.2 by discussing our addressing scheme of pixels using words over a four letter alphabet. Infinite languages define infinitely sharp images, so we next briefly discuss the concepts of multi-resolution and infinite resolution images. In Section 7.3 we discuss bilevel images defined by regular languages, and specifically using deterministic finite automata (DFA). This serves as an introduction to the more general idea of using finite automata to define gray-scale images. In Section 7.4 we introduce weighted finite automata (WFA). These are automata whose transitions are weighted using real numbers. Notice that weighted finite automata were studied extensively much earlier by M.P.Schützenberger in the more general case where the weights come from an arbitrary semiring [14]. Our contribution is a way to apply

* Research supported by the Academy of Finland grant 54102

J. Kari: *Image Processing Using Finite Automata*, Studies in Computational Intelligence (SCI) **25**, 171–208 (2006)
www.springerlink.com

these devices in image processing. In Section 7.5 we discuss how one can efficiently draw the image specified by a given WFA. Next we turn our attention to the converse problem: Section 7.6 concentrates on the problem of inferring a WFA that represents a given input image. We provide an efficient algorithm for this task. The algorithm is even guaranteed to produce the minimum state WFA. In the following Section 7.7 we outline the ideas used to transform the theoretical inference algorithm into a practical image compression technique. The details of these ideas are skipped and the interested reader is referred to the more technical descriptions [6, 11]. We compare our algorithm with the image compression standard JPEG that is based on the discrete cosine transform. Using several test images we show the difference in performance, and we demonstrate that the WFA technique perform especially well on images with sharp edges. Finally, in Section 7.8 we briefly discuss one nice aspect of the WFA representation of images: several natural image transformations can be implemented directly in the compressed form, without the need to decode the image first. We show several examples of image operations that can be defined using a weighted analogy of finite state transducers, called weighted finite transducers (WFT).

7.2 Image Types

A *finite resolution* image is a $w \times h$ rectangular array of pixels, each with its own color. In the context of this work, squares of dimensions that are powers of two come up naturally, so that

$$w = h = 2^k \text{ for some } k \in \mathbb{Z}_+.$$

The image is then just a function

$$\{1, 2, \dots w\} \times \{1, 2, \dots, h\} \longrightarrow \mathcal{C}$$

that assigns each pixel a color from a set $\mathcal{C}$ of possible colors. The color set depends on the type of the image. It can be (see Figure 7.1)

- $\mathcal{C} = \{\text{Black}, \text{White}\}$ on *bilevel* or *binary* images,
- $\mathcal{C} = \mathbb{R}$, the set of reals, on *greyscale* images. The color value represents the intensity of the pixel. Sometimes only non-negative real numbers $\mathcal{C} = [0, \infty)$ are allowed, sometimes the values are restricted to an interval, e.g. $\mathcal{C} = [0, 1]$. In digital images the intensity values are quantized in a finite number of intensity levels. It is common to use 256 possible intensity values, or 8 bits per pixel (bpp).
- $\mathcal{C} = \mathbb{R}^3$, vectors of three real numbers, on color images. Three values represent the intensities of three color components. Instead of $\mathbb{R}$ also $[0, \infty)$ or $[0, 1]$ may be used.

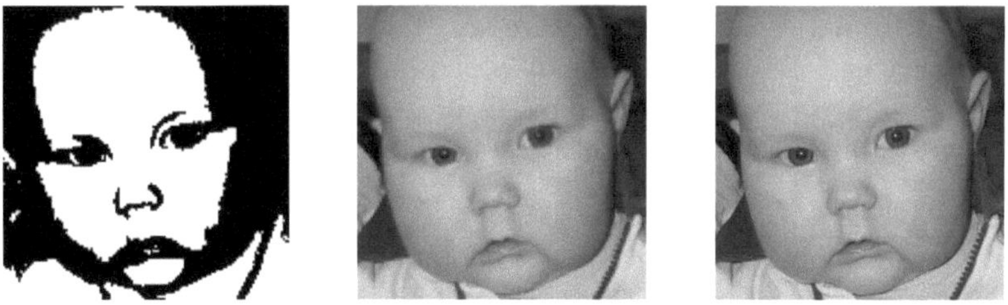

Fig. 7.1. Bilevel, grey-scale and color images.

Rather than addressing the pixels by their x- and y-coordinates, let us address the pixels using words over the four letter alphabet $\Sigma = \{0, 1, 2, 3\}$. Pixels of a $2^k \times 2^k$ image are addressed by words of length k as follows. The image is first divided into four quadrants. The first letter of the address of a pixel is determined by the quadrant that contains the pixel:

$2\ldots$	$3\ldots$
$0\ldots$	$1\ldots$

The rest of the address is defined recursively as the word of length $k-1$ that is the address of the pixel in the quadrant when the quadrant is viewed as a $2^{k-1} \times 2^{k-1}$ image. The only pixel of a 1×1 resolution "image" has the empty word ε as the address. For example, the addresses of pixels in 4×4 images are shown in Figure 7.2(a), and Figure 7.2(b) shows the location of the pixel addressed by 2131 at resolution 16×16.

22	23	32	33
20	21	30	31
02	03	12	13
00	01	10	11

(a)

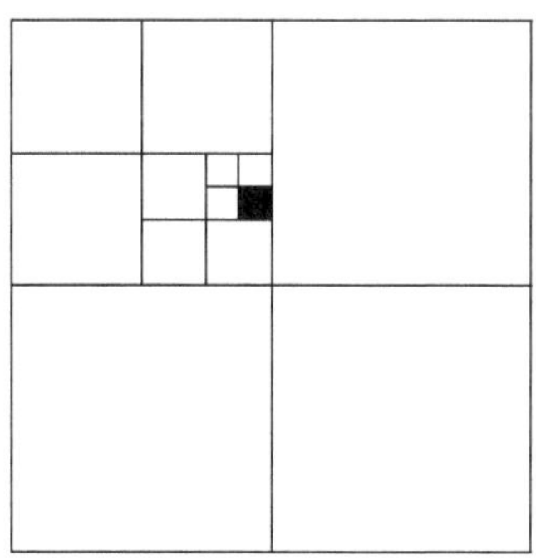

(b)

Fig. 7.2. (a) Addresses of pixels in 4×4 images, (b) The pixel with address 2131.

Viewing from the end of the addresses the addressing scheme looks as follows: Every $w \in \Sigma^k$ is the address of a pixel in a $2^k \times 2^k$ image. To make a $2^{k+1} \times 2^{k+1}$ image we subdivide all pixels into 2×2 blocks of smaller pixels. The addresses in the subdivision of pixel w are words $w0, w1, w2$ and $w3$:

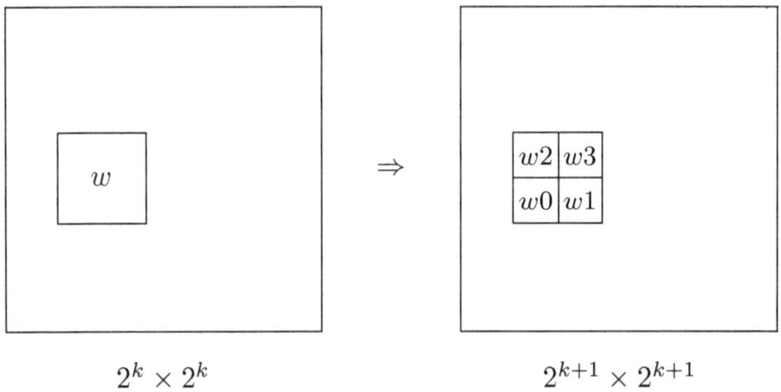

A $2^k \times 2^k$ resolution image is now a function

$$f_k : \Sigma^k \longrightarrow \mathcal{C}$$

that assigns a color to each pixel, where $\mathcal{C} = \{\text{Black}, \text{White}\}, \mathbb{R}$ or $\mathbb{R}^3$ depending on the type of the image.

7.2.1 Multi-resolution Images

A *multi-resolution* image is a function

$$f : \Sigma^* \longrightarrow \mathcal{C}$$

that assigns a color to each pixel in resolutions $2^k \times 2^k$ for all $k \geq 0$. A multi-resolution image can be viewed as a sequence $f_0, f_1, f_2, \ldots$ of images where $f_k = f_{|_{\Sigma^k}}$, the restriction of f to words of length k, is a $2^k \times 2^k$ image. Our addressing scheme defines a quad-tree structure on the pixels of the images:

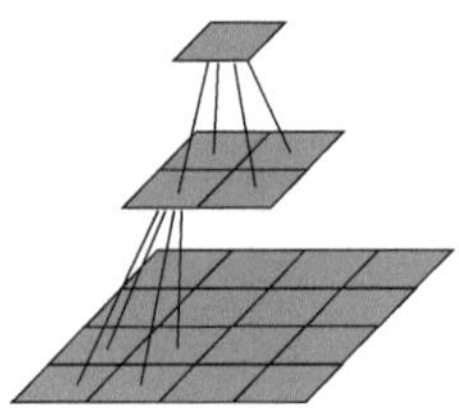

or, more abstractly,

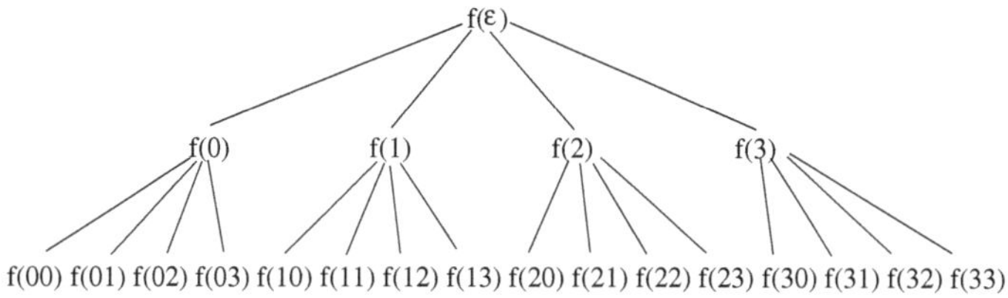

The definition of a multi-resolution image f does not require any similarity between the finite resolution images $f_0, f_1, f_2, \ldots$ it contains. However, it would be useful if each f_k is a finite resolution approximation of the same infinitely sharp image. For this purpose we introduce the concept of average preserving multi-resolution images.

Consider first gray-scale images, i.e. the case $\mathcal{C} = \mathbb{R}$, $\mathcal{C} = [0, \infty)$ or $\mathcal{C} = [0, 1]$. A simple way to interpolate a $2^k \times 2^k$ resolution gray-scale image into a $2^{k-1} \times 2^{k-1}$ resolution image is to average the intensity values on 2×2 blocks of pixels. Let f be a multi-resolution grey-scale image. If

$$f(w) = \frac{f(w0) + f(w1) + f(w2) + f(w3)}{4}$$

for every $w \in \Sigma^*$ then f_{k-1} is the interpolation of the next sharper image f_k, for every $k = 1, 2, \ldots$. In this case we say that f is *average preserving*, or *ap* for short. In the quad-tree form this property simply states that each node is the average of its four children.

If f is average preserving the images in the sequence $f_0, f_1, f_2, \ldots$ form sharper and sharper approximations of some infinitely sharp gray-scale image. For example, Figure 7.3 shows six images that are the finite resolution members $f_2, f_3, f_4, f_5, f_6, f_7$ of an ap multi-resolution image.

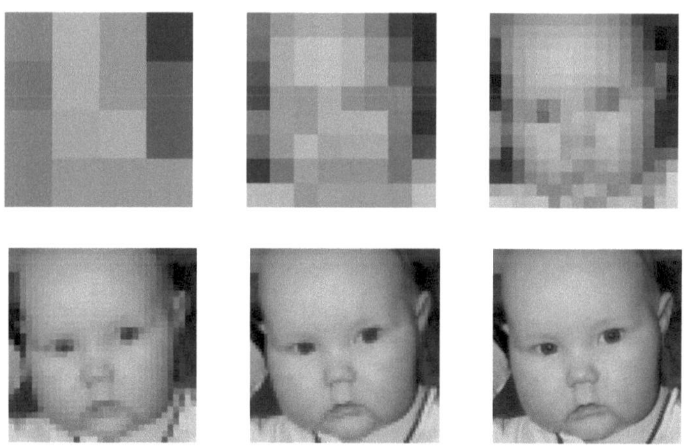

Fig. 7.3. Finite resolution images $f_2 \longrightarrow f_7$.

More precisely, an *infinite resolution* gray-scale image can be defined as a Borel measure on the unit square. The measure of a set represents the total amount of color needed to color the set. Note that this definition implies that we only use non-negative intensities, i.e. $\mathcal{C} = [0, \infty)$, but this is not a significant restriction. Average preserving multi-resolution images f correspond to such infinite resolution images where the correspondence is given by the fact that the measure of a pixel addressed by word w (and viewed as a half open square region of the unit square) has to be

$$f(w)/4^{|w|},$$

that is, the average intensity $f(w)$ of the pixel multiplied by the area $1/4^{|w|}$ of the pixel. This uniquely determines the measure. The ap-property of f guarantees that the measure is well-defined and additive.

Some infinite resolution gray-scale images can be expressed as an integrable function

$$g : [0,1] \times [0,1] \longrightarrow \mathcal{C}$$

that assigns greyness values to all points of the unit square. The measure is then obtained by integrating g over the area in question. Note, however, that not all measures (i.e. infinite resolution images) can be expressed in such a form. For example, the multi-resolution

$$\begin{aligned} f(0^n) &= 4^n \text{ for every } n \geq 0, \\ f(w) &= 0 \text{ for all } w \notin 0^* \end{aligned}$$

is average preserving. The measure is concentrated in one point at the lower left corner of the square, and it can not be defined as the integral of any function g.

Example 1. Let us define a multi-resolution image f recursively as follows:

$$\begin{cases} f(\varepsilon) = \frac{1}{2}, \\ f(w0) = f(w) - \left(\frac{1}{2}\right)^{|w|+2}, \\ f(w1) = f(w), \\ f(w2) = f(w), \\ f(w3) = f(w) + \left(\frac{1}{2}\right)^{|w|+2}. \end{cases}$$

Then, for example,

$$f(0232) = f(023) = f(02) + \frac{1}{16} = \frac{1}{2} - \frac{1}{4} + \frac{1}{16} = \frac{5}{16}.$$

It is immediately clear from the definition that this f is average preserving. The finite resolution images $f_0, f_1, \ldots, f_5$ are shown in Figure 7.4 at increasing resolutions. It is easy to see that the infinite resolution image defined by f is the measure obtained by integrating the linear function $g(x, y) = \frac{x+y}{2}$. □

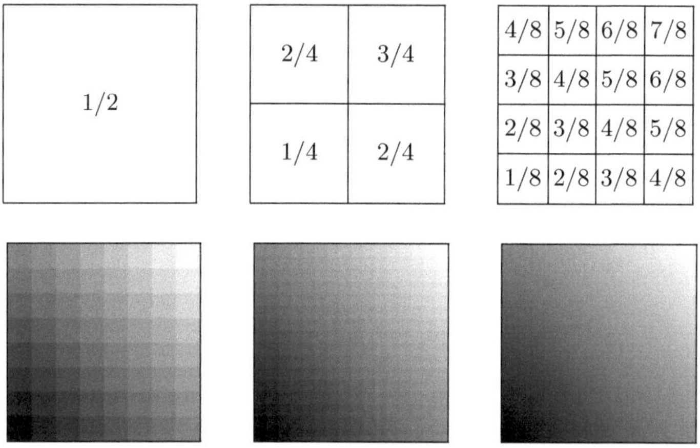

Fig. 7.4. First six finite resolution images of the multiresolution f from Example 1.

The set

$$\{f \mid f : \Sigma^\star \longrightarrow \mathbb{R}\}$$

of multi-resolution images is a linear space, where the operations of addition and multiplication with a real number are defined pixel-wise:

$$(f+g)(w) = f(w) + g(w) \quad \forall w \in \Sigma^\star, and$$
$$(r \cdot f)(w) = r \cdot f(w) \quad \forall w \in \Sigma^\star, r \in \mathbb{R}.$$

The set of average preserving multi-resolution images is a linear subspace.

In an ideal situation multi-resolution images are used. In practical applications in digital image processing one only has finite resolution $2^k \times 2^k$ images available. They also form a linear space that is isomorphic to $\mathbb{R}^{4^k}$. We naturally have standard algorithms of linear algebra available. In particular, we can find an expression of a given finite resolution image f as a linear combination of finite resolution images $\psi_1, \ldots, \psi_n$, if such an expression

$$f = c_1 \cdot \psi_1 + c_2 \cdot \psi_2 + \ldots c_n \cdot \psi_n$$

with $c_1, c_2, \ldots, c_n \in \mathbb{R}$ exists.

7.2.2 Multi-resolution Bilevel Images

Next, let us consider multi-resolution bilevel images. Function

$$f : \Sigma^* \longrightarrow \{\text{Black}, \text{White}\}$$

can be viewed as a language over alphabet $\Sigma = \{0, 1, 2, 3\}$, where word w is in the language if and only if $f(w) = \text{Black}$. The finite resolution image f_k

provides the words of length k of the language. Conversely, every language $L \subseteq \Sigma^*$ corresponds to a multi-resolution image.

Language L is called *prefix closed*, iff every prefix of every element of the language is also in the language, that is,

$$\forall u, v \in \Sigma^* : uv \in L \Longrightarrow u \in L.$$

We call language L *prefix trimmed* iff every element of L is a proper prefix of another element of L, that is,

$$\forall u \in L \ : uv \in L \text{ for some } v \neq \varepsilon.$$

Prefix closed and trimmed languages correspond to multi-resolution images where finite resolution images are obtained from the next higher resolution using the max-interpolation: in this interpolation each 2×2 block of the higher resolution image is replaced by

- a black pixel if the 2×2 block contains a black pixel,
- a white pixel if the block does not contain a black pixel.

Prefix closed and trimmed languages define multi-resolution images where the finite resolution images approach some infinitely sharp image as the resolution is increased. In this respect they correspond to average preserving multi-resolutions in the gray-scale case.

Example 2. Language $L = (0+1+2)^*$ is prefix closed and trimmed. The first few finite resolution images are shown in Figure 7.5.

□

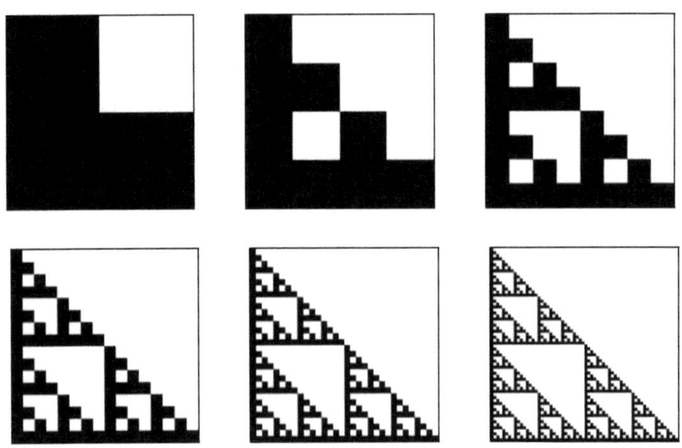

Fig. 7.5. Finite resolution images of the language $(0+1+2)^*$.

Infinite resolution bilevel images are defined as compact (i.e. topologically closed) subsets of the unit square $[0,1]^2$. Every prefix closed language defines an infinite resolution image which is obtained as the intersection

$$\bigcap_{k=0}^{\infty} f_k$$

of the finite resolution images f_k, each interpreted as a compact subset of the unit square. The prefix property simply means that we have a decreasing sequence of sets:

$$f_1 \supseteq f_2 \supseteq f_3 \supseteq \dots$$

so the intersection is guaranteed to be non-empty if L is an infinite language.

For example, the infinite resolution image defined by $(0+1+2)^*$ is the Sierpinski triangle

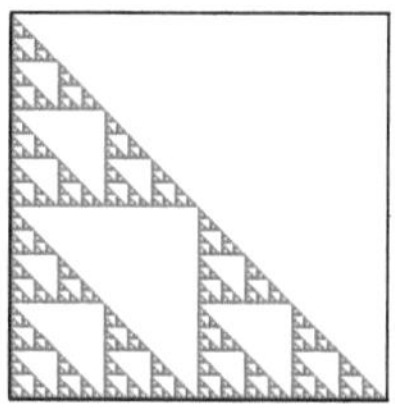

Every compact set $C \subseteq [0,1]^2$ is specified by the prefix closed and trimmed language L where $w \in L$ if and only if the sub-square addressed by w contains at least one element of C. Indeed, each finite resolution image f_k specified by this L covers C, so that

$$C \subseteq \bigcap_{k=0}^{\infty} f_k.$$

And conversely, if $x \notin C$ then some open neighborhood U of x does not intersect C, so for all sufficiently large k we have $x \notin f_k$.

Notice, however, that the same compact set may be specified by several prefix closed and trimmed languages. For example, languages $\varepsilon + 03^*$ and $\varepsilon + 30^*$ both specify the single point at the center of the unit square.

The prefix property is a natural requirement corresponding to the ap property of gray-scale images. Languages that are not prefix closed do not necessarily define any infinite resolution image under any reasonable definition. For example, the regular language $\Sigma^*(1+2)$ defines at finite resolutions the $2^k \times 2^k$ checker boards shown in Figure 7.6, and there is no natural infinite resolution limit to this sequence.

7.3 Regular Languages as Bilevel Images

All our examples have been regular languages. Every such language is accepted by a deterministic finite automaton (DFA). Prefix closed languages

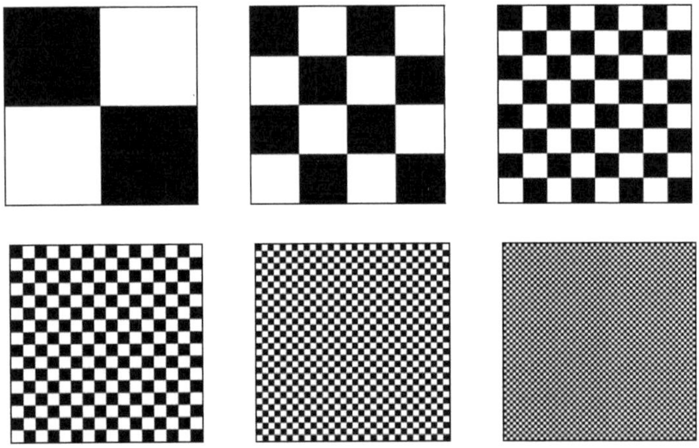

Fig. 7.6. Finite resolution images of the language $(0+1+2+3)^*(1+2)$.

are accepted by DFA all of whose states are accepting. The property of being prefix trimmed then means that there is at least one transition from every state. Note that the transition rule of a DFA is not required to be a complete function.

In a DFA, it is natural to assign to each state s its own language that consists of those words that are accepted starting from state s. Languages correspond to multi-resolution images, which means that each state specifies its own multi-resolution image. Then for every letter $a = 0, 1, 2, 3$ the transition

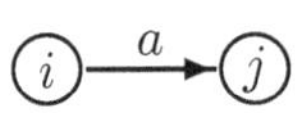

from state i to state j with label a means that the quadrant a of the image of state i is identical to the image of state j. Hence the transitions of the automaton give mutually recursive definitions of the state images.

Example 3. In the five state DFA of Figure 7.7 the transitions specify which state images are identical to the various quadrants of the state images. All states are final states.

□

For every multi-resolution function $f : \Sigma^* \longrightarrow \mathcal{C}$ and every $w \in \Sigma^*$, let

$$w^{-1}f : \Sigma^* \longrightarrow \mathcal{C}$$

be the multi-resolution function

$$w^{-1}f(u) = f(wu), \ \forall u \in \Sigma^*.$$

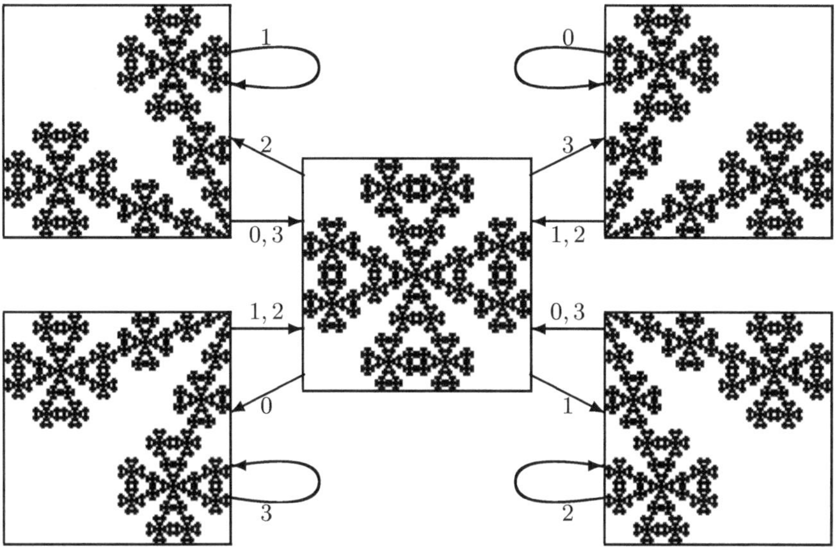

Fig. 7.7. A five state DFA with the state images drawn inside the states.

- As a quad-tree, $w^{-1}f$ is the sub-tree of f rooted at the node with address w.
- As a multi-resolution image, $w^{-1}f$ is the image obtained by zooming into the sub-square with address w.
- In the bilevel case, if f corresponds to the language L then $w^{-1}f$ corresponds to its left quotient $w^{-1}L$ with respect to word w, that is, the language
$$w^{-1}L = \{u \in \Sigma^* \mid wu \in L\}.$$

Let A be a DFA accepting language L, and let f be the corresponding multi-resolution bilevel image. Language $w^{-1}L$ is accepted by the DFA A' that is obtained from A by changing the initial state to be the state j where input w takes the original automaton A.

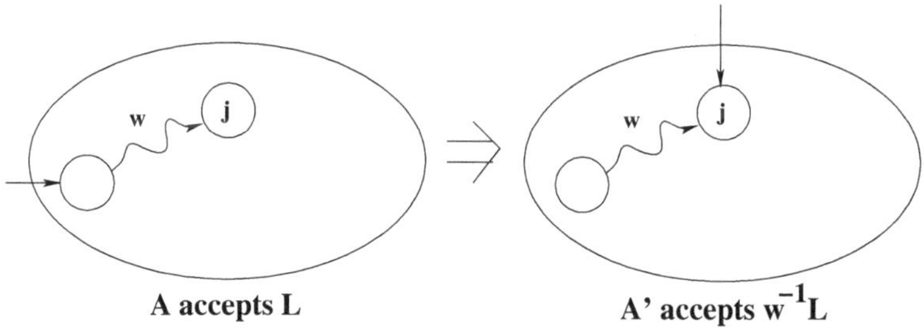

Number of different left quotients of L is the same as the number of states in the minimum state DFA that accepts language L. From this we conclude that the number of different subtrees of the quad-tree f is the same as the number of states in the smallest DFA accepting the corresponding language L. This is also the same as the number of different images that one can obtain by zooming into sub-squares addressed by Σ^* in the corresponding bilevel image.

The previous observation suggest the following well known algorithm to construct the minimum state DFA for a given bilevel multi-resolution image (i.e. for a given language L):

Encoding algorithm for bilevel images

Input: Language L
Variables: n : number of states so far
i : first non-processed state
L_j : Language accepted from state j

1. $n \leftarrow 1, i \leftarrow 1, L_1 \leftarrow L$
2. For $a = 0, 1, 2, 3$ do
   a) If $L_j = a^{-1}L_i$ for some $j = 1, 2, \ldots, n$ then add the transition

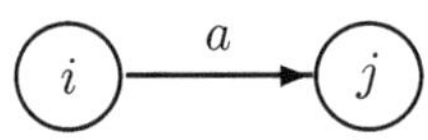

   b) else create a new state: Set $n \leftarrow n + 1$, $L_n \leftarrow a^{-1}L_i$, and add the transition

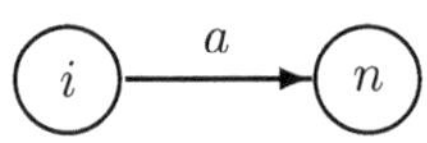

3. $i \leftarrow i + 1$. If $i \leq n$ then goto 2
4. State 1 is the initial state,
   State i is a final state iff $\varepsilon \in L_i$

7.4 Weighted Finite Automata (WFA)

WFA are finite automata whose transitions are weighted by real weights, and whose states have two associated weights called the initial and the final distribution values. In our applications the input alphabet is $\Sigma = \{0, 1, 2, 3\}$. More precisely, a WFA is specified by

- the number of states $n \in \mathbb{Z}_+$,
- four transition matrices $A_0, A_1, A_2, A_3 \in \mathbb{R}^{n \times n}$,
- a final distribution vector $F \in \mathbb{R}^{n \times 1}$, and
- an initial distribution vector $I \in \mathbb{R}^{1 \times n}$.

The WFA defines a multi-resolution image f as follows:

$$f(a_1 a_2 \ldots a_k) = I A_{a_1} A_{a_2} \ldots A_{a_k} F$$

for all $a_1a_2\dots a_k \in \Sigma^*$. Let us use the following shorthand notation: For every $w = a_1a_2\dots a_k \in \Sigma^*$ let

$$A_w = A_{a_1}A_{a_2}\dots A_{a_k}$$

be the product of the matrices corresponding to the letters of the word w. Then $f(w) = IA_wF$.

Usually a WFA is given as a labeled, weighted directed graph: There are n nodes, and there is an edge from node i to node j with label $a \in \Sigma$ and weight $r \neq 0$ iff $(A_a)_{ij} = r$. Transitions with weight 0 are not drawn. The initial and final distribution values are marked inside the nodes. For example,

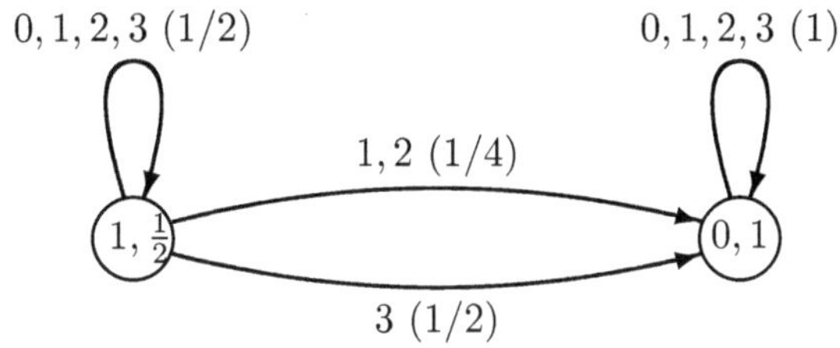

is the WFA whose matrix representation is

$$I = \begin{pmatrix} 1 & 0 \end{pmatrix} \qquad A_1 = \begin{pmatrix} 1/2 & 1/4 \\ 0 & 1 \end{pmatrix}$$

$$F = \begin{pmatrix} 1/2 \\ 1 \end{pmatrix} \qquad A_2 = \begin{pmatrix} 1/2 & 1/4 \\ 0 & 1 \end{pmatrix}$$

$$A_0 = \begin{pmatrix} 1/2 & 0 \\ 0 & 1 \end{pmatrix} \quad A_3 = \begin{pmatrix} 1/2 & 1/2 \\ 0 & 1 \end{pmatrix}$$

From the graph the multi-resolution image f can be read as follows: The weight of a path in the graph is obtained by multiplying the weights of the transitions on the path, the initial distribution value of the first node and the final distribution value of the last node on the path. The value $f(w)$ is the sum of the weights of all paths whose labels read word w. For example, in the WFA above, the matrix representation gives

$$f(03) = IA_0A_3F = \frac{3}{8},$$

but the same value is obtained by adding up the weights of the three paths whose labels read $w = 03$:

$$f(03) = 1 \cdot \frac{1}{2} \cdot \frac{1}{2} \cdot \frac{1}{2} + 1 \cdot \frac{1}{2} \cdot \frac{1}{2} \cdot 1 + 0 \cdot 1 \cdot 1 \cdot 1 = \frac{3}{8}.$$

If

$$(A_0 + A_1 + A_2 + A_3) \cdot F = 4F \tag{7.1}$$

holds then

$$f(w0) + f(w1) + f(w2) + f(w3) = 4 \cdot f(w)$$

for all $w \in \Sigma^\star$. In other words, the multi-resolution image specified by the automaton is average preserving. Therefore, a WFA is called *average preserving* if (7.1) is satisfied. Note that condition (7.1) states that number 4 is an eigenvalue of the matrix $A_0 + A_1 + A_2 + A_3$, and F is a corresponding eigenvector.

Given an n-state WFA A and an m-state WFA B that compute multi-resolution functions f and g, respectively, it is easy construct

- an $n+m$-state WFA that computes the point-wise sum $f+g$ where $(f+g)(w) = f(w)+g(w)$ for all $w \in \Sigma^*$. This WFA is just the union $A \cup B$ of the automata A and B:

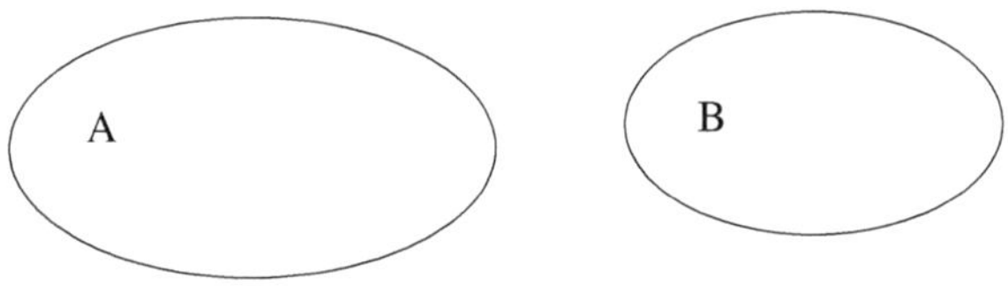

- an n-state WFA that computes the point-wise scalar multiple rf of f by $r \in \mathbb{R}$, where $(rf)(w) = rf(w)$ for all $w \in \Sigma^*$. In this construction one just replaces the initial distribution I of A by rI.
- an nm-state WFA that computes the point-wise product fg where for all $w \in \Sigma^*$ we have $(fg)(w) = f(w)g(w)$. This is the usual cartesian product $A \times B$ of the two automata:

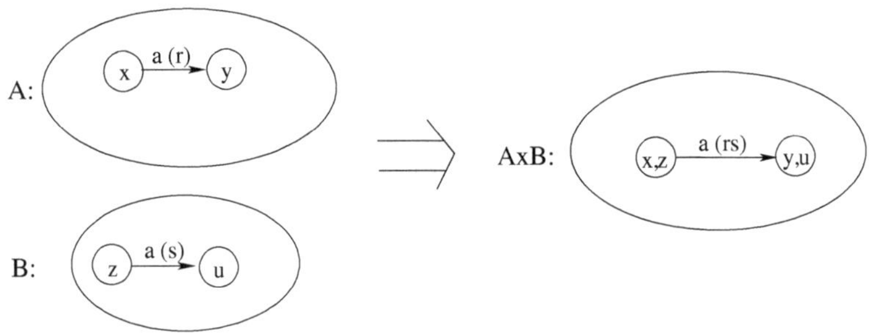

Example 4. The 2-state WFAs

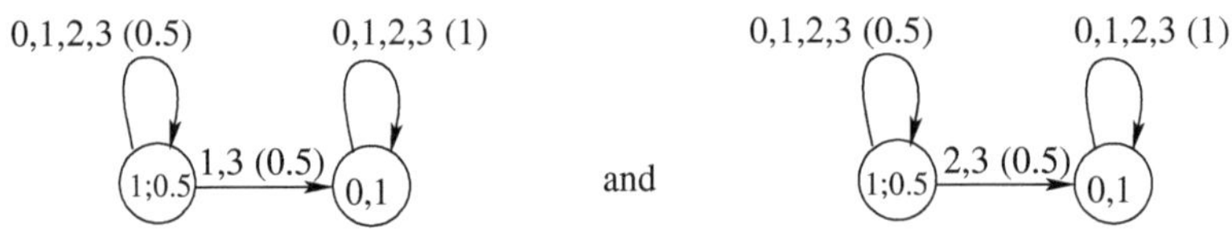

generate the linear functions of Figure 7.8. From these and the constant function $f(x, y) = 1$ one can build a WFA for any polynomial $p(x, y)$ of variables x and y using the operations above.

□

Fig. 7.8. Images generated by the two state WFA's of example 4.

Color images consist of three color layers, each of which is a gray-scale image. A common color representation is the RGB representation where red, green and blue color components are used to specify the colors. To define color images by WFA we use three initial distributions I_R, I_G and I_B, one for each color layer. The color layers are then generated by the WFA using the same transition weights and final distributions but different initial distributions. For example, the color WFA

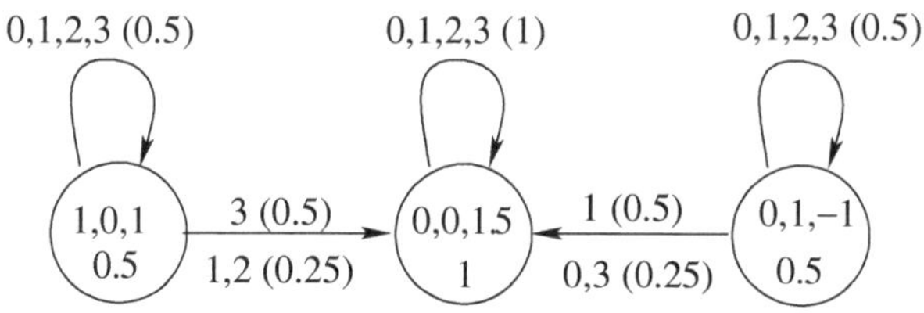

defines three linear slopes in three different orientations:

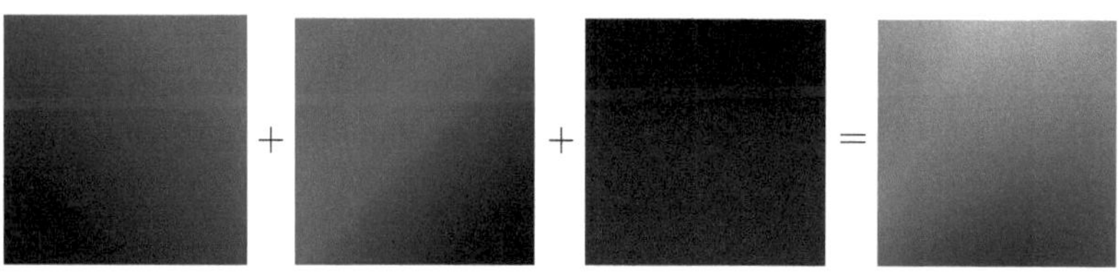

The fourth image is the color image defined by the color WFA.

The product of the 3-state color WFA above and the 5-state DFA of example 3 (interpreted as a WFA with weights 1) gives a 15-state color WFA that draws Figure 7.9.

Fig. 7.9. Image defined by a 15 state color WFA.

7.5 Drawing WFA Images

In this section we consider the following *decoding problem:* Given a WFA, draw the corresponding image at some specified finite resolution $2^k \times 2^k$. Decoding at resolution $2^k \times 2^k$ involves forming the matrix products IA_wF for all $w \in \Sigma^k$. Note that the number of multiplications (and additions) required by the trivial algorithm to compute the product of

- two $n \times n$ matrices is n^3,
- an $n \times n$ matrix and an n-vector is n^2,
- two n-vectors is n.

Observe that it is naturally better to multiply vectors than matrices.

Decoding algorithm #1

1. Form the products IA_w for all non-empty words $w \in \Sigma^{\leq k}$ in the order of increasing length of w. Because $IA_{ua} = IA_uA_a$, we need one product of a vector and a matrix for each word $w = ua$.

2. In the end, for every $w \in \Sigma^k$, multiply vectors IA_w and F.

Let us analyze the complexity of the algorithm. Let $N = 4^k$ be the number of pixels, and let n be the number of states. The number of non-empty words of length $\leq k$ is

$$4 + 16 + \ldots + 4^k \leq N(1 + \frac{1}{4} + \frac{1}{16} + \ldots) = \frac{4}{3}N.$$

Step 1 of the algorithm requires then at most $\frac{4}{3}Nn^2$ multiplications. Step 2 requires Nn multiplications so the total number of multiplications is at most $Nn(1 + \frac{4n}{3})$.

Let us consider then a more efficient alternative:

Decoding algorithm #2

1. Use the first step of the decoding algorithm #1 to compute the products IA_u for words u of length $k/2$ and products A_vF for words v of length $k/2$. If k is odd then we round the lengths so that u has length $\lfloor k/2 \rfloor$ and v has length $\lceil k/2 \rceil$.

2. Form all possible products $(IA_u)(A_vF)$ for all $u, v \in \Sigma^{k/2}$.

Step 1 requires at most $2 \times \frac{4}{3}4^{k/2}n^2 = \frac{8}{3}\sqrt{N}n^2$ multiplications. Step 2 requires Nn multiplications. Now the total is $Nn(1 + \frac{8n}{3\sqrt{N}})$ multiplications. This is considerably better than algorithm #1, especially when N, the number of pixels, is very large.

7.6 An Encoding Algorithm

Analogously to DFA, in WFA A the transitions A_0, A_1, A_2 and A_3 and the final distribution F define a multi-resolution image ψ_i for every state i: Image ψ_i is the multi-resolution image computed by the WFA that is obtained from A by changing the initial distribution so that the initial distribution value of state i is 1, and all other states have initial distribution 0. In other words, for every $w \in \Sigma^*$ we have

$$\psi_i(w) = (A_wF)_i$$

where we use the notation that $(A_wF)_i$ is the i'th component of vector A_wF. Multi-resolution ψ_i is called the image of state i. It is average preserving if the WFA is.

The WFA gives a mutually recursive definition of the ψ_i multi-resolutions:

- $\psi_i(\varepsilon) = F_i$, that is, the final distribution values are the average intensities of the state images.
- For every $a \in \Sigma, w \in \Sigma^\star$

$$a^{-1}\psi_i(w) = \psi_i(aw) = (A_{aw}F)_i = [A_a(A_wF)]_i = \sum_{j=1}^{n}(A_a)_{ij}(A_wF)_j$$
$$= r_1\psi_1(w) + r_2\psi_2(w) + \ldots + r_n\psi_n(w)$$

  where $r_j = (A_a)_{ij}$ is the weight of the transition from state i into state j with label a. Since the coefficients $r_j = (A_a)_{ij}$ are independent of w, we have

$$a^{-1}\psi_i = r_1\psi_1 + r_2\psi_2 + \ldots + r_n\psi_n$$

  in the vector space of multi-resolution images. In other words, the i'th row of the transition matrix A_a tells how the quadrant a of the state image ψ_i is expressed as a linear combination of state images $\psi_1, \ldots, \psi_n$. See Figure 7.10 for an illustration.

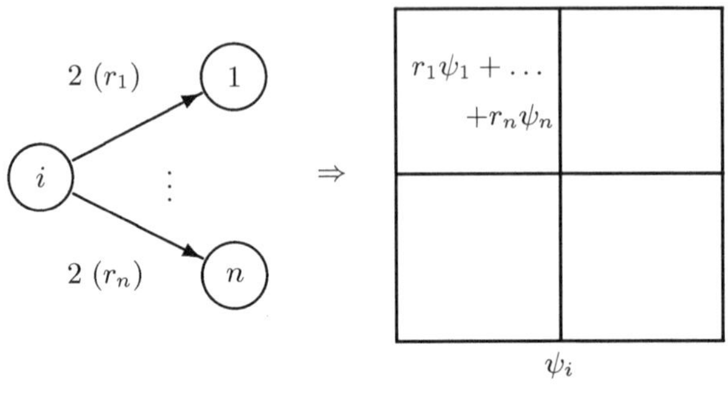

Fig. 7.10. The transitions from state i with label a specify how the quadrant a of the state image ψ_i is expressed as a linear combination of the state images $\psi_1, \ldots, \psi_n$.

- The initial distribution I tells how the multi-resolution image f computed by the WFA is composed of state images $\psi_1, \psi_2, \ldots, \psi_n$:

$$f = I_1 \cdot \psi_1 + I_2 \cdot \psi_2 + \ldots + I_n \cdot \psi_n$$

The final distribution values and the transition matrices define state images $\psi_1, \ldots, \psi_n$ uniquely.

Example 5. In our sample WFA the images ψ_1 and ψ_2 of the two states are shown in Figure 7.11. Here $\mathcal{C} = [0, 1]$ where 0 is black and 1 is white. □

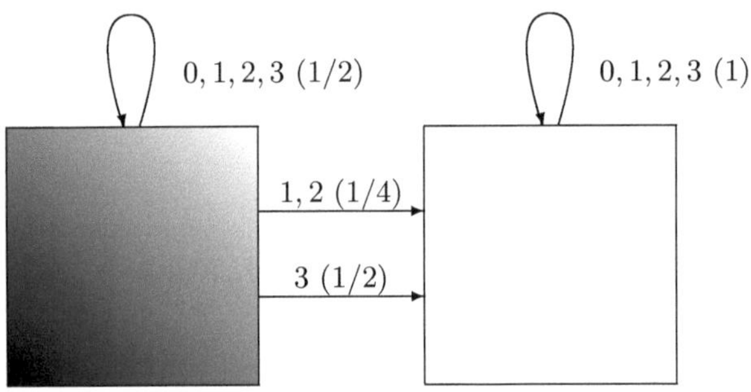

Fig. 7.11. A two state WFA with the state images drawn inside the states.

Analogously to DFA, we have the following algorithm to infer a minimum state WFA for a given multi-resolution function f:

Encoding algorithm for grey-scale images

Input: multi-resolution image ψ
Variables: n : number of states so far
i : first non-processed state
ψ_j : Image of state j

1. $n \leftarrow 1, i \leftarrow 1, \psi_1 \leftarrow \psi$
2. For quadrants $a = 0, 1, 2, 3$ do
   a) If $\exists r_1, \ldots, r_n \in \mathbb{R}$ such that $a^{-1}\psi_i = r_1\psi_1 + \ldots + r_n\psi_n$ then add the transitions

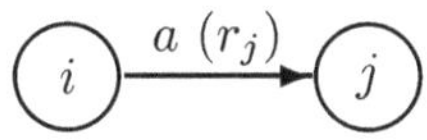

   for all $j = 1, 2, \ldots n$.
   b) else create a new state: Set $n \leftarrow n + 1$, $\psi_n \leftarrow a^{-1}\psi_i$, and add the transition

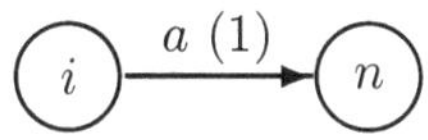

3. $i \leftarrow i + 1$. If $i \leq n$ then goto 2
4. Initial distribution: $I_1 = 1, I_i = 0$ for $i = 2, 3, \ldots, n$.
   Final distribution: $F_i = \psi_i(\epsilon)$ for $i = 1, 2, \ldots n$.

Example 6. Let us find a WFA that generates the image

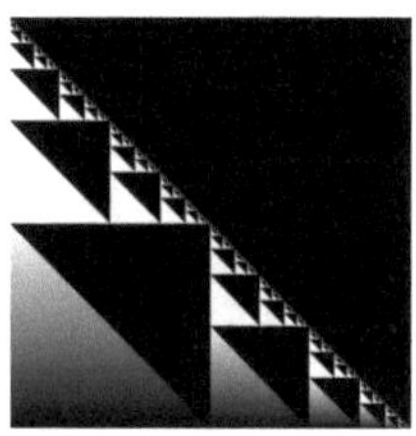

First we create one state whose image is the given input. We then process the four quadrants. Quadrants 0 and 2 can not be expressed as linear combinations, and new states are created for them. Quadrant 1 is identical to the input image, and quadrant 3 is the zero image. At this stage we have three states, one of which has been completely processed, as shown in Figure 7.12(a). Processing the four quadrants of the newly created two states yields two more states, see Figure 7.12(b). Notice that several quadrants were expressed as non-trivial linear combinations.

We continue in this fashion. After six states the process stops as all quadrants of all six state images have been expressed as linear combinations of the state images. We have found the WFA representation shown in Figure 7.12(c).
□

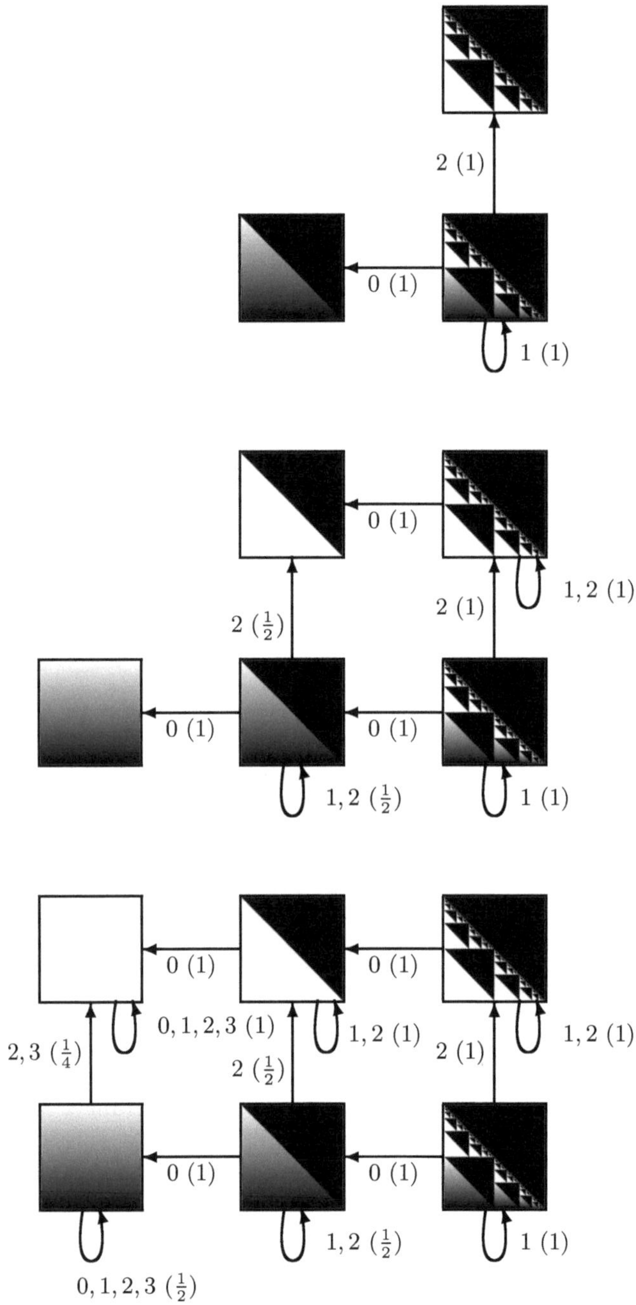

Fig. 7.12. The construction of a WFA from Example 6. The final result is the six state WFA shown in (c).

Theorem 1. *[7]*

- *ψ can be generated by a WFA if and only if the functions*

$$u^{-1}\psi, \text{ for all } u \in \Sigma^\star,$$

generate a finite dimensional vector space, where

$$u^{-1}\psi(w) = \psi(uw)\ \forall w \in \Sigma^\star$$

is the image in the sub-square addressed by u. The dimension of the vector space is the same as the smallest possible number of states in any WFA that generates ψ.
- *If ψ can be generated by a WFA, then the algorithm above produces a WFA with the minimum number of states.*
- *If ψ is average preserving then the algorithm produces an average preserving WFA.*

In practice, the input to the algorithm is a finite resolution image. When encoding a $2^k \times 2^k$ size image, the subtrees $w^{-1}\psi$ are only known up to depth $k - |w|$. When forming linear combinations the values deeper in the tree are "don't care" nodes, that is, it is enough to express the known part of each subtree as a linear combination of other trees and the don't care nodes may get arbitrary values. Fortunately, as the process is done in the breadth-first order, the sub-images are processed in the decreasing order of size. This means that all previously created states have trees assigned to them that are known at least to the depth of the current subtree. Hence the don't care values or prior images only can affect the don't care values of the present image, and the linear expressions are precise at the known part of the tree.

Note also that linear algebra can be used to minimize the number of states of a given WFA. This state minimization algorithm is due to the classical work by M.P.Shützenberger [14]. If the linear sub-space

$$V_L = \langle IA_w \mid w \in \Sigma^* \rangle$$

of $\mathbb{R}^n$ has dimension m that is strictly smaller than n then an equivalent WFA with m states can be formed as follows. There are $n \times m$ and $m \times n$ matrices X and X^{-1} such that

$$IA_w XX^{-1} = IA_w \text{ for every } w \in \Sigma^*.$$

Then the new m-state WFA has initial and final distributions $I' = IX$ and $F' = X^{-1}F$, and the transition matrices are $A'_i = X^{-1}A_iX$ for $i = 0, 1, 2, 3$. Then clearly $I'A'_wF' = IA_wF$ for every $w \in \Sigma^*$, so the new WFA defines the same multi-resolution image as the original WFA.

Analogously, if the dimension of

$$V_R = \langle A_wF \mid w \in \Sigma^* \rangle$$

is smaller than n then the number of states can be reduced. It is not difficult to see that if $\dim V_R = \dim V_L = n$ then the WFA has the minimum number of states.

7.7 Image Compression Using WFA

In the previous section we saw how we can find a minimum state WFA for a given grey-scale image. If the WFA is small in size, then the WFA can be used as a compressed representation of the image. However, the encoding algorithm of the previous section as such is ill suited for image compression purposes. Even though the resulting WFA is minimal with respect to the number of states, it may have a very large number of edges. The algorithm also represents the image exactly, and the last details of the image often require a very large increase in the size of the WFA. This is called *lossless* image compression. More often one is interested in *lossy* compression where small errors are allowed in the regenerated image, if this admits sufficient decrease in the size of the compressed image file. In this section we outline how the encoding algorithm can be modified into a practical image compression method. Note that the techniques are heuristic in nature and no guarantees of optimality exist. We can only use compression experiments and comparisons with other algorithms to show their usefulness.

Recall the step 2 of the encoding algorithm from the previous section:

a) If $\exists r_1, \ldots, r_n \in \mathbb{R}$ such that $a^{-1}\psi_i = r_1\psi_1 + \ldots + r_n\psi_n$ then add the transitions

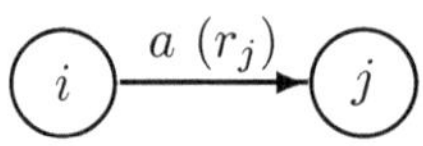

for all $j = 1, 2, \ldots n$.

b) else create a new state: Set $n \leftarrow n+1$, $\psi_n \leftarrow a^{-1}\psi_i$, and add the transition

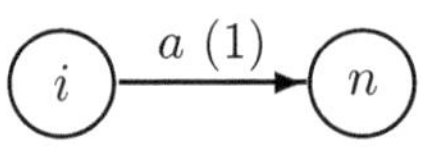

It turns out that better results are obtained in practice if we try both alternatives (a) and (b), and choose the one that gives smaller automaton. It may namely be that the linear combination in (a) contains so many coefficients that it is better to create a new state in (b) and process its quadrants.

In order to make a fair comparison between (a) and (b) we need to process the new state created in (b) completely before we can determine which choice to make. Therefore we change the order of processing the states from the breadth-first order into the depth-first order: instead of processing all four quadrants of ψ_i before advancing to the next state image, all new states created by a quadrant are processed before advancing to the next quadrant.

Let us first measure the size of the automaton by the quantity

$$|E| + P \cdot |V|$$

where $P \in \mathbb{R}_+$ is a given constant, $|E|$ is the number of transitions and $|V|$ is the number of states of the automaton. Constant P is a Lagrange multiplier that formulates the relative cost of states vs. edges in the automaton. If we want to minimize the number of edges we set $P = 0$. The goal of the next inference algorithm is to find for a given multi-resolution image ψ a small WFA in the sense that the value $|E| + P \cdot |V|$ is small.

Because we now process the quad-tree in the depth-first order, it is natural to make the algorithm recursive. The new encoding algorithm consists of a recursive routine *make_wfa*(ψ_i,*max*) that adds new states and edges to the WFA constructed so far, with the goal of representing the state image ψ_i in such a way that the value $\Delta_E + P \cdot \Delta_V$ is small, where Δ_E and Δ_V are the numbers of new transitions and states added by the recursive call. If $\Delta_E + P \cdot \Delta_S \leq max$ then the routine returns the value $\Delta_E + P \cdot \Delta_V$, otherwise it returns ∞ to indicate that no improvement over the target value max was obtained.

Encoding algorithm #2

Input: Multi-resolution image ψ and a positive real number P
Global variables used: n : number of states
ψ_j : image of state j

1. $n \leftarrow 1, \psi_1 \leftarrow \psi$
2. *make_wfa*(ψ_1, ∞)

make_wfa(ψ_i,*max*) :

1. If $max < 0$ then return(∞)
2. Set $cost \leftarrow 0$
3. For quadrants $a = 0, 1, 2, 3$ do
   a) If $\exists r_1, \ldots, r_n \in \mathbb{R}$ such that $a^{-1}\psi_i = r_1\psi_1 + \ldots + r_n\psi_n$ then

      cost1 $\leftarrow$ number of non-zero coefficients r_j

      else

      cost1 $\leftarrow \infty$

   b) Set $n_0 \leftarrow n$, $n \leftarrow n + 1$, $\psi_n \leftarrow a^{-1}\psi_i$ and add the transition

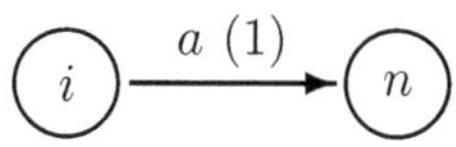

   c) Set *cost2* $\leftarrow P +$ *make_wfa*(ψ_n,*min*$\{max - cost - P, cost1 - P\}$)
   d) If *cost2* $\leq$ *cost1* then
      - *cost* $\leftarrow$ *cost* + *cost2*

      else

- *cost* $\leftarrow$ *cost* + *cost1*,
- remove all transitions from states $n_0 + 1, \dots n$, and set $n \leftarrow n_0$.
- remove the transition

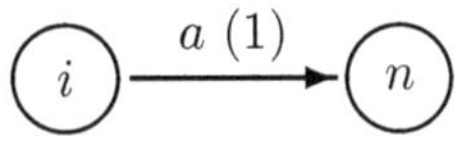

- add transitions

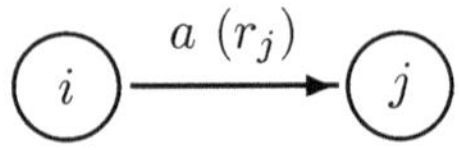

  for $r_j \neq 0$

4. If *cost* $\leq$ *max* then return(*cost*) else return(∞)

A few words to explain the algorithm: The main step is line 3 where we try to find a WFA representation for each of the four quadrants of ψ_i. For each quadrant we try two alternatives: (a) to express the quadrant as a linear combination of existing states, and (b) to create a new state whose state image is the quadrant, and to recursively process the new state. In variable *cost1* we store the cost of alternative (a), i.e. the number of transitions created in the automaton, and in *cost2* we store the cost of alternative (b), i.e. the sum of P (for the new state created) and the cost returned from the recursive call to process the new state. In step 3(d) the algorithm chooses the better of the two alternatives.

The algorithm above is still lossless. The WFA represents the input image precisely, without any loss of detail. Much better compression is obtained if we allow small errors in the regenerated image whenever that helps to compress the image more. There is a trade-off between the amount of image degradation and the size of the compressed image. Let us measure the amount of degradation by the square difference metric $d(\cdot,\cdot)$. For two images ψ and ϕ at resolution $2^k \times 2^k$ this metric defines the image difference value as

$$d(\psi, \phi) = \sum_{w \in \Sigma^k} (\psi(w) - \phi(w))^2.$$

This is a reasonable measure for the reconstruction error. It is also convenient for our linear combinations approach since it is the square of the normal Euclidean distance.

We also introduce a new Lagrange multiplier G that controls the trade-off between the image size and the reconstruction error. Parameter G is given as an input by the user, and the algorithm will produce a WFA that generates an image ψ' such that the value of

$$d(\psi, \psi') + G \cdot S$$

is small, where ψ is the image to be compressed and S is the size of the WFA constructed by the algorithm. We may continue using

$$S = |E| + P \cdot |V|,$$

but in practical image compression it is better to define S as the actual number of bits required to store the WFA in a file. The WFA is stored using suitable entropy coder, e.g. an arithmetic coder. See [11] for details on how the different items such that states, transitions and weights are stored using arithmetic coding.

The Lagrange multiplier G is the parameter that the user can change to adjust the file size and the image quality:

Small G $\Rightarrow$ big automaton, small error,
Large G $\Rightarrow$ small automaton, big error.

The following modifications to the algorithm were also made:

- Edges back to states that have not yet been completely processed are problematic in lossy coding, as the actual image of those states is not yet precisely known. Therefore, we opted to only allow edges to states already completely processed. Note that this prevents the creation of any loops in the WFA.
- In order to have at least one processed state to begin with, we introduce an initial base: Before calling *make_wfa* the first time we set $n \leftarrow N$ with some fixed images $\psi_1, \ldots, \psi_N$. In our tests below we used $N = 6$ base images that were linearly independent quadratic polynomials, i.e. functions 1, x, y, x^2, y^2 and xy.

With these modifications we have a practical encoding algorithm that compares favourably with other compression techniques. Let us compare WFA compression with the JPEG image compression standard using the color image of Figure 7.13. This test image contains large, smooth areas, and is therefore well suited for JPEG compression.

A color image consists of three color layers, each of which is compressed as a grayscale image. However, only one automaton is build, that is, different color components can refer to the sub-images of other color layers. In the RGB color representation each color is formed as a weighted sum of red, green and blue basic colors. These color components are typically strongly correlated, so most compression algorithms that treat the color components separately first apply a linear transformation of $\mathbb{R}^3$ to the color space to remove the correlation. Good color representations are YUV and YIQ representations, where the Y-component is luminance and contains most of the image information. UV- or IQ-components are chrominance components and they get compressed easily. Figure 7.14 shows the RGB and YUV components of the test image.
In WFA compression the color representation has little or no effect since the linear combinations are allowed to span across different color components. Hence the algorithm automatically exploits the correlations between the RGB color components.

In the following compression examples the reconstruction errors are reported as the peak signal-to-noise ratio (PSNR). The units of this measure

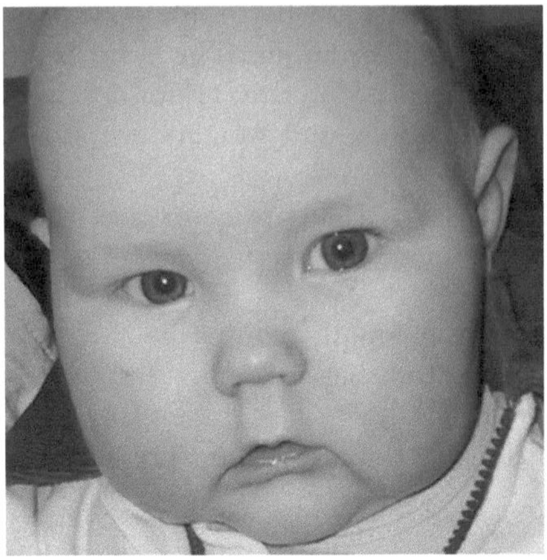

Fig. 7.13. The first test image.

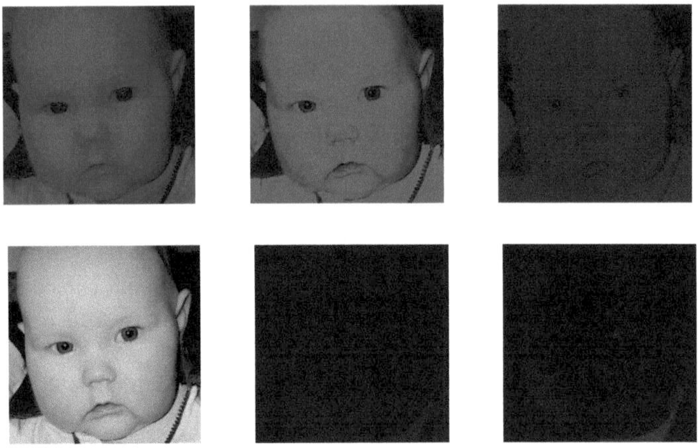

Fig. 7.14. The RGB and the YUV color components of the test image.

are decibels (dB). The PSNR value is directly obtained from the square difference $d(\cdot,\cdot)$ as follows:

$$\mathrm{PSNR}(\cdot,\cdot) = 10\log_{10}\left(\frac{A^2}{\sigma}\right)$$

where A is the maximum intensity value ($A = 255$ in our 8 bits-per-pixel image) and

$$\sigma = \frac{d(\cdot,\cdot)}{\#\text{ of pixels}}$$

is the average square difference between the pixel values of the two images. Notice that larger PSNR values mean better quality. Typically, PSNR of 35 dB is considered reasonable image quality, and a 40 dB reconstruction is already of very high quality.

Our first comparison is at the very low bitrate of 7.2 kbytes. The results are shown in Figure 7.15. The WFA compressed image is more than 4 dB better than the JPEG compressed image at the same bitrate. However, this comparison is not fair to JPEG since the JPEG algorithm is not meant to be used at such low bitrates.

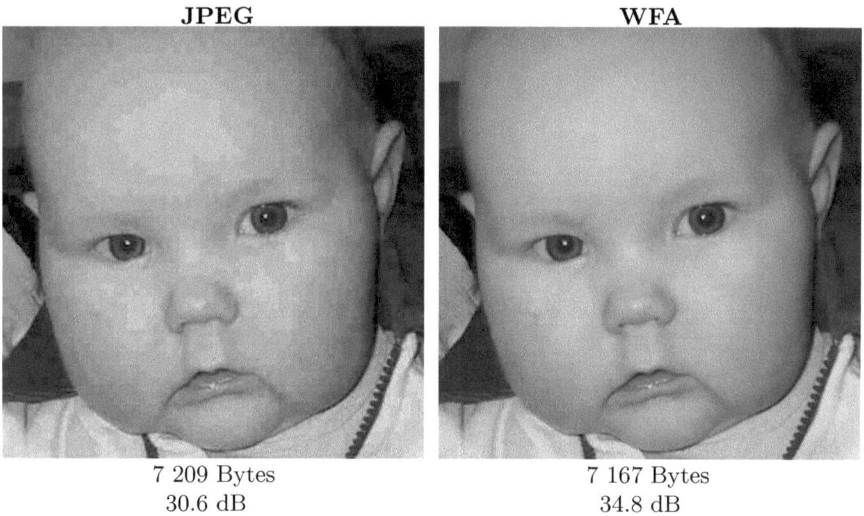

Fig. 7.15. The comparison of JPEG and the WFA compression at a low bitrate.

As we increase the bitrate, the JPEG algorithm catches up with WFA. At 15.1 kbytes the images are of the same quality, see Figure 7.16. The next Figure 7.17 summarizes the numerical results. Both algorithms were used at various bitrates, and the bitrate versus image quality values were plotted to obtain the rate-distortion curves of the two algorithms. Note how JPEG surpasses WFA at 16 kbyte compression.

Very good compression is obtained if the WFA encoding algorithm is applied to an image composed of wavelet coefficients instead of the original image. In this way the algorithm becomes a fancy way of storing the result of the wavelet transformation.

In our tests the Daubechies W6 wavelets are used. The sub-bands obtained from the wavelet transformation are arranged into a Mallat pyramid (see Figure 7.18), and this is compressed as an image using the WFA encoding algorithm. Because the transformation is an orthogonal linear transformation, the Euclidean distances between the original and regenerated images before and after the wavelet transformation are the same. In other words, the error

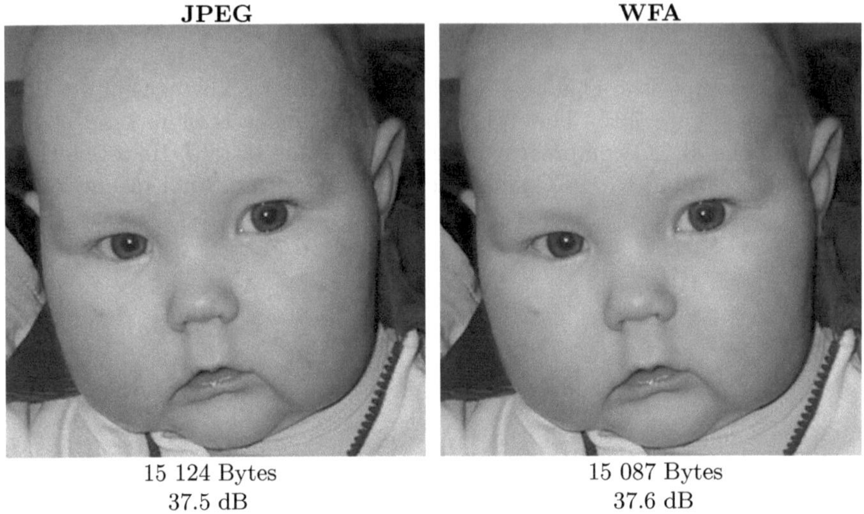

Fig. 7.16. The comparison of JPEG and the WFA compression at 15 kbytes.

done in the image consisting of the wavelet coefficients is the same as the error in the final regenerated image.

The WFA algorithm is able to take advantage of the self-similarity of different sub-bands in the wavelet transformation. The subdivision into quadrants used by our algorithm matches with the organization of the sub-bands in the Mallat form. This yields better compression results. The chart in Figure 7.19 shows the rate-distortion results.

WFA (without wavelets) compress very well images with sharp edges. Let us compare WFA compression and JPEG on the test image of Figure 7.20. The difference between JPEG and WFA is clear even at high quality setting, as seen in Figure 7.21. Numerical comparisons in Figure 7.22 indicate that WFA compression remains superior through all bitrates.

As our final test image, consider the tiling by decorated Penrose kites and darts shown in Figure 7.23. The image contains sharp edges along the boundaries of the tiles and smooth color changes everywhere. WFA compression outperforms JPEG. The magnification in Figure 7.24 shows the differences between the two algorithms. The magnification is from the result when the image was compressed into 41 kbytes. Numerically the error to the original image is still large: 22.3 dB and 28.3 dB with JPEG and WFA, respectively.

The magnification shows how the colors bleed into the white boundaries in the JPEG image, while the boundaries remain sharp in the WFA compressed image. In contrast, the WFA image suffers from blockiness in smooth regions.

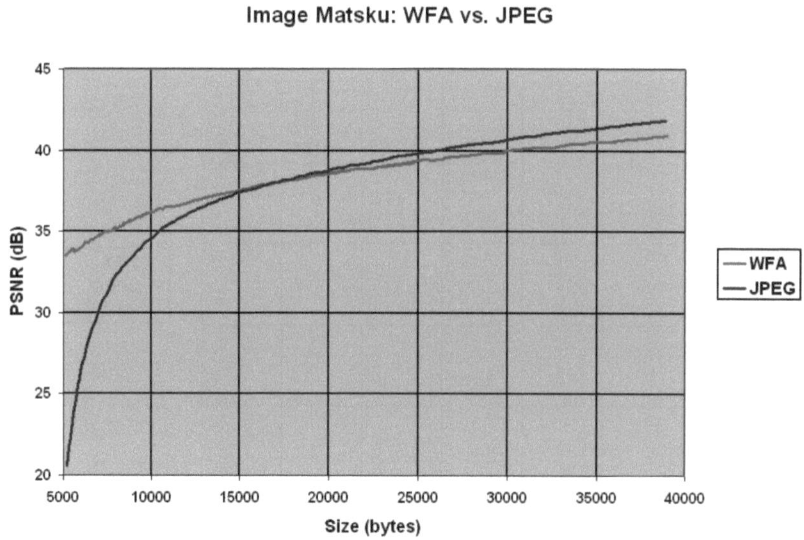

Fig. 7.17. Rate-distortion comparison of JPEG and WFA compression. The test image of Figure 7.13 was used.

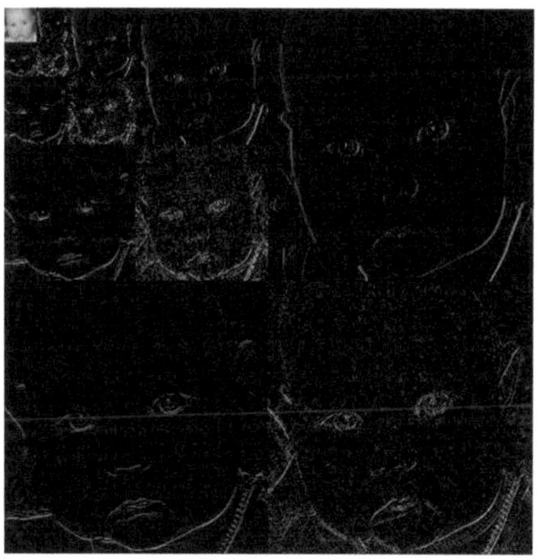

Fig. 7.18. The Mallat pyramid of the wavelet coefficients.

7.8 Weighted Finite Transducers (WFT)

A nice feature of WFA image representations is the property that one can perform several interesting and useful image operations directly in the WFA form. Bilevel images and regular languages can be transformed using finite state transducers. Analogously, grey-scale images and WFA are transformed

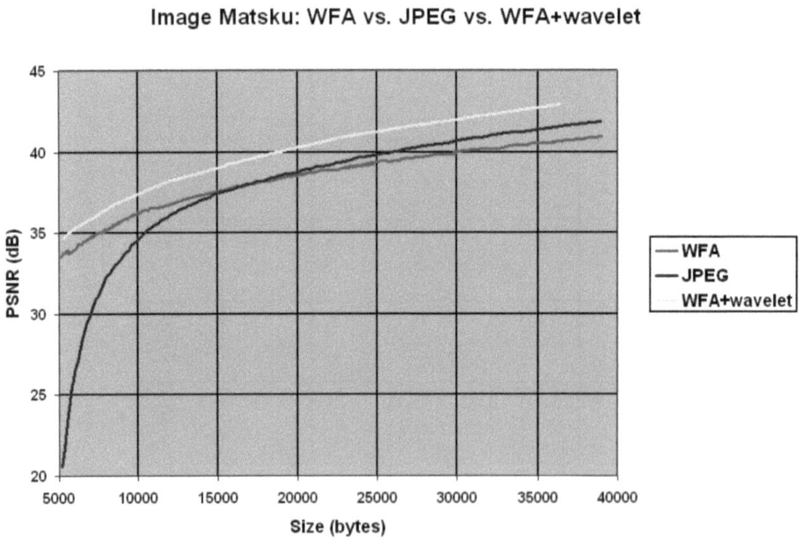

Fig. 7.19. The rate-distortion comparison of JPEG, WFA compression and WFA compression of the wavelet coefficients.

Fig. 7.20. The second test image that contains sharp edges.

using finite state transducers with real weights. More details of the examples and results in this section can be found in [4].

A weighted finite transducer (WFT) is obtained by introducing edge weights and initial and final distribution values to an ordinary finite state

JPEG **WFA**

5 064 Bytes
28.9 dB

5 015 Bytes
35.2 dB

Fig. 7.21. The compression results using JPEG and WFA compression at 5 kbytes.

transducer. The transitions are labeled by pairs a/b where $a, b \in \Sigma \cup \{\varepsilon\}$. More precisely, a WFT is specified by

- number of states $n \in \mathbb{Z}_+$,
- weight matrices $A_{a,b} \in \mathbb{R}^{n\times n}$ for all $a, b \in \Sigma \cup \{\varepsilon\}$,
- final distribution vector $F \in \mathbb{R}^{n\times 1}$, and
- initial distribution vector $I \in \mathbb{R}^{1\times n}$.

The WFT is called ε-*free* if the weight matrices $A_{a,\varepsilon}$, $A_{\varepsilon,b}$ and $A_{\varepsilon,\varepsilon}$ are zero matrices for all $a, b \in \Sigma$. The WFT defines a function

$$\rho : \Sigma^* \times \Sigma^* \longrightarrow \mathbb{R}$$

called a *weighted relation* as follows: For every $u, v \in \Sigma^*$ we have

$$\rho(u, v) = IA_{u,v}F$$

where

$$A_{u,v} = \sum_{\substack{a_1 \ldots a_m = u \\ b_1 \ldots b_m = v}} A_{a_1,b_1} \ldots A_{a_m,b_m}$$

if the sum converges. The sum is over all decompositions of u and v into symbols $a_i, b_i \in \Sigma \cup \{\varepsilon\}$. Note that the sum is finite (and hence converges) if the WFT does not contain any cycles that read ε/ε.

If the WFT is ε-free then

$$\rho(a_1 \ldots a_k, b_1 \ldots b_k) = IA_{a_1,b_1} \ldots A_{a_k,b_k}F$$

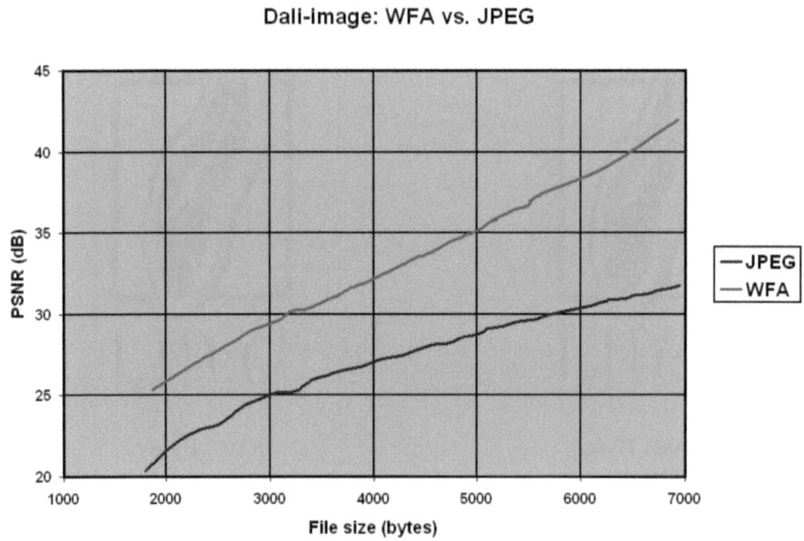

Fig. 7.22. The rate-distortion comparison of JPEG and WFA compression on the second test image.

Fig. 7.23. The third test image.

where all $a_i, b_i \in \Sigma$, and $\rho(u,v) = 0$ when $|u| \neq |v|$.

Next we define the action of a weighted relation ρ on a multi-resolution image f. The result is a new multi-resolution function $g = \rho(f)$, defined by

$$g(w) = \sum_{u \in \Sigma^*} f(u)\rho(u,w), \text{ for all } w \in \Sigma^*,$$

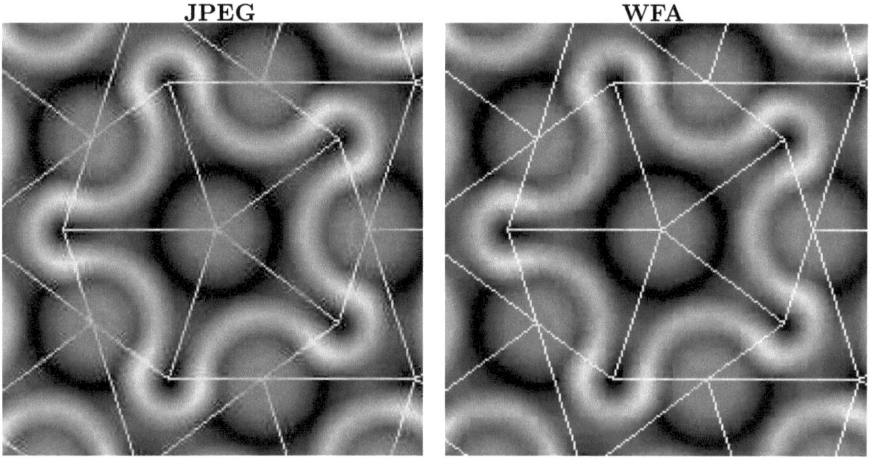

Fig. 7.24. Magnifications of the compressed image using JPEG and WFA compression at 41 kbytes.

provided the sum converges. The sum is finite if the WFT is ε-free, or more generally, if the weight matrices $A_{a,\varepsilon}$ are zero, for all $a \in \Sigma \cup \{\varepsilon\}$. In this case the sum is over all words u whose length is not greater than the length of w.

It is easy to see that the operator

$$\rho : \mathbb{R}^{\Sigma^*} \longrightarrow \mathbb{R}^{\Sigma^*}$$

is linear, that is, for arbitrary multi-resolution functions f and g, and arbitrary real numbers x and y we have

$$\rho(xf + yg) = x\rho(f) + y\rho(g).$$

Many interesting and natural linear image transformations can be implemented as a WFT. In the following we see several examples.

Example 7. Let $w \in \Sigma^*$ be a fixed word. The WFT

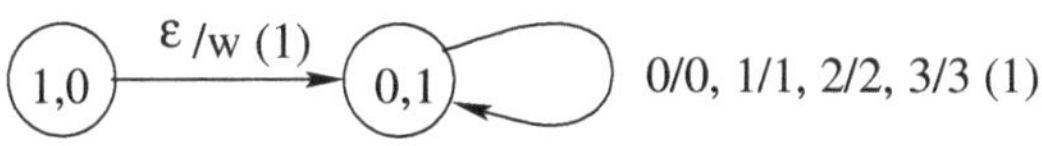

computes the weighted relation $\rho(u, wu) = 1$ for every $u \in \Sigma^*$, and $\rho(u, v) = 0$ if $v \neq wu$. The effect is to shrink the input image and place it at the sub-square addressed by w. For example, with $w = 21$ our test image becomes

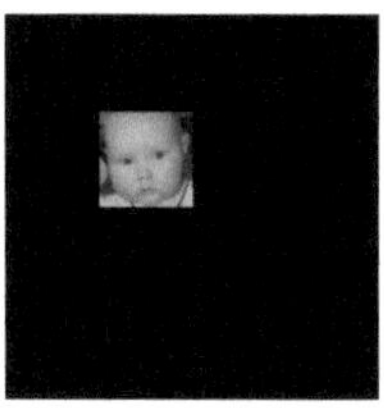

Consider then the WFT

It computes the weighted relation $\rho(wu, u) = 1$ for every $u \in \Sigma^*$, and $\rho(v, u) = 0$ if $v \neq wu$. Now the effect is to zoom the sub-square of the input image whose address is w. For example, with $w = 30$ our test image is mapped into the image

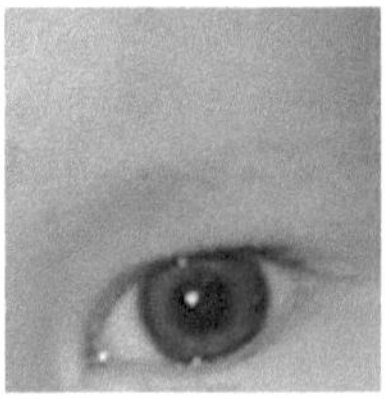

□

Example 8. The WFT

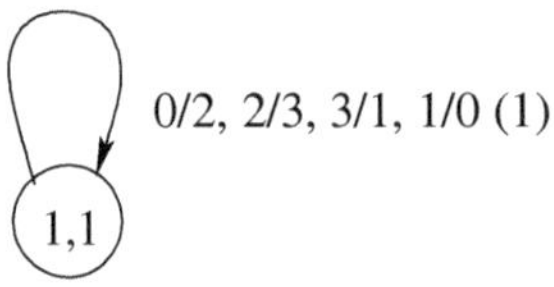

rotates the image 90°:

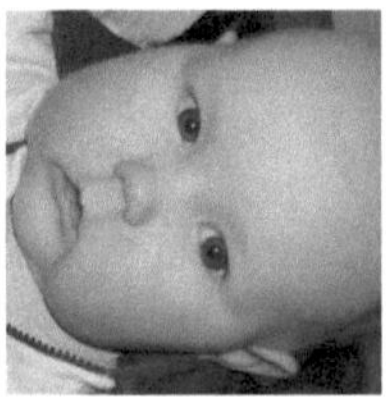

□

Also fractal-like transformations can be defined. For example, 11 states are enough to implement a transformation that produces

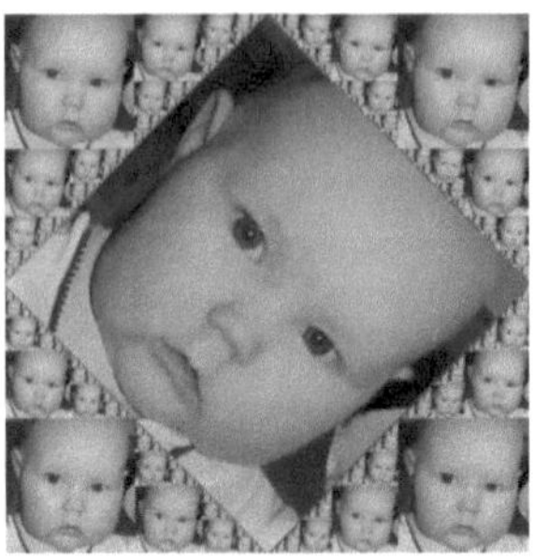

Example 9. The circular right shift σ_R by a single pixel:

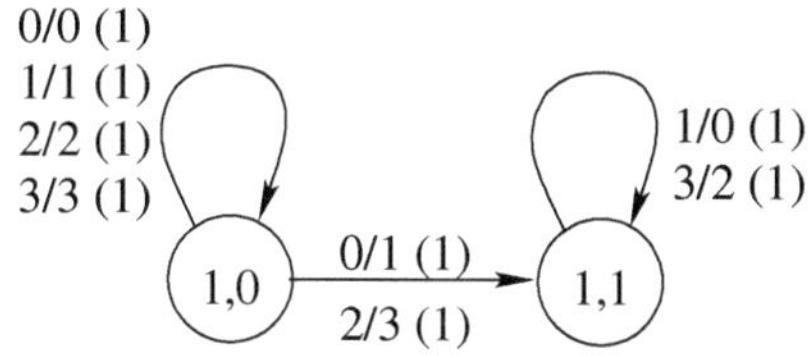

By taking the union with the WFT

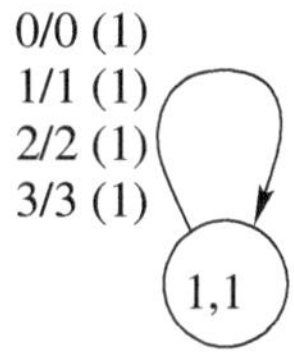

that computes the negative of an image, we obtain a WFT that computes the differences in consecutive pixels in the horizontal direction. The same difference transformation is computed also by the two state transducer

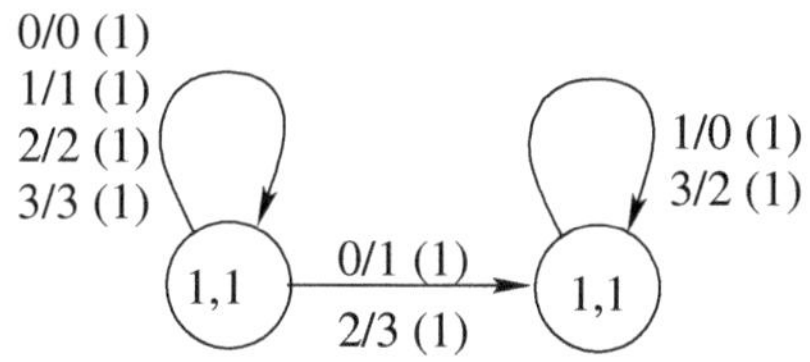

The horizontal derivative of an image is defined as the limit of the horizontal pixel differences divided by the width of the pixels. The width of the pixel at the resolution given by words of length k is $1/2^k$, so the horizontal derivative is computed by the WFT

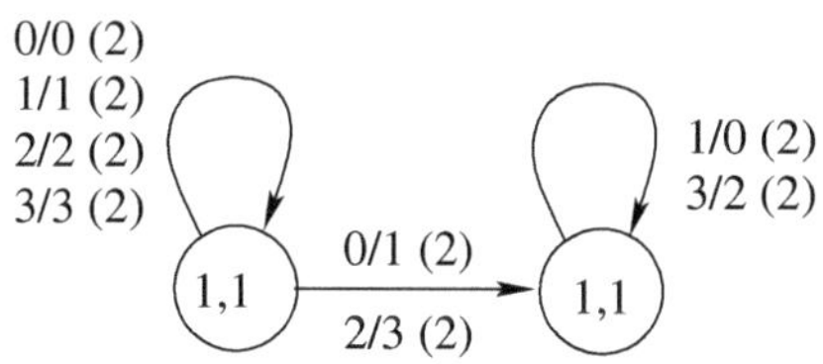

that is obtained from the difference WFT by changing all transition weights to 2.

□

Example 10. The sum of all pixels to the left of a given pixel is computed by

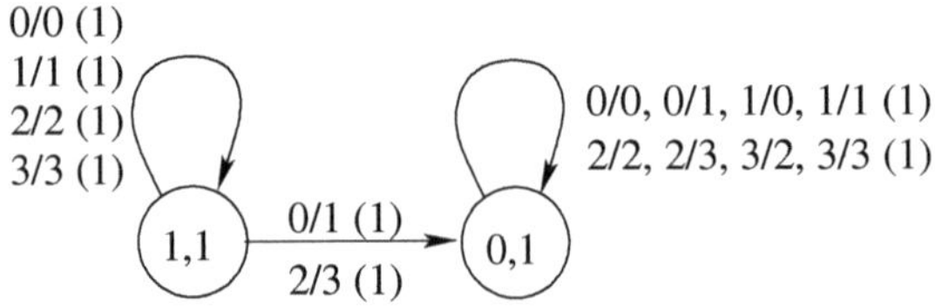

The integral of the image in the horizontal direction is then the sum scaled by multiplying the sum by $1/2^k$, where k is the length of the input. This is established by changing all weights from 1 to $\frac{1}{2}$:

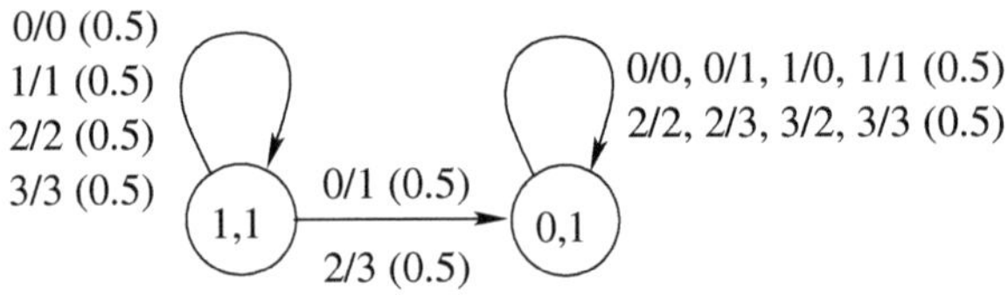

□

The examples above show several natural image operations defined by small WFT's. Next we define the action of WFT's on WFA's. An application of an ε-free n-state WFT M to an m-state WFA A is the mn state WFA $M(A)$ whose states are the pairs (p, q) of states of A and M, the initial and final distribution values are obtained by multiplying the corresponding distributions of A and M, and the weight of the transition

$$(p, q) \xrightarrow{a} (s, t)$$

is

$$\sum_{x \in \Sigma} (A_x)_{p,s} (M_{x,a})_{q,t}$$

where A_x and $M_{x,a}$ are the weight matrices of A and M, respectively. This is a straightforward generalization of the usual applications of a finite letter-to-letter transducer on a finite automaton. It is easy to see that $M(A)$ generates the multi-resolution function $\rho(f)$ where ρ is the weighted relation of M and f is the multi-resolution determined by A.

Applying WFT M directly to a WFA A has the advantages that

- it is fast if A and M are small. There is no need to decode the image into the usual pixel form before applying the operation.
- the result is correct at every resolution.

Let us call an ε-free WFT *average preserving* if for every $a \in \Sigma$ we have

$$\sum_{b \in \Sigma} A_{a,b} F = F,$$

that is, if F is the eigenvector corresponding to eigenvalue 1 in each of the sums

$$\sum_{b\in\Sigma} A_{a,b}.$$

The following result is easy to establish [4]. It justifies the use of term average preserving.

Theorem 2. *Let M be an ε-free WFT. Then*

- *$M(A)$ is an average preserving WFA for every average preserving A iff M is average preserving.*
- *if M is average preserving then $\rho(f)$ is average preserving for every average preserving multi-resolution function, where ρ is the weighted relation specified by M.*

As a final topic we mention some closure properties of weighted relations defined by WFT's. Let ρ and σ be two weighted relations, and $r \in \mathbb{R}$. Let us define the weighted relations $\rho+\sigma$, $r\rho$ and $\rho\circ\sigma$ as follows:

- $(\rho+\sigma)(u,v) = \rho(u,v) + \sigma(u,v)$,
- $(r\rho)(u,v) = r\rho(u,v)$,
- $\rho\circ\sigma(u,v) = \sum_{w\in\Sigma^*} \rho(u,w)\sigma(w,v)$.

Let f be an arbitrary multi-resolution function. Then we have

- $(\rho+\sigma)(f) = \rho(f) + \sigma(f)$,
- $(r\rho)(f) = r\rho(f)$,
- $(\rho\circ\sigma)(f) = \rho(\sigma(f))$.

Theorem 3. *wft] If ρ and σ are computed by an n-state and an m-state WFT, respectively, then $\rho+\sigma$ is computed by an $(n+m)$-state WFT, $r\rho$ is computed by an n-state WFT, and $\rho\circ\sigma$ is computed by an nm-state WFT.*

Using the operations above we can build more complicated operations from simple ones.

7.9 Conclusion

This tutorial aimed to be an easy to read overview of a way to apply automata theory in image processing. Using finite automata we can define a wide range of artificial images at infinite resolution. The definition uses mutual recursion between a finite number of state images. Hence the images obtained are self-similar, i.e. fractal. With weighted finite automata we can also define smooth grey-scale images.

We have demonstrated through several examples that weighted finite automata provide a compact representation of real images. They are especially good in compressing images with sharp edges. Finally, we have shown how finite state transducers define interesting and natural image transformations. These transformations can be applied directly in the finite automata, hence obtaining infinitely sharp results.

References

1. S. Bader, S. Holldobler and A. Scalzitti. Semiring Artificial Neural Networks and Weighted Automata – And an Application to Digital Image Encoding, *Lecture Notes in Computer Science* 3238, Springer-Verlag, Berlin, 2004, 281–294.
2. P. Bao and X. L. Wu. L-infinity-constrained Near-lossless Image Compression Using Weighted Finite Automata Encoding, *Computers & Graphics,* 22 (1998), 217–223.
3. K. Culik and J. Kari. Digital Images and Formal languages, *Handbook of Formal Languages, vol.3,* (A. Salomaa, G. Rozenberg, eds.), Springer-Verlag, 1997, 599–616.
4. K. Culik and J. Kari. Finite State Transformations of Images, *Computers & Graphics,* 20 (1996), 125–135.
5. K. Culik and J. Kari. Finite State Methods for Compression and Manipulation of Images, *Proceedings of the Data Compression Conference,* (J. A. Storer, M. Cohen, eds.), IEEE Computer Society Press, 1995, 142–151.
6. K. Culik and J. Kari. Image-data Compression Using Edge-optimizing Algorithm for WFA Inference, *Information Processing & Management,* 30 (1994), 829–838.
7. K. Culik and J. Kari. Image Compression Using Weighted Finite Automata, *Computers & Graphics,* 17 (1993), 305–313.
8. U. Hafner, J. Albert, S. Frank and M. Unger. Weighted Finite Automata for Video Compression, *IEEE Journal on selected areas in communication,* 16 (1998), 108–119.
9. Z. H. Jiang, O. de Vel and B. Litow. Unification and Extension of Weighted Finite Automata Applicable to Image Compression, *Theoretical Computer Science,* 302 (2003), 275–294.
10. Z. H. Jiang, B. Litow and O. de Vel. Similarity Enrichment in Image Compression Through Weighted Finite Automata, *Lecture Notes in Computer Science* 1858, Springer-Verlag, Berlin, 2000, 447–456.
11. J. Kari and P. Fränti. Arithmetic Coding of Weighted Finite Automata, *RAIRO Informatique théorique et Applications,* 28 (1994), 343–360.
12. F. Katritzke, W. Merzenich and M. Thomas. Enhancements of Partitioning Techniques for Image Compression Using Weighted Finite Automata, *Theoretical Computer Science,* 313 (2004), 133–144.
13. S. V. Ramasubramanian and K. Krithivasan. Finite Automata and Digital Images, *International Journal of Pattern Recognition and Artificial Intelligence,* 14 (2000), 501–524.
14. M. P. Schützenberger. On the Definition of a Family of Automata, *Information and Control* 4 (1961), 245–270.

8

Mathematical Foundations of Learning Theory*

Satoshi Kobayashi

Department of Computer Science
University of Electro-Communications
1-5-1, Chofugaoka, Chofu, Tokyo 182-8585, Japan
E-mail: satoshi@cs.uec.ac.jp

Summary. This chapter gives a mathematical foundation of learning theory, especially focusing on learnability theory based on identification in the limit model, proposed by Gold ([5]). This framework is one of the most important and fundamental basis for learning theory, where we can obtain various theoretical insights into learning, which leads us to an appropriate research direction.

8.1 Introduction

This chapter gives a mathematical foundation of learning theory, especially focusing on learnability theory based on identification in the limit model, proposed by Gold ([5]). Gold obtained an insight into the learning activity as an infinite inference process, and proposed the first mathematically well-defined learning framework, called *identification in the limit*. This framework is one of the most important and fundamental basis for learning theory, where we can obtain various theoretical insights into learning, which leads us to an appropriate research direction.

We first introduce the identification in the limit model, and give some theoretical results concerning the ability of learning machines which can use both positive and negative example information, what is called a *complete data*, about the target concept. An important result says that any indexable family of recursive languages can be identified in the limit from complete data. On the other hand, we will show that there is no learning machines which can identify the class $\mathcal{RC}$ of all recursive languages. The class $\mathcal{RC}$ is shown to be learnable from complete data when we relax the learning criterion so that a learning machine may waive its conjecture if it *behaviorally correctly* identifies the target concept ([4]).

Second, we will focus on learning machines which can use only positive examples drawn from the target concept. We will show that such learning

* The author is sincerely grateful to variable comments by Prof. Takashi Yokomori on the contents of this chapter.

S. Kobayashi: *Mathematical Foundations of Learning Theory*, Studies in Computational Intelligence (SCI) **25**, 209–228 (2006)
www.springerlink.com

machines have strictly weaker learning capability than those which can use complete data. An elegant characterization of learning from positive data will be presented based on the formulation by Angluin ([1]). We will also give sufficient conditions for learning from positive data, which are often useful for showing the learnability of concept classes from positive data. We then introduce *length-bounded* elementary formal systems, a rich hierarchy of language classes each of which is identifiable in the limit from positive data.

Finally we will consider a learning model where a target concept may be outside the hypothesis space. In the framework of identification in the limit, the theory of approximately identifying a target concept outside the hypothesis space was first proposed by Mukouchi ([11]). Inspired from Mukouchi's work, we will introduce an approximate learning framework, called *upper-best approximate learning* ([9]). An upper-best approximation of a concept is defined as the *minimum* concept in the hypothesis space that contains the concept, and the objective of the upper-best approximate learning is to identify the upper-best approximation of a target concept in the limit. In this framework, we can completely characterize the condition for a concept class in order to upper-best approximately identify another concept class. In particular, surprisingly enough, we will show that when a learning machine should learn *every* concept approximately, negative data is not useful. Furthermore, we will show the existence of rich families of languages which can approximately identify every concept.

8.2 Preliminaries

Let us consider a recursively enumerable set U, called a *universal set*. By $\subseteq$ ($\subset$), we denote the inclusion (proper inclusion) relation. A subset of U is called a *concept*, and a set of concepts is called a *concept class*. In case of $U = \Sigma^*$ for some finite alphabet Σ, a concept and a concept class correspond to a *language* and a *language class* (*language family*), respectively. For any (possibly infinite) concept class $\mathcal{C}$, by $\cap\mathcal{C}$ and $\cup\mathcal{C}$, we denote the sets $\{x \mid \forall L \in \mathcal{C}\ x \in L\}$ and $\{x \mid \exists L \in \mathcal{C}\ x \in L\}$, respectively. By $\mathcal{F}in$, we denote the class of all finite concepts. For any concept class $\mathcal{C}$, by $\mathcal{I}nt(\mathcal{C})$, we denote the smallest class containing $\mathcal{C}$ and closed under *finite* intersections.

A class $\mathcal{C} = \{L_i\}_{i\in\mathbf{N}}$ of concepts is *an indexed family of recursive concepts* (or, *indexed family*, for short), iff there exists a recursive function $f : \mathbf{N} \times U \to \{0, 1\}$ such that

$$f(i, w) = \begin{cases} 1 & \mathit{if}\ w \in L_i, \\ 0 & \mathit{otherwise}. \end{cases}$$

where $\mathbf{N}$ denotes the set of positive integers.

8.3 Identification in the Limit

For a given $L \in \mathcal{C}$, *a complete presentation of* L is any infinite sequence $(w_1, l_1), (w_2, l_2), (w_3, l_3), \ldots$ of elements in $U \times \{0,1\}$ such that $U = \{w_1, w_2, \ldots\}$ and $\forall i\ l_i = 1$ iff $w_i \in L$. An algorithmic device M is said to be a learning machine if it received an infinite sequence of elements in $U \times \{0,1\}$ and outputs an infinite sequence of integers. Let L be a given concept. We say that a learning machine M *identifies L in the limit from complete data* iff for any complete presentation of L, the infinite sequence, $g_1, g_2, g_3, \ldots$, produced by M converges to an integer g such that $L = L_g$. An indexed family $\mathcal{C}$ of recursive concepts is said to be *identifiable in the limit from complete data* iff there exists a learning machine M such that M identifies every concept in $\mathcal{C}$ in the limit from complete data.

For integers i, j with $i \leq j$, by $\sigma[i..j]$ we denote the set of elements of σ located from the ith to jth position. In case of $i = j$, we simply write $\sigma[i]$. In case of $i > j$, $\sigma[i..j]$ denotes the empty set. Let S be a set of elements in $U \times \{0,1\}$. A concept L is *consistent with* S iff for any $(w, l) \in S$, $w \in L \Leftrightarrow l = 1$.

Let us consider the following algorithm.

Algorithm A_1
input : a complete presentation σ of a concept $L \in \mathcal{C}$
begin
  initialize $i := 1$;
  initialize $S := \emptyset$;
  repeat forever
    $S := S \cup \{\sigma[i]\}$;
    $k := \min\{j \mid L_j \text{ is consistent with } S\}$;
    output $g_i = k$;
    $i := i + 1$;

Then, we can show the following theorem.

Theorem 1. *Any indexed family of recursive concepts is identifiable in the limit from complete data.*

Proof. We can prove that the algorithm A_1 identifies $\mathcal{C}$ in the limit from complete data. □

Let $w_1, w_2, \ldots$ be an enumeration of all elements of Σ^* such that $w_i \neq w_j$ for any $i \neq j$. By $< l_1, \ldots, l_n >$ ($l_i \in \{0,1\}, i = 1, \ldots, n$), we denote a sequence (w_1, l_1), ..., (w_1, l_n) of elements in $\Sigma^* \times \{0,1\}$. For $t_1 =< l_1, \ldots, l_m >$ and $t_2 =< l'_1, \ldots, l'_n >$, by $t_1 t_2$ we denote the sequence $< l_1, \ldots, l_m, l'_1, \ldots, l'_n >$. For (possibly infinite) sequence $t =< l_1, l_2, \ldots >$, by $L(t)$ we denote a language $\{w_i \mid l_i = 1\}$. Let $\mathcal{RC}$ be the class of all recursive languages. Then, we have the following negative result concerning the identification in the limit model.

Theorem 2. *There exists no learning machine which identifies $\mathcal{RC}$ in the limit.*

Proof. Assume that there exists a learning machine M which identifies $\mathcal{RC}$ in the limit.

Consider a language $L(t)$ for $t_* = t_0 t_1 t_2 \cdots$ defined as follows :

Step 0 : $t_0 =< 0 >$;
Step k ($k \geq 1$) :
for each $t =< 0 >, < 1 >, < 00 >, < 11 >, < 000 >, < 111 >, \cdots$
test whether the following inequality holds or not :
$M(< t_0 t_1 \cdots t_{k-1} >) \neq M(< t_0 t_1 \cdots t_{k-1} t >)$
If you find such t set $t_k = t$ and goto Step k+1;

Assume that at some step k, t_k can not be defined. Then, it is clear that the following equality holds :

$$\begin{aligned} M(< t_0 t_1 \cdots t_{k-1} 000 \cdots >) &= M(< t_0 t_1 \cdots t_{k-1} >) \\ &= M(< t_0 t_1 \cdots t_{k-1} 111 \cdots >) \end{aligned} \quad \cdots (a)$$

Let $u =< t_0 t_1 \cdots t_{k-1} 0000000 \cdots >$ and $v =< t_0 t_1 \cdots t_{k-1} 1111111 \cdots >$. Then, $L(u)$ and $L(v)$ are both recursive languages, since k is a fixed constant. Then, the above equality (a) implies that M converges to the same guess with inputs of distinct language presentations. This is a contradiction. Therefore, at each step k, t_k can be defined.

Furthermore, it is clear that $L(t_*)$ is a recursive language. Therefore, the learning machine M should learn $L(t_*)$ in the limit.

Note that t_* is a complete presentation of $L(t_*)$. Give t_* to M, then M changes its conjecture infinitely many times by definition of t_*. This is a contradiction. □

In order to identify the class $\mathcal{RC}$, we need to relax the learning criterion of identification in the limit. Case and Smith introduced a theoretically interesting learning criterion, called *behaviorally correct identification in the limit.*

We say that a learning machine M *behaviorally correctly (BC) identifies L in the limit from complete data* iff for any complete presentation of L, the output $g_1, g_2, ...$ of M satisfies the following condition:

$$\exists n_0 > 0 \ \forall i \geq 0 \ \ L = L(g_{n_0+i}).$$

In this BC-learning framework, the learning machine is not required to converge to a fixed conjecture. The learning process is considered to be successful if it outputs a *behaviorally correct* answer.

By $L_1 \Delta L_2$ we denote the symmetric difference of the sets L_1 and L_2. We write $L_1 =^* L_2$ iff $L_1 \Delta L_2$ is finite. We say that a learning machine M *BC*-identifies L in the limit from complete data* iff for any complete presentation of L, the output $g_1, g_2, ...$ of M satisfies the following condition:

$$\exists n_0 > 0 \; \forall i \geq 0 \;\; L =^* L(g_{n_0+i}).$$

Then, we have the following theorem:

Theorem 3. *There exists a learning machine which BC*-identifies $\mathcal{RC}$ in the limit from complete data.*

Proof. Let $p_1, p_2, ...$ be an enumeration of codes of Turing machines. Consider a learning algorithm M described bellow:

```
Initialize i := 1;
Initialize S := ∅;
Repeat forever
   S := S ∪ {σ[i]};
   output the following program (Turing machine (TM) code) ;

      main(x){
          for each p ∈ {p_1, ..., p_i} do
             run TM(p) at most lg(x) steps and checks
               the consistency of TM(p) w.r.t. S;
             if TM(p) is found to be consistent then
                run TM(p) on x and output the result;
             end if
          end
          output 'NO';
      }

   i := i + 1;
```

Let z be the minimum integer such that p_z is a Turing machine code representing a target language L. Let $k_i \stackrel{def}{=} \mathbf{min}\{\, j \mid w_j \in L \Delta L(p_i) \,\}$, $k_* \stackrel{def}{=} \mathbf{max}\{\, k_i \mid i = 1, ..., z-1 \,\}$, and $m_* = \mathbf{max}\{\, k_*, z \,\}$.

Let $pg(i)$ be the program (TM code) which is produced by the learning machine at step i. Then, we will show the claim: $L(pg(i)) =^* L$ holds for $i \geq m_*$.

The code p_z is contained in $\{p_1, ..., p_i\}$ since $i \geq m_* \geq z$. Define $s(j)$ to be the time steps required for $TM(p_z)$ to compute the result on input w_j. Define $s_* = \mathbf{max}\{s(j) \mid 1 \leq j \leq i\,\}$.

For $x \in U$ such that $lg(x) \geq s_*$, p_j ($j = 1, ..., z-1$) is ruled out from the hypotheses, since $i \geq m_* \geq k_*$ holds and therefore S contains w_{k_j} for each $j = 1, ..., z-1$. Then, $pg(i)$ selects p_z as a hypothesis since $s_* \leq lg(x)$. Thus, $pg(i)$ executes $TM(p_z)$ on x and outputs the correct answer. □

8.4 Identification from Positive Data

For a given $L \in \mathcal{C}$, *a positive presentation of* L is any infinite sequence $w_1, w_2, w_3, ...$ of elements in U such that $\{w_1, w_2, w_3, ...\} = L$. We say that a learning algorithm M *identifies* L *in the limit from positive data* iff for any positive presentation of L, the infinite sequence, $g_1, g_2, g_3, ...$, of conjectures produced by M converges to g such that $L = L_g$. An indexed family $\mathcal{C}$ of recursive concepts is said to be *identifiable in the limit from positive data* iff there exists an algorithm M such that M identifies every concept in $\mathcal{C}$ in the limit from positive data.

8.4.1 Characterization of Identification from Positive Data

Let $\mathcal{C}$ be an indexed family of recursive concepts. For $L \in \mathcal{C}$, a finite subset F of L is said to be a *finite tell-tale* (ft) of L w.r.t. $\mathcal{C}$ iff for any $L' \in \mathcal{C}$, $S \subseteq L'$ implies $L' \not\subset L$. An indexed family $\mathcal{C}$ has the *finite tell-tale property* (ftp) iff for any $L \in \mathcal{C}$, there exists a finite tell-tale F of L w.r.t. $\mathcal{C}$.

Condition EC1: There exists an effective procedure that, given an input i, enumerates a finite tell-tale S of L_i w.r.t. $\mathcal{C}$.

Note that the above enumeration procedure may not halt after producing a finite tell-tale of L_i.

For an infinite sequence $L_1, L_2, ...$ of concepts, we define $\lim_{i\to\infty} L_i \stackrel{def}{=} \cup_{i\geq 1} L_i$. Then, we have the following lemma:

Lemma 1. *For two infinite sequences* $L_1, L_2, L_3, ...$ *and* $K_1, K_2, K_3, ...$ *of concepts, and a concept* T, *we have:*

(1) if for any $i \geq 1$, $L_i \subseteq T$, *then* $\lim_{i\to\infty} L_i \subseteq T$,
(2) if for any $i \geq 1$, $T \subseteq L_i$, *then* $T \subseteq \lim_{i\to\infty} L_i$,
(3) if for any $i \geq 1$, $L_i \subseteq K_i$, *then* $\lim_{i\to\infty} L_i \subseteq \lim_{i\to\infty} K_i$.

Theorem 4. *An indexed family* $\mathcal{C}$ *is identifiable in the limit from positive data iff* $\mathcal{C}$ *satisfies the condition* **EC1**.

Proof. **if direction :** By T_i, we denote a finite tell-tale of L_i w.r.t. $\mathcal{C}$. We first assume that there is a procedure P which for a given input i, output T_i and *halts*. (This halting assumption will be removed away later.) Consider the following learning algorithm M using P:

Initialize $i = 1$;
$S := \emptyset$;
Repeat forever
  $S := S \cup \{\sigma[i]\}$;
  find $j_* =$ **min**$\{j \mid 1 \leq j \leq i,\ T_i \subseteq S \subseteq L_j\}$ and output $g_i = j_*$;
  if such j_* does not exist, find $j'_* =$ **min**$\{j \mid S \subseteq L_j\}$ and output $g_i = j'_*$;
  $i := i + 1$;

Let $L \in \mathcal{C}$ and t be the minimum integer such that $L = L_t$. Since T_t is finite, there exists some n such that $T_t \subseteq \sigma[1..n]$. By $T_t \subseteq S \subseteq L_t$, t can be a candidate of the output of M at stage $N = \mathbf{max}\{n, t\}$.

Suppose that there exists some j such that $j < t$ and for each stage $i \geq 1$, $T_j \subseteq S \subseteq L_j$. Then, S converges to L as $i \to \infty$. Therefore, by Lemma 1, we have $T_j \subseteq L \subseteq L_j$. Since T_j is a finite tell-tale of L_j, $L \not\subset L_j$. Thus, we have $L = L_j$, which contradicts to the minimality of t. Therefore, for all $j < t$, there is a stage i such that $T_j \subseteq S \subseteq L_j$ does not hold. This implies that at the second line of some stage of repeat loop, the algorithm output t.

In order to remove the halting assumption, we define $T_i^{(j)}$ as the set of strings enumerated by P up to j-th time step. Then, replace the condition $T_i \subseteq S \subset L_j$ by $T_i^{(i)} \subseteq S \subset L_j$ in the definition of j_*. Note that this modified algorithm P' can be implemented without the halting assumption. It is not difficult to see that P' also enumerates finite tell-tales.

only if direction : Assume that there is a learning algorithm M which identifies $\mathcal{C}$ in the limit from positive data. Then, we can show that the following procedure with input i effectively enumerates a finite tell-tale of L_i:

(Let $s_1, s_2, ...$ be an enumeration of L_i.)
(Let $\gamma_1, \gamma_2, ...$ be an enumeration of all finite sequences consisting of elements of L_i)
(For a finite sequence τ of strings, $M(\tau)$ denotes a conjecture of M on input τ.)
1. Initialize $n := 1$;
2. $\tau_1 =< s_1 >$;
3. output τ_n;
4. find j_*=$\mathbf{min}\{j \mid M(\tau_n) \neq M(\tau_n \gamma_j)\}$;
5. if such j_* exists, then $\tau_{n+1} = \tau_n \gamma_{j_*} < s_{n+1} >, \; n := n+1$ and goto step 3;

The correctness of this finite tell-tale enumeration procedure is omitted because of space constraint. □

8.4.2 Some Sufficient Conditions for Identification from Positive Data

An indexed family $\mathcal{C}$ of recursive concepts is said to have infinite elasticity if there exist an infinite sequence $L_1, L_2, ... \in \mathcal{C}$ and an infinite sequence $w_0, w_1, ... \in U$ such that for any $n \geq 1$, $\{w_0, ..., w_{n-1}\} \subseteq L_n$ and $w_n \notin L_n$. $\mathcal{C}$ has *finite elasticity* (fe), if $\mathcal{C}$ does not have *infinite elasticity*.

The finite subset T of L is called a *characteristic sample* (cs) of L w.r.t. $\mathcal{C}$ if for any $L' \in \mathcal{C}$, $T \subseteq L'$ implies $L \subseteq L'$. $\mathcal{C}$ has *characteristic sample property* (csp) if for any $L \in \mathcal{C}$, there exists a characteristic sample of L w.r.t. $\mathcal{C}$.

We will consider the following four conditions for an indexed family $\mathcal{C}$ of recursive languages:

Condition C1 : $\mathcal{C}$ has finite tell-tale property (ftp): i.e., for any $L \in \mathcal{C}$, there exists a finite tell-tale T of L.

Condition C2 : $\mathcal{C}$ has characteristic sample property

Condition C3 : $\mathcal{C}$ has finite elasticity.

Condition C4 : $\mathcal{C}$ has finite thickness, i.e., for any $w \in U$, $\#\{L \in \mathcal{C} \mid w \in L\}$ is finite.

Theorem 5. *We have the following implications:*
(1) **C4** *implies* **C3**.
(2) **C3** *implies* **C2**.
(3) **C2** *implies* **EC1**.
(4) **EC1** *implies* **C1**.

Proof. (1) Assume that $\mathcal{C}$ has infinite elasticity. Then, there exist an infinite sequence $L_1, L_2, ... \in \mathcal{C}$ and an infinite sequence $w_0, w_1, ... \in U$ such that for any $n \geq 1$, $\{w_0, ..., w_{n-1}\} \subseteq L_n$ and $w_n \notin L_n$. Then, $\#\{L \in \mathcal{C} \mid w_0 \in L\}$ is infinite.

(2) Assume that $\mathcal{C}$ does not have csp. Then, there exists $L \in \mathcal{C}$ such that L has no characteristic sample. Then, construct a sequence $w_0, w_1, ... \in U$ and a sequence $L_1, L_2, ... \in \mathcal{C}$ as follows:

Step 1 : Choose any $w \in L$ and set $w_0 = w$;
Step i **(**$i \geq 1$**):**
  Select $L_i \in \mathcal{C}$ such that $\{w_0, ..., w_{i-1}\} \subseteq L_i$ and $L \not\subseteq L_i$;
  Choose any $w_i \in L - L_i$;

Since $\{w_0, ..., w_{i-1}\}$ is not a cs of L, at each step i ($i \geq 1$), there is $L_i \in \mathcal{C}$ such that $\{w_0, ..., w_{i-1}\} \subseteq L_i$ and $L \not\subseteq L_i$. Then, these sequences show the infinite elasticity of $\mathcal{C}$.

(3) Let $s_1, s_2, ...$ be an enumeration of U. For a language L, let $L^{(n)} = L \cap \{s_1, ..., s_n\}$. Using the characteristic sample property of $\mathcal{C}$, we can prove that the following procedure with input i effectively enumerates a finite tell-tale of $L_i \in \mathcal{C}$

1. Initialize $n := 1$;
2. Initialize $T := \emptyset$;
3. output $T = L_i^{(n)}$ if there is a j ($1 \leq j \leq n$) such that $T \subseteq L_j^{(n)} \subset L_i^{(n)}$;
4. set $n := n + 1$ and go to step **3**;

(4) Immediate from the definitions. □

It is known that there exists an indexed family of recursive languages which satisfy the condition **C1** but not **EC1** ([1]).

8.4.3 Identifying Elementary Formal Systems from Positive Data

We show that there exist rich classes which are identifiable in the limit from positive data ([18]). Elementary formal systems (EFS's, for short), originally introduced by Smullyan [19], are a kind of logic system where we use string patterns instead of terms in first order logic. For more detailed definition or theoretical results on learning EFS's, refer to [3],[18].

Let Σ be a finite alphabet, whose elements are called *constants*. Let X and D be countable alphabets, whose elements are called *variables* and *predicates*, respectively, We note that Σ, X, and D are mutually disjoint. A *term* is a string over $\Sigma \cup X$. An *atomic formula* (*atom*) is an expression $P(t_1, ..., t_n)$, where t_i are terms for $i = 1, ..., n$. A *definite clause*, or a *clause* for short, is an expression of the form : $A \leftarrow B_1 \& ... \& B_n$., where A, B_1, ..., B_n are atoms. (n could be 0.) The atom A is called a head, and the sequence B_1, ..., B_n is called a *body*, of the clause. Elementary Formal System (EFS) is a finite set of definite clauses, where each definite clause is called an *axiom* of the EFS.

A substitution θ is a mapping from X to $(\Sigma \cup X)^+$. For a term t, by $\theta(t)$ we denote the term obtained by replacing every occurrence of a variable $x \in X$ in t by $\theta(x)$. For a substitution θ and an atom $p(t_1, , , t_n)$, we define $\theta(p(t_1, ..., t_n)) = p(\theta(t_1), ..., \theta(t_n))$. For a clause $C = A \leftarrow B_1, ..., B_n$, we define $\theta(C) = \theta(A) \leftarrow \theta(B_1), ..., \theta(B_n)$.

A clause C is *provable from* an EFS Γ, written $\Gamma \vdash C$, if C is obtained from Γ by applying substitutions and modus ponens finitely many times. More formally, we define the relation $\Gamma \vdash C$ inductively as follows:

1. If $C \in \Gamma$ then $\Gamma \vdash C$.
2. If $\Gamma \vdash C$ then $\Gamma \vdash \theta(C)$
 for any substitution.
3. If $\Gamma \vdash A \leftarrow B_1, ..., B_{n+1}$ and $\Gamma \vdash B_{n+1}$
 then $\Gamma \vdash A \leftarrow B_1, ..., B_n$.

For an EFS Γ and $p \in D$ with arity n, we define $L(\Gamma, p) = \{(w_1, ..., w_n) \in (\Sigma^+)^n \mid \Gamma \vdash p(w_1, ..., w_n)\}$. If p is unary, $L(\Gamma, p)$ is called an EFS language if such Γ and p exist.

By $|t|$ we denote the length of a term t. We define $|P(t_1, ..., t_n)| = |t_1| + \cdots + |t_n|$. By $v(A)$, we denote the set of all variables in an atom A. By $o(x, A)$, we denote the number of all occurrences of a variable x.

A clause $A \leftarrow B_1, ..., B_n$ is *variable-bounded* if $v(A) \supseteq v(B_i)$ for $i = 1, ..., n$. An EFS Γ is *variable-bounded* if each axiom of Γ is variable-bounded. A clause $A \leftarrow B_1, ..., B_n$ is *length-bounded* if $|\theta(A)| \geq |\theta(B_1)| + \cdots + |\theta(B_n)|$ for any substitution θ. An EFS Γ is *length-bounded* if each axiom of Γ is length-bounded.

The following lemmas are straightforward.

Lemma 2. *A clause* $A \leftarrow B_1, ..., B_n$ *is length-bounded if and only if* $|A| \geq |B_1| + \cdots + |B_n|$ *and* $o(x, A) \geq o(x, B_1) + \cdots + o(x, B_n)$ *for any* x.

Lemma 3. *If a clause C is provable from a length-bounded EFS, then C is length-bounded.*

Proof. By induction on the number of inference steps of a length-bounded EFS. □

Theorem 6. *A language is definable by a variable-bounded EFS if and only if it is recursively enumerable. A language is definable by a length-bounded EFS if and only if it is context-sensitive.*

Proof. By comparing the computational capacity with Turing machines and linear bounded automata. □

An atom $A(t_1, ..., t_n)$ is said to be *ground* if $t_i \in \Sigma^*$ holds for $i = 1, ..., n$. *Herbrand base*, written HB, is the set of all ground atoms. We define:

$M(\Gamma) \stackrel{def}{=} \{ A \in HB \mid \Gamma \vdash A \}.$

HB can be regarded as a universal set. Any subset of HB can be regarded as a concept. We consider the learning problem of the class:

$\mathcal{M}^n = \{ M(\Gamma) \mid \Gamma \text{ is length-bounded EFS}, \#\Gamma \leq n, \ M(\Gamma) \neq \emptyset \}.$

Lemma 4. *Let Γ be a length-bounded EFS and $C = A \leftarrow B_1, ..., B_n$ is provable from Γ. Then, the head of every axiom used to prove C is not longer than the head A of C.*

Proof. By induction on the number of inference steps for proving C. □

An EFS Γ is *reduced with respect to* a finite subset I of HB if $I \subseteq M(\Gamma)$ but $I \not\subseteq M(\Gamma')$ for any $\Gamma' \subset \Gamma$.

Lemma 5. *Let $I = \{A_1, ..., A_k\}$ be a nonempty finite subset of HB and Γ be a length-bounded EFS that is reduced with respect to I. Then, for any axiom $A \leftarrow B_1, ..., B_n$ of Γ, $|A| \leq$ **max**$\{ |A_1|, ..., |A_k| \}$.*

Proof. Assume that there is an axiom $C = A \leftarrow B_1, ..., B_n$ in Γ such that $|A| >$ **max**$\{ |A_1|, ..., |A_k| \}$ holds. Then, by Lemma 2.4, C is not used to prove A_i $(i = 1, ..., k)$, which contradicts to the assumption that Γ is reduced with respect to I. □

Lemma 6. *Given a nonempty finite subset I of HB, there exist only finitely many length-bounded EFS's that are reduced with respect to I.*

Proof. Let Γ be a reduced length-bounded EFS with respect to I. By Lemma 2.5, the head of every axiom in Γ is not longer than **max**$\{|A| \mid A \in I\}$. Note that the number of patterns t such that $|t| \leq k$ for some constant k is finite. Therefore, the number of such axioms is bounded by some constant, since Γ is length-bounded. □

Theorem 7. *For any $n \geq 1$, the class $\mathcal{M}^n$ has finite elasticity, therefore, is learnable in the limit from positive data.*

Proof. We will prove the claim by induction on n.

In the case of $n = 1$, consider an atom $A \in HB$. Any $\Gamma \in \mathcal{M}^1$ satisfying $A \in M(\Gamma)$ should be of the form $\{B\}$ for some B such that $A = \theta(B)$ for some θ. Therefore, we have $|B| \leq |A|$. Thus, the number of such Γ is finite except for renaming of variables. This implies that $\mathcal{M}^1$ has finite thickness, therefore, finite elasticity.

We assume that $\mathcal{M}^n$ has finite elasticity for any $n < k$. Suppose that $\mathcal{M}^k$ has infinite elasticity, then we will derive a contradiction as follows. There exists an infinite sequence $A_0, A_1, A_2, ...$ of ground atoms and an infinite sequence $\Gamma_1, \Gamma_2, ...$ of length-bounded EFS's such that $\#\Gamma_i \leq k$ and $\{A_0, ..., A_{i-1}\} \subseteq M(\Gamma_i)$ but $A_i \notin M(\Gamma_i)$ for any $i \geq 1$.
We define:
$h(i) = \mathbf{min}\{\, j \mid \Gamma_i \text{ is reduced with respect to } \{A_0, ..., A_j\} \text{ or } j = i\,\}$

We have two cases: (1) h has a finite upper-bound, and (2) no upper-bound.
Case 1. If there is j such that $h(i) \leq j$ for any $i \geq 1$. Then, for any $i > j$, Γ_i should be reduced with respect to $\{A_0, ..., A_j\}$. However, Lemma 2.5 claims that the number of such EFS's is finite, a contradiction.
Case 2. If $h(i)$ $(i = 1, 2, ...)$ does not have any finite upper-bound, there exists an infinite sequence $i_1 < i_2 < ...$ such that $1 < h(i_1) < h(i_2) < ....$

(Case 2.1) If Γ_{i_j} is reduced with respect to $\{A_0, ..., A_{h(i_j)}\}$, then $\{A_0, ..., A_{h(i_j)}\} \not\subseteq M(\Gamma'_{i_j})$ for any $\Gamma'_{i_j} \subset \Gamma_{i_j}$. However, $\{A_0, ..., A_{h(i_j)-1}\} \subseteq M(\Gamma'_{i_j})$ for some $\Gamma'_{i_j} \subset \Gamma_{i_j}$ since Γ_{i_j} is not reduced with respect to $\{A_0, ..., A_{h(i_j)-1}\}$. Therefore, there exists $\Gamma'_{i_j} \subset \Gamma_{i_j}$ such that $\{A_0, ..., A_{h(i_j)-1}\} \subseteq M(\Gamma'_{i_j})$ but $A_{h(i_j)} \notin M(\Gamma'_{i_j})$.

(Case 2.2) If Γ_{i_j} is not reduced with respect to $\{A_0, ..., A_{h(i_j)}\}$, then $h(i_j) = i_j$. Since M^k has infinite elasticity, $A_{h(i_j)} = A_{i_j} \notin M(\Gamma_{i_j})$ holds. Since Γ_{i_j} is not reduced with respect to $\{A_0, ..., A_{h(i_j)-1}\}$, $\{A_0, ..., A_{h(i_j)-1}\} \subseteq M(\Gamma'_{i_j})$ for some $\Gamma'_{i_j} \subset \Gamma_{i_j}$. Furthermore, for this Γ'_{i_j}, we have $A_{i_j} \notin M(\Gamma'_{i_j})$ since $A_{i_j} \notin M(\Gamma_{i_j})$.

Therefore, in both subcases **(2.1)** and **(2.2)**, there exists $\Gamma'_{i_j} \subset \Gamma_{i_j}$ such that
$\{A_0, ..., A_{h(i_j)-1}\} \subseteq M(\Gamma'_{i_j})$ but $A_{h(i_j)} \notin M(\Gamma'_{i_j})$.

Note that $\#\Gamma'_{i_j} \leq k - 1$. Therefore, the existence of two infinite sequences:

$$A_0, A_{h(i_1)}, A_{h(i_2)}, ... \text{ and } \Gamma'_{i_1}, \Gamma'_{i_2}, \Gamma'_{i_3}, ...$$

show the infinite elasticity of $\mathcal{M}^{k-1}$. This is a contradiction.

In conclusion, $\mathcal{M}^n$ has finite elasticity for any $n \geq 1$. Therefore, $\mathcal{M}^n$ is identifiable in the limit from positive data. □

Theorem 8. *For any $n \geq 1$, the class*
$\mathcal{L}^n = \{L(\Gamma, p) \mid \Gamma$ *is length-bounded EFS*, p *is unary*, $\#\Gamma \leq n$, $L(\Gamma, p) \neq \emptyset\}$
has finite elasticity, therefore, is learnable in the limit from positive data.

Proof. Assume that $\mathcal{L}^n$ has infinite elasticity. Then, there exist an infinite sequence $w_0, w_1, w_2, ...$ of words and an infinite sequence $(\Gamma_1, p_1), (\Gamma_2, p_2), ...$ of pairs of length bounded EFS's and unary predicates such that $\#\Gamma \leq n$ and $\{w_0, ..., w_{k-1}\} \subseteq L(\Gamma_k, p_k)$ but $w_k \notin L(\Gamma_k, p_k)$.

We should note that without loss of generality, we may assume that all the predicate symbols $p_k' s$ are the same predicate symbol p. Therefore, the infinite sequence $p(w_0), p(w_1), p(w_2), ...$ of atoms and an infinite sequence $\Gamma_1, \Gamma_2, ...$ of length bounded EFS's show the infinite elasticity of the class $\mathcal{M}^n$, which contradicts to Theorem 7. □

8.5 Approximate Identification in the Limit

Let U be the universal set. Let $\mathcal{C}$ be a concept class and X be a concept (not always in $\mathcal{C}$). A concept $Y \in \mathcal{C}$ is called a *$\mathcal{C}$-upper-best approximation of a concept X* iff $X \subseteq Y$ and for any concept $C \in \mathcal{C}$ such that $X \subseteq C$, $Y \subseteq C$ holds. By $\overline{\mathcal{C}}X$ we denote the $\mathcal{C}$-upper-best approximation of a concept X. $\mathcal{C}_1$ has *upper-best approximation property (u.b.a.p)* relative to $\mathcal{C}_2$, written $\mathcal{C}_1 \stackrel{u.b.}{\rightarrow} \mathcal{C}_2$, iff for any concept X in $\mathcal{C}_2$, there exists a $\mathcal{C}_1$-upper-best approximation of X.

Proposition 1 *Let $\mathcal{C}_1, \mathcal{C}_2, \mathcal{C}_3$ be concept classes.*
(1) $\mathcal{C}_1 \stackrel{u.b.}{\rightarrow} \mathcal{C}_2$ and $\mathcal{C}_1 \stackrel{u.b.}{\rightarrow} \mathcal{C}_3$ imply $\mathcal{C}_1 \stackrel{u.b.}{\rightarrow} \mathcal{C}_2 \cup \mathcal{C}_3$.
(2) $\mathcal{C}_1 \stackrel{u.b.}{\rightarrow} \mathcal{C}_3$ and $\mathcal{C}_2 \stackrel{u.b.}{\rightarrow} \mathcal{C}_3$ imply $Int(\mathcal{C}_1 \cup \mathcal{C}_2) \stackrel{u.b.}{\rightarrow} \mathcal{C}_3$.

Proof. **(1)** Immediate from the definition.
(2) Let L be any concept in $\mathcal{C}_3$, and define $L_1 = \overline{\mathcal{C}}_1 L$, $L_2 = \overline{\mathcal{C}}_2 L$, and $L_* = L_1 \cap L_2$. Note that $L_* \in \mathcal{I}nt(\mathcal{C}_1 \cup \mathcal{C}_2)$. We will show that L_* is $\mathcal{I}nt(\mathcal{C}_1 \cup \mathcal{C}_2)$-upper-best approximation of L. Note that $L \subseteq L_1$ and $L \subseteq L_2$ hold. Therefore, we have $L \subseteq L_*$.

Let us consider any concept $A \in \mathcal{I}nt(\mathcal{C}_1 \cup \mathcal{C}_2)$ such that $L \subseteq A$. Then, there are a finite sequence $L_{p_1}, ..., L_{p_k}$ of concepts in $\mathcal{C}_1$ and a finite sequence $L_{q_1}, ..., L_{q_l}$ of concepts in $\mathcal{C}_2$ such that $A = L_{p_1} \cap \cdots L_{p_k} \cap L_{q_1} \cap \cdots L_{q_l}$. We have that $L \subseteq L_{p_i}$ $(1 \leq i \leq k)$ and $L \subseteq L_{q_i}$ $(1 \leq i \leq l)$, which implies that $L_1 = \overline{\mathcal{C}}_1 L \subseteq L_{p_i}$ $(1 \leq i \leq k)$ and $L_2 = \overline{\mathcal{C}}_2 L \subseteq L_{q_i}$ $(1 \leq i \leq l)$. Therefore, we have $L_* = L_1 \cap L_2 \subseteq A$. This implies that $L_* \in \mathcal{I}nt(\mathcal{C}_1 \cup \mathcal{C}_2)$ is a $\mathcal{I}nt(\mathcal{C}_1 \cup \mathcal{C}_2)$-upper-best approximation of L. □

Let $\mathcal{C}$ be an indexed family. A learning algorithm M *$\mathcal{C}$-upper-best approximately identifies* a concept L in the limit from positive data iff for any positive presentation σ of L, M with σ outputs infinite sequence of integers,

$g_1, g_2, ...,$ such that $\exists n \geq 1\ (\forall j \geq n\ g_j = g_n) \wedge (L_{g_n} = \overline{\mathcal{C}}L)$. A concept class $\mathcal{C}_1$ is *upper-best approximately identifiable in the limit* from positive data by an indexed family $\mathcal{C}_2$ iff there is a learning algorithm M which $\mathcal{C}_2$-upper-best approximately identifies every concept in $\mathcal{C}_1$ in the limit from positive data. The original definition of the identifiability from positive data ([1]) just corresponds to the case of $\mathcal{C}_1 = \mathcal{C}_2$.

8.5.1 Approximate Identification and Finite Elasticity

We introduce an extended notion of the finite elasticity ([20][10]). Let $\mathcal{C}$ be a concept class and L be any concept. $\mathcal{C}$ has *infinite elasticity* at L iff there are an infinite sequence $F_0, F_1, ...$ of finite subsets of L and an infinite sequence $L_1, L_2, ...$ of concepts in $\mathcal{C}$ such that (1) $\bigcup\{F_i \mid i \geq 1\} = L$ and (2) for any $i \geq 1$, $F_{i-1} \subset F_i$, $F_{i-1} \subseteq L_i$ and $F_i \not\subseteq L_i$ hold. A concept class $\mathcal{C}_1$ has *finite elasticity* relative to $\mathcal{C}_2$, written $\mathcal{C}_1 \overset{f.e.}{\rightarrow} \mathcal{C}_2$, iff for every concept $L \in \mathcal{C}_2$, $\mathcal{C}_1$ does not have infinite elasticity at L. In case of $\mathcal{C}_2 = 2^U$, this corresponds to the original definition of the finite elasticity, and we simply say that $\mathcal{C}_1$ has finite elasticity.

Further, we must note that the above notion of the infinite elasticity of $\mathcal{C}$ at L was first equivalently introduced by [15] as the notion of the *infinite cross property* of L within $\mathcal{C}$, and this equivalence has been proved in [17].

We have the following interesting proposition.

Proposition 2 *Let $\mathcal{C}_1$, $\mathcal{C}_2$, and $\mathcal{C}_3$ be concept classes.*
(1) $\mathcal{C}_1 \overset{f.e.}{\rightarrow} \mathcal{C}_2$ and $\mathcal{C}_1 \overset{f.e.}{\rightarrow} \mathcal{C}_3$ imply $\mathcal{C}_1 \overset{f.e.}{\rightarrow} \mathcal{C}_2 \cup \mathcal{C}_3$.
(2) $\mathcal{C}_1 \overset{f.e.}{\rightarrow} \mathcal{C}_3$ and $\mathcal{C}_2 \overset{f.e.}{\rightarrow} \mathcal{C}_3$ imply $Int(\mathcal{C}_1 \cup \mathcal{C}_2) \overset{f.e.}{\rightarrow} \mathcal{C}_3$.

Proof. (1) Assume $\mathcal{C}_1 \overset{f.e.}{\not\rightarrow} \mathcal{C}_2 \cup \mathcal{C}_3$. Then, $\mathcal{C}_1$ has infinite elasticity at some concept $L \in \mathcal{C}_2 \cup \mathcal{C}_3$. Therefore, there are an infinite sequence $F_0, F_1, ...$ of finite subsets of L and an infinite sequence $L_1, L_2, ...$ of concepts in $\mathcal{C}_1$ satisfying the conditions of infinite elasticity at L. Since $L \in \mathcal{C}_2 \cup \mathcal{C}_3$, this implies that $\mathcal{C}_1$ does not have finite elasticity relative to either $\mathcal{C}_2$ or $\mathcal{C}_3$, which is a contradiction.
(2) Assume $Int(\mathcal{C}_1 \cup \mathcal{C}_2) \overset{f.e.}{\not\rightarrow} \mathcal{C}_3$. Then, $Int(\mathcal{C}_1 \cup \mathcal{C}_2)$ has infinite elasticity at some concept $L \in \mathcal{C}_3$. Therefore, there are an infinite sequence $F_0, F_1, ...$ of finite subsets of L and an infinite sequence $L_1, L_2, ...$ of concepts in $Int(\mathcal{C}_1 \cup \mathcal{C}_2)$ satisfying the conditions of infinite elasticity at L. We can write $L_i = A_1^i \cap \cdots \cap A_{k(i)}^i$ for each $i \geq 1$, where $A_j^i \in \mathcal{C}_1 \cup \mathcal{C}_2$ $(1 \leq j \leq k(i))$. By $F_{i-1} \subseteq L_i$ $(i \geq 1)$, we have $F_{i-1} \subseteq A_j^i$ $(i \geq 1, 1 \leq j \leq k(i))$. By $F_i \not\subseteq L_i$ $(i \geq 1)$, we have for each $i \geq 1$, that some A_j^i does not contain F_i (for convenience, we say that A_j^i *breaks the elasticity*). Since $A_j^i \in \mathcal{C}_1 \cup \mathcal{C}_2$ $(i \geq 1, 1 \leq j \leq k(i))$, either $\mathcal{C}_1$ or $\mathcal{C}_2$ contains an infinite number of concepts A_j^i which breaks the elasticity. Without loss of generality, we can assume that $\mathcal{C}_1$ contains infinite number of such concepts, $A_{l_1}^{n_1}, A_{l_2}^{n_2}, ...,$ where $n_i < n_{i+1}$ $(i \geq 1)$. Then, the sequences $F_0, F_{n_1}, F_{n_2}, ...$ and $A_{l_1}^{n_1}, A_{l_2}^{n_2}, ...$ lead to $\mathcal{C}_1 \overset{f.e.}{\not\rightarrow} \mathcal{C}_3$, which is a contradiction. □

Next, we introduce an extended notion of characteristic samples ([9]).

Let $\mathcal{C}$ be a concept class and consider a concept $L \in 2^U$. A finite subset F of L is called a *characteristic sample* of L relative to $\mathcal{C}$ iff for any $A \in \mathcal{C}$, $F \subseteq A$ implies $L \subseteq A$. Note that when we restrict L to a concept in $\mathcal{C}$, this definition coincides with the original notion of the characteristic sample ([1]).

Let $\mathcal{C}_1$ and $\mathcal{C}_2$ be concept classes. We say that $\mathcal{C}_1$ *has characteristic samples* relative to $\mathcal{C}_2$ iff for any $L \in \mathcal{C}_1$, there exists a characteristic sample of L relative to $\mathcal{C}_2$. The following lemma is useful.

Lemma 7. *$\mathcal{C}_1$ has characteristic samples relative to $\mathcal{C}_2$ iff $\mathcal{C}_2 \overset{f.e.}{\rightarrow} \mathcal{C}_1$.*

Proof. We first prove the only if direction of the claim. Suppose that there is a concept $L \in \mathcal{C}_1$ such that $\mathcal{C}_2$ has infinite elasticity at L. Let $F_0, F_1, ...$ be an infinite sequence of finite subsets of L and $L_1, L_2, ...$ be an infinite sequence of concepts in $\mathcal{C}_2$ satisfying the conditions of infinite elasticity at L. By the assumption, L has a characteristic sample $T \subseteq L$ relative to $\mathcal{C}_2$. Let n be the minimum integer j such that $T \subseteq F_j$. Then, we have for each $j > n$, $L \subseteq L_j$. This implies that for each $j > n$, $F_j \subseteq L \subseteq L_j$, which contradicts the condition of infinite elasticity at L.

For proving the converse direction, assume that there is a concept $L \in \mathcal{C}_1$ such that L does not have characteristic samples relative to $\mathcal{C}_2$. Let $w_0, w_1, ...$ be an enumeration of elements of L such that $w_i \neq w_j$ whenever $i \neq j$. Consider the following procedure:

stage 0 : $F_0 = \{w_0\}$;
stage i $(i \geq 1)$**:**
  Find a concept $L_i \in \mathcal{C}_2$ such that $F_{i-1} \subseteq L_i$ and $L \not\subseteq L_i$;
  Select an element e_i from $L - L_i$;
  Set $F_i = F_{i-1} \cup \{w_i, e_i\}$;

Since L does not have characteristic samples, it holds that at the first step of each stage i $(i \geq 1)$, we can find a concept L_i satisfying $F_{i-1} \subseteq L_i$ and $L \not\subseteq L_i$. (Otherwise, F_{i-1} should be a characteristic sample, a contradiction.) By $F_i - L_i \supseteq \{e_i\} \neq \emptyset$, we have $F_i \not\subseteq L_i$. Further, it is clear that $\bigcup\{F_i \mid i \geq 1\} = L$ holds. Thus, the sequences $F_0, F_1, ...$ and $L_1, L_2, ...$ satisfy the conditions of infinite elasticity at L. This is a contradiction. □

Let $e_1, e_2, ...$ be a fixed recursive enumeration of the universal set U. For any concept L, by $L^{(n)}$, we denote the set $L \cap \{e_1, ..., e_n\}$. It is clear that for any recursive concepts L_1 and L_2, and for any positive integer n, the inclusion $L_1^{(n)} \subseteq L_2^{(n)}$ is decidable.

Lemma 8. *Let $\mathcal{C}_1 = \{L_i\}_{i \in \mathbf{N}}$ be an indexed family and $\mathcal{C}_2$ be a concept class. $\mathcal{C}_2$ is upper-best approximately identifiable in the limit from positive data by $\mathcal{C}_1$ if $\mathcal{C}_1 \overset{u.a.}{\rightarrow} \mathcal{C}_2$ and $\mathcal{C}_1 \overset{f.e.}{\rightarrow} \mathcal{C}_2$.*

Proof. Let $w_1, w_2, ...$ be a positive presentation of a target concept $L_* \in \mathcal{C}_2$ and consider the following learning algorithm:

stage 0 : $P_0 = \emptyset$;
stage n $(n \geq 1)$**:**
$P_n = P_{n-1} \cup \{w_n\}$;
Select all concepts L_{p_i} in $\{L_1, ..., L_n\}$ such that $P_n \subseteq L_{p_i}$ and construct the set $Cns_n = \{L_{p_1}, ..., L_{p_{k_n}}\}$;
Select and output the minimum index g_n of concepts in Cns_n such that $L_{g_n}^{(n)}$ is a minimal concept in $\{L_{p_1}^{(n)}, ..., L_{p_{k_n}}^{(n)}\}$ with respect to $\subseteq$;

Claim A: There is a stage n_1 such that at any stage $n \geq n_1$, $L_* \subseteq L_{p_i}$ holds for any $L_{p_i} \in Cns_n$.
Proof of the Claim
By Lemma 7, L_* has a characteristic sample T relative to $\mathcal{C}_1$. Let n_1 be the minimum integer n such that $T \subseteq P_n$. Then, the claim holds. □

Let g_* be the minimum integer j of the concepts in $\mathcal{C}_1$ such that $\overline{\mathcal{C}_1} L_* = L_j$.
Claim B: There is a stage n_2 such that at any stage $n \geq n_2$, the learning algorithm always outputs g_*.
Proof of the Claim
Select all concepts L_{r_i} from $\{L_1, ..., L_{g_*-1}\}$ such that $L_* \subseteq L_{r_i}$ and construct the set $\{L_{r_1}, ..., L_{r_m}\}$. Then, for each L_{r_i} $(1 \leq i \leq m)$, $L_{g_*} = \overline{\mathcal{C}_1} L_* \subseteq L_{r_i}$ holds. By the minimality of g_*, for each L_{r_i} $(1 \leq i \leq m)$, $L_{g_*} \subset L_{r_i}$ holds. Therefore, for each i $(1 \leq i \leq m)$, we can choose the minimum integer s_i such that $L_{g_*}^{(s_i)} \subset L_{r_i}^{(s_i)}$. Let $n_2 = max\{s_1, ..., s_m, g_*, n_1\}$.

At any stage $n \geq n_2$, $L_{g_*} \in Cns_n$ holds since $n_2 \geq g_*$ and $L_* \subseteq L_{g_*}$. Further, at any stage $n \geq n_2$, $L_{g_*}^{(n)}$ is a minimal concept in $\{L_{p_1}^{(n)}, ..., L_{p_{k_n}}^{(n)}\}$, since for each L_{p_i} $(1 \leq i \leq k_n)$, $L_* \subseteq L_{p_i}$ (by $n_2 \geq n_1$) and therefore $L_{g_*} = \overline{\mathcal{C}_1} L_* \subseteq L_{p_i}$ hold.

It is only left to show that every $L_{p_i}^{(n)}$ with $p_i < g_*$ is not a minimal concept in $\{L_{p_1}^{(n)}, ..., L_{p_{k_n}}^{(n)}\}$. For any concept $L_{p_i}^{(n)}$ with $p_i < g_*$, by $n_2 \geq max\{s_1, ..., s_m\}$, $L_{g_*}^{(n)} \subset L_{p_i}^{(n)}$ holds. Therefore, every $L_{p_i}^{(n)}$ with $p_i < g_*$ is not a minimal concept. □

This completes the proof of the lemma. □

Lemma 9. *Let $\mathcal{C}_1$ be an indexed family and $\mathcal{C}_2$ be a concept class such that $\mathcal{F}in \subseteq \mathcal{C}_2$. $\mathcal{C}_2$ is upper-best approximately identifiable in the limit from positive data by $\mathcal{C}_1$ only if $\mathcal{C}_1 \overset{u.a.}{\rightarrow} \mathcal{C}_2$ and $\mathcal{C}_1 \overset{f.e.}{\rightarrow} \mathcal{C}_2$.*

Proof. By definition, it is clear that $\mathcal{C}_1 \overset{u.a.}{\rightarrow} \mathcal{C}_2$ is necessary for the upper-best approximate identification. Assume $\mathcal{C}_1 \overset{f.e.}{\not\rightarrow} \mathcal{C}_2$. Then, $\mathcal{C}_1$ has infinite elasticity at some concept $L \in \mathcal{C}_2$. Let $F_0, F_1, ...$ be an infinite sequence of finite subsets of L and $L_1, L_2, ...$ be an infinite sequence of concepts in $\mathcal{C}_1$ satisfying the conditions of infinite elasticity at L. Let $w_0, w_1, ...$ be a recursive enumeration of elements of L. Let M be a learning algorithm which $\mathcal{C}_1$-upper-best approximately identifies every concept in $\mathcal{C}_2$ in the limit from positive data. For any finite sequence σ of elements of U, by $M(\sigma)$, we denote the last conjecture (integer) output by M when it is fed with σ. Consider the following procedure:

stage 0 : $\sigma_0 = w_0$;
stage i $(i \geq 1)$**:**
Find a finite sequence σ_f of elements in L such that $M(\sigma_{i-1}) \neq M(\sigma_{i-1} \cdot \sigma_f)$;
$\sigma_i = \sigma_{i-1} \cdot \sigma_f \cdot w_i$;

At the first step of some stage i $(i \geq 1)$, the procedure above can not find any finite sequence σ_f satisfying the condition, and does not stop forever. Since otherwise, σ_∞ could be a positive presentation of L, and M with input σ_∞ does not converge to any integer, which is a contradiction.

Let n be the stage where the procedure can not find any finite sequence σ_f satisfying the condition. Then, $L_{M(\sigma_n)}$ should be equivalent to $\overline{\mathcal{C}_1}L$ since M $\mathcal{C}_1$-upper-best approximately identifies L in the limit.

There is an F_m such that F_m contains all elements of σ_n, since σ_n is finite. Let δ be any positive presentation of the concept F_m. Then, M with input $\sigma_n \cdot \delta$ should converge to the index of the concept $\overline{\mathcal{C}_1}F_m$ since $F_m \in \mathcal{C}_2$ holds and $\sigma_n \cdot \delta$ is a positive presentation of F_m. Then, by definition of σ_n, $F_m \in \mathcal{C}_2$ should be equivalent to $L_{M(\sigma_n)} = \overline{\mathcal{C}_1}L$.

Note that by the conditions of infinite elasticity, we have:

$$F_m \subseteq L_{m+1}, \tag{8.1}$$
$$F_{m+1} \not\subseteq L_{m+1}, \tag{8.2}$$
$$F_{m+1} \subseteq L. \tag{8.3}$$

By (8.1) and the definition of $\overline{\mathcal{C}}_1 F_m$, we have:

$$\overline{\mathcal{C}}_1 F_m \subseteq L_{m+1} \tag{8.4}$$

Therefore,

$$\begin{aligned}
\overline{\mathcal{C}}_1 L - \overline{\mathcal{C}}_1 F_m &\supseteq L - \overline{\mathcal{C}}_1 F_m && (\text{ by definition of } \overline{\mathcal{C}}_1 L\ ) \\
&\supseteq F_{m+1} - \overline{\mathcal{C}}_1 F_m && (\text{ by (8.3) }) \\
&\supseteq F_{m+1} - L_{m+1} && (\text{ by (8.4) }) \\
&\neq \emptyset && (\text{ by (8.2) })
\end{aligned}$$

holds, which implies $\overline{\mathcal{C}}_1 L \neq \overline{\mathcal{C}}_1 F_m$, a contradiction. □

Theorem 9. *Let $\mathcal{C}_1$ be an indexed family and $\mathcal{C}_2$ be a concept class such that $\mathcal{F}in \subseteq \mathcal{C}_2$. Then, the following conditions are equivalent:*
(1) $\mathcal{C}_2$ is upper-best approximately identifiable in the limit from positive data by $\mathcal{C}_1$,
(2) $\mathcal{C}_1 \overset{u.a.}{\rightarrow} \mathcal{C}_2$ and $\mathcal{C}_1 \overset{f.e.}{\rightarrow} \mathcal{C}_2$.

Proof. By Lemma 8 and Lemma 9. □

Theorem 10. *Let $\mathcal{C}_1$ and $\mathcal{C}_2$ be indexed families and let $\mathcal{C}_3$ and $\mathcal{C}_4$ be concept classes such that $\mathcal{F}in \subseteq \mathcal{C}_3$ and $\mathcal{F}in \subseteq \mathcal{C}_4$. Consider any indexed family $\mathcal{C}_5$ for the concept class $\mathcal{I}nt(\mathcal{C}_1 \cup \mathcal{C}_2)$. Then, the followings hold.*
(1) Assume that both $\mathcal{C}_3$ and $\mathcal{C}_4$ are upper-best approximately identifiable in the limit from positive data by $\mathcal{C}_1$. Then, $\mathcal{C}_3 \cup \mathcal{C}_4$ is upper-best approximately identifiable in the limit from positive data by $\mathcal{C}_1$.
(2) Assume that $\mathcal{C}_3$ is upper-best approximately identifiable in the limit from positive data by both $\mathcal{C}_1$ and $\mathcal{C}_2$. Then, $\mathcal{C}_3$ is upper-best approximately identifiable in the limit from positive data by $\mathcal{C}_5$.

Proof. By Proposition 1, Proposition 2 and Theorem 9 □

This theorem is important in the sense that it implicitly provides us with a method for enlarging the target concept class and for refining the mesh of the hypothesis space.

8.5.2 A Characterization of Upper-Best Approximate Identifiability from Positive Data

In the rest of the chapter, we argue on the issue of upper-best approximately identifying the class 2^U in the limit by some indexed family. In this section, we present a characterization theorem for such an approximate identifiability from positive data.

We will first prove the following easy lemma:

Lemma 10. *A concept class $\mathcal{C}$ has u.b.a.p. iff $\mathcal{C}$ is closed under infinite intersections.*

Proof. For proving the if part, assume that $\mathcal{C}$ is closed under infinite intersections. Let S be any concept. We have $\mathcal{F} = \{C \in \mathcal{C} \mid S \subseteq C\} \neq \emptyset$, since $U = \bigcap \emptyset \in \mathcal{C}$. Let $X = \bigcap \mathcal{F}$. Then, we have $S \subseteq X$. Further, for any concept $C' \in \mathcal{C}$ such that $S \subseteq C'$, $X \subseteq C'$ holds, since $C' \in \mathcal{F}$. Therefore, $X \in \mathcal{C}$ is $\mathcal{C}$-upper best approximation of S. Hence, $\mathcal{C}$ has u.b.a.p.

Conversely, let us assume $\mathcal{C}$ has u.b.a.p., and consider any subclass $\mathcal{F}$ of $\mathcal{C}$. Let $X = \bigcap \mathcal{F}$ and define $Y = \overline{\mathcal{C}}X$. From the assumption, we have $Y \in \mathcal{C}$.

By definition, $X \subseteq Y$ holds. We will show that $Y \subseteq X$ as follows.

Assume that there exists an element u such that $u \in Y$ and $u \notin X$. Then, by the definition of X, we have that there exists a concept $A \in \mathcal{F}$ such that $u \notin A$. For the concept A, we have $X \subseteq A$ and $Y \not\subseteq A$, which contradicts the fact that Y is the $\mathcal{C}$-upper-best approximation of X. Therefore, we have $Y \subseteq X$.

Hence, $X = Y \in \mathcal{C}$ holds. This completes the proof. □

The following is a characterization for the upper-best approximate identifiability from positive data.

Theorem 11. *Let $\mathcal{C}$ be an indexed family with u.b.a.p. Then, the following statements are equivalent.*
(1) 2^U is upper-best approximately identifiable in the limit from positive data by $\mathcal{C}$.
(2) 2^U is upper-best approximately identifiable in the limit from complete data by $\mathcal{C}$.
(3) $\mathcal{C}$ has finite elasticity.
(4) $\mathcal{C}$ has no infinite ascending sequences.

Proof. **(1)⇒(2) :** Immediately from the definition.
(2)⇒(4) : Assume that $\mathcal{C}$ has an infinite ascending sequence $\tilde{L}_1 \subset \tilde{L}_2 \subset \tilde{L}_3 \subset \cdots$ and let $\mathcal{F} = \{\tilde{L}_i \mid i \geq 1\}$. The condition (2) implies that there exists an algorithm M which upper-best approximately identifies 2^U in the limit from complete data with respect to $\mathcal{C}$. In the following, $\sigma = w_1, w_2, ...$ is a positive presentation of U such that for any i, j (≥ 1) with $i \neq j$, $w_i \neq w_j$ holds.

We will define an infinite sequence $\tilde{L}_{p_0}, \tilde{L}_{p_1}, ...$ of concepts in $\mathcal{F}$ and an infinite sequence $(w_1, l_1), (w_2, l_2), ...$ of pairs on $U \times \{0,1\}$ fed to M as follows:

Stage 0 **:** $n_0 = 0$; $p_0 = 0$; $\tilde{L}_{p_0} = \emptyset$; initialize M;
Stage i $(i \geq 1)$ **:**
(i) Let $\tilde{L}_{p_i}$ be a concept in $\mathcal{F}$ such that $\tilde{L}_{p_i} - (\tilde{L}_{p_{i-1}} \cup \{w_1, ..., w_{n_{i-1}}\}) \neq \emptyset$;
(ii) Additionally feed M a sequence, $(w_{n_{i-1}+1}, \tilde{L}_{p_i}(w_{n_{i-1}+1})), (w_{n_{i-1}+2}, \tilde{L}_{p_i}(w_{n_{i-1}+2})), (w_{n_{i-1}+3}, \tilde{L}_{p_i}(w_{n_{i-1}+3})), ...$ of pairs on $U \times \{0,1\}$ until the outputs of M converges to some index g_i;
(iii) Let n_i be the total number of pairs fed to M up to this point;
(iv) Go to Stage $i+1$;

At the first step of each ith ($i \geq 1$) stage, $\tilde{L}_{p_i}$ can be defined, since $\tilde{L}_1 \subset \tilde{L}_2 \subset \tilde{L}_3 \subset \cdots$ is an infinite sequence and the cardinality of $\{w_1, ..., w_{n_{i-1}}\}$ is finite. At the second step of each ith ($i \geq 1$) stage, the output of M converges to some index g_i, since for any concept L, M can identify a $\mathcal{C}$-upper-best approximation of L in the limit. Further, $\tilde{L}_{p_i} - \{w_1, ..., w_{n_{i-1}}\} \subseteq L_{g_i} \subseteq \tilde{L}_{p_i}$ holds for each $i \geq 1$, since at the ith stage, the set of all positive examples in a complete presentation fed to M contains $\tilde{L}_{p_i} - \{w_1, ..., w_{n_{i-1}}\}$ and is contained in $\tilde{L}_{p_i}$. Therefore, we have for each $i \geq 1$, $L_{g_{i+1}} - L_{g_i} \supseteq \tilde{L}_{p_{i+1}} - \{w_1, ..., w_{n_i}\} - \tilde{L}_{p_i} \neq \emptyset$. Thus, M changes its conjectures infinitely many times.

Then, we define a language L_* using the infinite sequence τ_∞ which is fed to M in the definition above: $w \in L_*$ iff $(w, 1)$ belongs to τ_∞. (Recall that L_* may not be recursively enumerable.) We have that the $\mathcal{C}$-upper-best approximation of L_* can not be identifiable in the limit from the complete presentation τ_∞ by M, since M on input τ_∞ changes its conjectures infinitely many times. This is a contradiction.
(4)⇒(3) : Assume that $\mathcal{C}$ has infinite elasticity. Then there exists an infinite sequence $w_0, w_1, w_2, ...$ of elements in U and an infinite sequence $L_1, L_2, ...$ of

concepts in $\mathcal{C}$ such that for any $k \geq 1$, $\{w_0, w_1, ..., w_{k-1}\} \subseteq L_k$ and $w_k \notin L_k$ hold. Let us define concepts $L'_i = \bigcap\{L_j \mid j \geq i\}$ $(i \geq 1)$. We have that for each $i \geq 1$, $L'_i \in \mathcal{C}$, since $\mathcal{C}$ is closed under infinite intersections by Lemma 10. It is easy to see $L'_1 \subset L'_2 \subset L'_3 \subset \cdots$, which implies that $\mathcal{C}$ has an infinite ascending sequence. This completes the proof.
(3)⇒(1) : By Theorem 9. □

Corollary 1. *2^U is upper-best approximately identifiable in the limit from complete data by an indexed family $\mathcal{C}$ iff 2^U is upper-best approximately identifiable in the limit from positive data by $\mathcal{C}$.*

Proof. Note that the upper-best approximate identifiability requires the u.b.a.p. of $\mathcal{C}$. □

Thus, it is remarkable that the upper-best approximate identifiability from *complete* data is collapsed into the upper-best approximate identifiability from *positive* data.

Consider once again length-bounded EFS's. We note the following lemma without proof.

Lemma 11. *$\mathcal{I}nt(\mathcal{L}^n)$ is closed under infinite intersections.*

Theorem 12. *For any $n \geq 1$, 2^U is upper-best approximately identifiable in the limit from positive data by $Intsct(\mathcal{L}^n)$).*

Proof. By Lemma 11 and Lemma 10, we have $Intsct(\mathcal{L}^n) \overset{u.b.}{\to} 2^U$. Recall that $\mathcal{L}^n$ has finite elasticity. Then, by Proposition 2, we have $Intsct(\mathcal{L}^n) \overset{f.e.}{\to} 2^U$. Therefore, we have the claim by Theorem 9. □

8.6 Concluding Remarks

In this chapter, we gave fundamental theoretical results concerning the identification in the limit model proposed by Gold. This learning framework is widely applied to various learning problems of grammatical devices, such as finite automata ([2][12]), linear grammars([6]), context free grammars ([13][14]), categorical grammars([7]), etc. Finite elasticity, proposed by Wright ([20]), is one of the most important notions in inductive inference from positive data. Theoretical important properties of finite elasticity is well studied by Sato ([17]). An interesting approximate learnability of regular languages by k-reversible languages is obtained by Kobayashi and Yokomori ([8]).

References

1. D. Angluin. Inductive Inference of Formal Languages From Positive Data, *Information and Control*, 45 (1980), 117-135.

2. D. Angluin. Inference of Reversible Languages, *Journal of the ACM*, 29 (1982), 741-765.
3. S. Arikawa, T. Shinohara and A. Yamamoto. Learning Elementary Formal Systems, *Theoretical Computer Science*, 95 (1992), 97-113.
4. J. Case and C. Smith. Anomaly Hierarchies of Mechanized Inductive Inference, *Proc. of STOC*, 1978, 314-319.
5. E. Mark Gold. Language Identification in the Limit. *Information and Control*, 10 (1967), 447-474.
6. C. Higuera, J. Oncina. Inferring Deterministic Linear Languages. *Proc. of COLT'02*, 2002, 185-200.
7. M. Kanazawa. Identification in the Limit of Categorial Grammars. *Journal of Logic, Language and Information*, 5, 2 (1996), 115-155.
8. S. Kobayashi and T. Yokomori. Learning Approximately Regular Languages with Reversible Languages, *Theor. Comput. Sci.* 174, 1-2 (1997), 251-257.
9. S. Kobayashi and T.Yokomori. On Approximately Identifying Concept Classes in the Limit. *Proc. of 6th International Workshop on Algorithmic Learning Theory*, (*Lecture Notes in Artificial Intelligence* 997), Springer-Verlag, 1995, 298-312.
10. T. Motoki, T. Shinohara and K. Wright. The Correct Definition of Finite Elasticity: Corrigendum to Identification of Unions. *Proc. of 4th Workshop on Computational Learning Theory*, 1991, 375-375.
11. Y. Mukouchi. Inductive Inference of an Approximate Concept from Positive Data. *Proc. of 5th International Workshop on Algorithmic Learning Theory*, Lecture Notes in Artificial Intelligence **872**, Springer-Verlag, Berlin, 1994, 484-499.
12. J. Oncina and P. Garcia. Inferring Regular Languages in Polynomial Update Time, *Pattern Recognition and Image Analysis*, World Scientific, 1991, 49-61.
13. Y. Sakakibara. Learning Context-Free Grammars from Structural Data in Polynomial Time, *Theor. Comput. Sci.*, 76, 2-3 (1990), 223-242.
14. Y. Sakakibara and M Kondo. GA-based Learning of Context-Free Grammars using Tabular Representations. *Proc. of International Conference on Machine Learning'99*, 1999, 354-360.
15. M. Sato and K. Umayahara. Inductive Inferability for Formal Languages from Positive Data. *IEICE Trans. Inf. & Syst.*, E75-D, 4 (1992), 415-419.
16. M. Sato and T. Moriyama. Inductive Inference of Length Bounded EFS's from Positive Data. DMSIS-RR-94-2, Department of Mathematical Sciences and Information Sciences, Univ. of Osaka Pref., Japan, April 1994
17. M. Sato. Inductive Inference of Formal Languages. *Bulletin of Informatics and Cybernetics*, 27, 1 (1995).
18. T. Shinohara. Rich Classes Inferable from Positive Data : Length Bounded Elementary Formal Systems. *Information and Computation*, 108 (1994), 175-186.
19. R. M. Smullyan. *Theory of Formal Systems*. Princeton Univ. Press, 1961.
20. K. Wright. Identification of Unions of Languages Drawn from an Identifiable Class. *Proc. of 2nd Workshop on Computational Learning Theory*, 1989, 328-333.

9

Some Essentials of Graph Transformation*

Hans-Jörg Kreowski, Renate Klempien-Hinrichs, and Sabine Kuske

Department of Computer Science
University of Bremen
P.O.Box 33 04 40, 28334 Bremen, Germany
E-mail: {kreo,rena,kuske}@informatik.uni-bremen.de

Summary. This chapter introduces rule-based graph transformation, which constitutes a well-studied research area in computer science. The chapter presents the most fundamental definitions and illustrates them with some selected examples. It presents also the concept of transformation units, which makes pure graph transformation more feasible for specification and modeling aspects. Moreover, a translation of Chomsky grammars into graph grammars is given and the main theorems concerning parallelism and concurrency are presented. Finally, an introduction to hyperedge replacement is given, a concept which has nice properties because it transforms hypergraphs in a context-free way.

9.1 Introduction

Graphs are a well-established means in computer science for representing data structures, states of concurrent and distributed systems, or more generally sets of objects with relations between them. Famous examples of graphs are Petri nets, flow diagrams, Entity-Relationship diagrams, finite automata, and UML diagrams.

In many situations one does not only want to employ graphs as a static structure, but also to transform them e.g. by firing transitions in the case of Petri nets or UML state diagrams, or by generating or deleting objects and links in the case of UML object diagrams. The area of graph transformation brings together the concepts of graphs and rules with various methods from the theory of formal languages and from the theory of concurrency, and with a spectrum of applications, see the three volumes of the Handbook of Graph

* Research partially supported by the EC Research Training Network SegraVis (Syntactic and Semantic Integration of Visual Modeling Techniques) and the Collaborative Research Centre 637 (Autonomous Cooperating Logistic Processes: A Paradigm Shift and Its Limitations) funded by the German Research Foundation (DFG).

H.-J. Kreowski, et al.: *Some Essentials of Graph Transformation*, Studies in Computational Intelligence (SCI) **25**, 229–254 (2006)
www.springerlink.com

Grammars and Computing by Graph Transformation as an overview [29, 7, 10].

In this chapter, we give a survey of some essentials of graph transformation including

- a translation of Chomsky grammars into graph grammars in Section 9.5 showing the computational completeness of graph transformation,
- the basic notions and results on parallelism and concurrency in graph transformation in Section 9.6, and
- a context-free model of graph transformation in Section 9.7.

Unfortunately, graphs are quite generic structures that can be encountered in many variants in the literature, and there are also many ways to apply rules to graphs. One cannot deal with all possibilities in an introductory survey. Therefore, we focus on directed, edge-labeled graphs (Section 9.2) and on rule application in the sense of the so-called double-pushout approach (Section 9.3). The directed, edge-labeled graphs can be specialized into many other types of graphs. And the double-pushout approach (which is introduced here by means of set-theoretic constructions on graphs without reference to categorical concepts) is one of the most frequently used approaches. In Section 9.4, we define graph grammars as a language-generating device and the more general notion of a transformation unit that models binary relations on graphs.

9.2 Graphs and the Need to Transform Them

Graphs are well-suited and frequently-used structures to represent complex relations between objects of various kinds. They are the central structures of interest in at least four areas of mathematics and computer science: graph theory (see, e.g., Harary [16]), graph algorithms (see, e.g., [13]), Petri nets (see, e.g., [27, 14]), and graph transformation (see, e.g., the Handbooks on Graph Grammars and Computing by Graph Transformation [29, 7, 10]). But they are also popular and useful in many other disciplines like biology, chemistry, economics, logistics, engineering, and many others.

Maps are typical examples of structures that are often represented by graphs. Already in 1736, Euler formulated the *Königsberger Brückenproblem* concerning the map of Königsberg, which consists of four areas that are separated from each other by the two arms of the river Pregel. There are seven bridges connecting two areas each, and the question is whether one can walk around passing each bridge exactly once. This becomes a graph problem if the areas are considered as nodes and the bridges as edges between the corresponding nodes. A sketch of the map and the respective graph are shown in the left side of Figure 9.1. For such graphs, the general question (known as the Eulerian Cycle Problem) is whether there is a cycle passing each edge exactly once. Similarly, maps of countries can be represented as graphs by considering the countries as nodes and by connecting each two nodes with an edge the corresponding countries of which share a borderline. In this way the famous Four-Color-Problem of maps becomes the Four-Color-Problem of graphs (see, e.g., [13, 2]). Finally, road maps are nicely represented as graphs by considering sites as nodes and a road that connects two sites directly as an edge that may be labeled with the distance. Such graphs are the basic data structures for various transportation and tour planning problems.

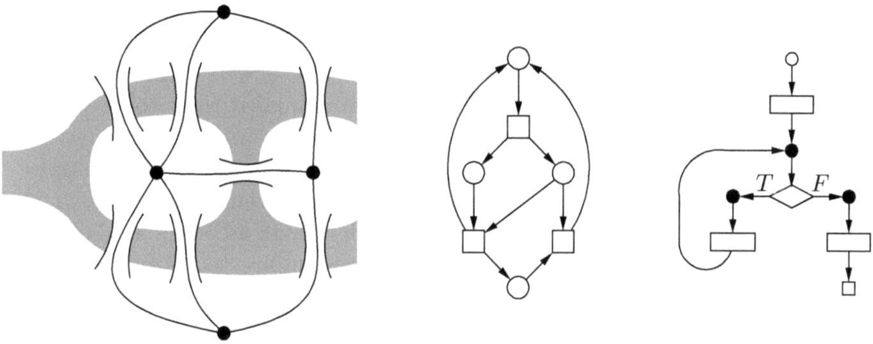

Fig. 9.1. Various graphs

Another typical example of graphs are Petri nets, which allow one to model concurrent and distributed systems (see, e.g., [28]). A Petri net is a simple bipartite graph meaning that there are two types of nodes, called conditions and events or places and transitions, and a set of edges, called flow relation, which connect nodes of distinct types only. The middle graph of Figure 9.1 shows a sample Petri net; as usual, round nodes represent conditions or places, and square nodes represent events or transitions.

A further example of graphs are well-structured flow diagrams such as the right graph in Figure 9.1. Such a graph has an entry node (the circle) and an exit node (the square), boxes representing statements, rhombs representing tests, and auxiliary nodes in between each two linked boxes or box and rhomb. The edges represent the control flow. No edge leaves the exit. Each rhomb is left by two edges representing the test results *TRUE* and *FALSE*, respectively. Each other node is left by a single edge. Each rhomb is the test of a *while*-loop meaning that the *TRUE*-edge starts a path that ends at the node immediately before the rhomb. Like well-structured flow diagrams, many other kinds of diagrams including all the UML diagrams may be represented by and seen as graphs.

Graphs are quite generic structures which can be encountered in the literature in many variants: directed and undirected, labeled and unlabeled, simple and multiple, with binary edges and hyperedges, etc. In this survey, we focus on directed, edge-labeled, and multiple graphs with binary edges.

9.2.1 Graphs

Let Σ be a set of labels. A (multiple directed edge-labeled) *graph* over Σ is a system $G = (V, E, s, t, l)$ where V is a finite set of *nodes*, E is a finite set of *edges*, $s, t \colon E \to V$ are mappings assigning a *source* $s(e)$ and a *target* $t(e)$ to every edge in E, and $l \colon E \to \Sigma$ is a mapping assigning a label to every edge in E. An edge e with $s(e) = t(e)$ is also called a *loop*. The components V, E, s, t, and l of G are also denoted by V_G, E_G, s_G, t_G, and l_G, respectively. The set of all graphs over Σ is denoted by $\mathcal{G}_\Sigma$.

The notion of multiple directed edge-labeled graphs with binary edges is flexible enough to cover other types of graphs. *Simple graphs* form a subclass consisting of those graphs two edges of which are equal if their sources and their targets are equal respectively. A label of a loop can be interpreted as a label of the node to which the loop is attached so that *node-labeled graphs* are covered. On the other hand, we assume a particular label $*$ which is omitted in drawings of graphs. In this way, graphs where all edges are labeled with $*$ may be seen as *unlabeled graphs*. Moreover, undirected graphs can be represented by directed graphs if one replaces each undirected edge by two directed edges attached to the same two nodes, but in opposite directions. Finally, even *hypergraphs* can be handled by the introduced type of graphs as done explicitly in Section 9.7.

If graphs are the structures of interest, it is rarely the case that just a single static graph is considered. Rather, graphs may be the inputs of algorithms and processes, so that means are needed to search and manipulate graphs. Graphs may represent states of systems, so that means for updates and state transitions are needed. Or graph languages are in the center of consideration like the set of all well-structured flow diagrams or all Petri nets or all connected and planar graphs. Like in the case of string languages, one needs means to generate and recognize graph languages. To meet all these needs, rule-based graph transformation is defined in the next section. This requires some prerequisites to deal with graphs, which are introduced in the following.

9.2.2 Subgraphs

A graph $G \in \mathcal{G}_\Sigma$ is a *subgraph* of a graph $H \in \mathcal{G}_\Sigma$, denoted by $G \subseteq H$, if $V_G \subseteq V_H$, $E_G \subseteq E_H$, $s_G(e) = s_H(e)$, $t_G(e) = t_H(e)$, and $l_G(e) = l_H(e)$ for all $e \in E_G$. In drawings of graphs and subgraphs, shapes, colors, and names will be used to indicate the identical nodes and edges.

Given a graph, a subgraph is obtained by removing some nodes and edges subject to the condition that the removal of a node is accompanied by the removal of all its incident edges. More formally, let $G = (V, E, s, t, l)$ be a graph and $X = (V_X, E_X) \subseteq (V, E)$ be a pair of sets of nodes and edges. Then $G - X = (V - V_X, E - E_X, s', t', l')$ with $s'(e) = s(e)$, $t'(e) = t(e)$, and $l'(e) = l(e)$ for all $e \in E - E_X$ is a subgraph of G if and only there is no

$e \in E - E_X$ with $s(e) \in V_X$ or $t(e) \in V_X$. This condition is called *contact condition* of X in G.

In other words, two subsets of nodes and edges $Y = (V_Y, E_Y) \subseteq (V, E)$ induce a subgraph $Y^\bullet = (V_Y, E_Y, s', t', l') \subseteq G$ with $s'(e) = s(e), t'(e) = t(e)$, and $l'(e) = l(e)$ for all $e \in E_Y$ if and only if $(V - V_Y, E - E_Y)$ satisfies the contact condition in G, i.e. there is no edge $e \in E_Y$ with $s(e) \in V - V_Y$ or $t(e) \in V - V_Y$.

9.2.3 Graph Morphisms

For graphs $G, H \in \mathcal{G}_\Sigma$ a *graph morphism* $g\colon G \to H$ is a pair of mappings $g_V\colon V_G \to V_H$ and $g_E\colon E_G \to E_H$ that are structure-preserving, i.e. $g_V(s_G(e)) = s_H(g_E(e))$, $g_V(t_G(e)) = t_H(g_E(e))$, and $l_H(g_E(e)) = l_G(e)$ for all $e \in E_G$. We will usually write $g(v)$ and $g(e)$ for nodes $v \in V_G$ and edges $e \in E_G$ since the indices V and E can be reconstructed easily from the type of the argument.

For a graph morphism $g\colon G \to H$ the image of G in H is called a *match* of G in H, i.e. the match of G with respect to the morphism g is the subgraph $g(G) \subseteq H$ which is induced by $(g(V), g(E))$. The corresponding contact condition is satisfied because g preserves the structure of graphs.

Given $F \subseteq G$, then the two inclusions of the sets of nodes and edges define a graph morphism. It is also easy to see that the (componentwise) sequential composition of two graph morphisms $f\colon F \to G$ and $g\colon G \to H$ yields a graph morphism $g \circ f\colon F \to H$. Consequently, if f is the inclusion w.r.t. $F \subseteq G$, $g(F)$ is the match of F in H w.r.t. g restricted to F.

9.2.4 Extension of Graphs

Instead of removing nodes and edges, one may add some nodes and edges to extend a graph such that the given graph is a subgraph of the extension. The addition of nodes causes no problem at all, whereas the addition of edges requires the specification of their labels, sources, and targets, where the latter two may be given or new nodes. Let $G = (V, E, s, t, l)$ be a graph and (V', E', s', t', l') be a structure consisting of two sets V' and E' and three mappings $s'\colon E' \to V \uplus V'$, $t'\colon E' \to V \uplus V'$, and $l'\colon E' \to \Sigma$ (where $\uplus$ denotes the disjoint union of sets). Then $H = G + (V', E', s', t', l') = (V \uplus V', E \uplus E', s'', t'', l'')$ is a graph with $G \subseteq H$ (which establishes the definition of the three mappings s'', t'', l'' on E) and $s''(e') = s'(e')$, $t''(e') = t'(e')$, and $l''(e') = l'(e')$ for all $e' \in E'$.

9.2.5 Disjoint Union

If G is extended by a full graph $G' = (V', E', s', t', l')$, the graph $G + G'$ is the *disjoint union* of G and G'. Note that in this case s' and t' map E' to V' rather than $V \uplus V'$, but V' is included in $V \uplus V'$ such that the extension works.

The disjoint union of graphs puts graphs together without any interconnection. If graphs are disjoint, their disjoint union is just the union. If they are not disjoint, the shared nodes and edges must be made different from each other. Because this part of the construction is not made explicit, the disjoint union is only unique up to isomorphism, i.e. up to naming. Nevertheless, the disjoint union of graphs has got some useful properties. It is associative and commutative. If $G_1 + G_2$ is the disjoint union of G_1 and G_2, there are two inclusions $incl_i \colon G_i \to G_1 + G_2$. And whenever one has two graph morphisms $g_i \colon G_i \to G$ into some graph G, there exists a unique graph morphism $g \colon G_1 + G_2 \to G$ with $g \circ incl_i = g_i$ for $i = 1, 2$. This property is the categorical characterization of the disjoint union up to isomorphism.

9.3 Rule-Based Transformation of Graphs

Graph transformation is a rule-based method that performs local changes on graphs. With graph transformation rules it is possible to specify formally and visually for instance the semantics of rule-based systems (like the firing of transitions in Petri nets or in state charts, or the semantics of functional languages), specific graph languages (like the set of all well-formed flow graphs), graph algorithms (like the search of all Eulerian cycles in a graph), and many more.

9.3.1 Graph Transformation Rule

The idea of a graph transformation rule is to express which part of a graph is to be replaced by another graph. Unlike strings, a subgraph to be replaced can be linked in many ways (i.e., by many edges) with the surrounding graph. Consequently, a rule also has to specify which kind of links are allowed; this is done with the help of a third graph that is common to the replaced and the replacing graph and requires that the surrounding graph may be linked to the replaced graph only with edges incident to this third graph.

Formally, a *rule* $r = (L \supseteq K \subseteq R)$ consists of three graphs $L, K, R \in \mathcal{G}_\Sigma$ such that K is a subgraph of L and R. The components L, K, and R of r are called *left-hand side*, *gluing graph*, and *right-hand side*, respectively.

Example 1 (flow diagrams). Figure 9.2 shows two rules representing the replacement of a single statement in a flow diagram by a more complex instruction: the statement is replaced by two consecutive statements with $r_{compound}$, and by a while-loop with $r_{while\text{-}do}$. For both rules, the gluing graph consists of two nodes that can be located in the respective left- and right-hand sides by their shape and color.

Example 2 (shortest paths). Figure 9.3 shows the two essential rules for the computation of shortest paths in distance graphs, that is graphs labeled with

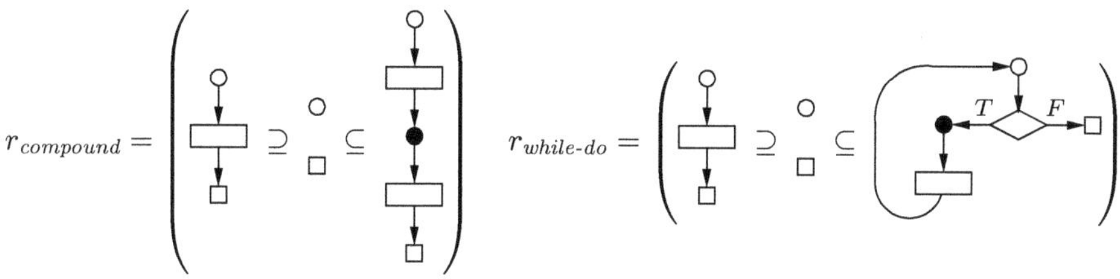

Fig. 9.2. Graph transformation rules for the construction of flow diagrams

non-negative integers. The first rule adds a direct connection between each two nodes that are connected by a path of length 2 and sums the distances up. Using this rule, one can compute the transitive closure of the given distance graph. If one applies the second rule, which chooses the shortest connection of two direct connections as long as possible, one ends up with shortest connections between each two nodes.

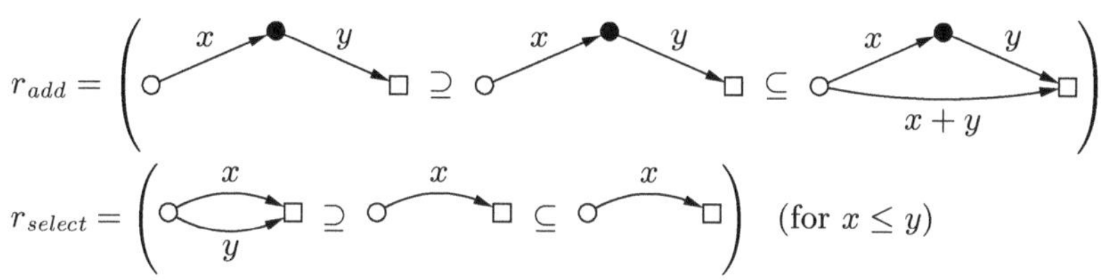

Fig. 9.3. Graph transformation rules for the computation of shortest paths

In practice one often has the special case where the gluing graph of a rule $r = (L \supseteq K \subseteq R)$ is a set of nodes. In this case the graphical representation of r may omit the gluing graph K by depicting only the graphs L and R, with numbers uniquely identifying the nodes in K. As an example, the rule $r_{\textit{while-do}}$ from Figure 9.2 is drawn in Figure 9.4 using this alternative representation.

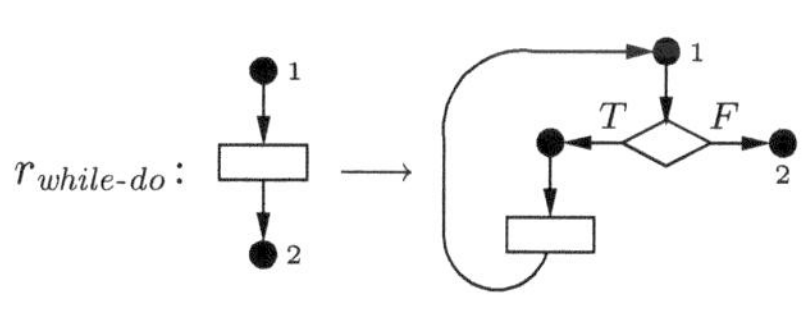

Fig. 9.4. Alternative representation of the rule $r_{\textit{while-do}}$

9.3.2 Application of a Graph Transformation Rule

The application of a graph transformation rule to a graph G consists of replacing a match of the left-hand side in G by the right-hand side such that the match of the gluing graph is kept. Hence, the application of $r = (L \supseteq K \subseteq R)$ to a graph $G = (V, E, s, t, l)$ comprises the following three steps.

1. A graph morphism $g\colon L \to G$ is chosen to establish a match of L in G subject to the following two *application conditions*:
   a) *Contact condition* of $g(L) - g(K) = (g(V_L) - g(V_K), g(E_L) - g(E_K))$ in G; and
   b) *Identification condition.* If two nodes or edges of L are identified in the match of L they must be in K.
2. Now the match of L up to $g(K)$ is removed from G, resulting in a new intermediate graph $Z = G - (g(L) - g(K))$.
3. Afterwards the right-hand side R is added to Z by gluing Z with R in $g(K)$ yielding the graph $H = Z + (R - K, g)$ where $(R - K, g) = (V_R - V_K, E_R - E_K, s', t', l')$ with $s'(e') = s_R(e')$ if $s_R(e') \in V_R - V_K$ and $s'(e') = g(s_R(e'))$ otherwise, $t'(e') = t_R(e')$ if $t_R(e') \in V_R - V_K$ and $t'(e') = g(t_R(e'))$ otherwise, and $l'(e') = l_R(e')$ for all $e' \in E_R - E_K$.

The contact condition guarantees that the removal of $g(L) - g(K)$, yields a subgraph of G. The identification condition is not needed for the construction of a direct derivation, but will be helpful in dealing with parallel rules as considered in Section 9.6. The extension of Z to H is properly defined because s' and t' map the edges of $E_R - E_K$ into nodes of $V_R - V_K$ or $g(V_K)$ which is part of V_Z.

Example 3 (flow diagrams). Figure 9.5 shows an application of the rule $r_{\mathit{while\text{-}do}}$ to a flow graph representing a sequence of three statements. The gray areas indicate the match (left), its parts belonging to the image of the gluing graph (middle), and the right-hand side (right).

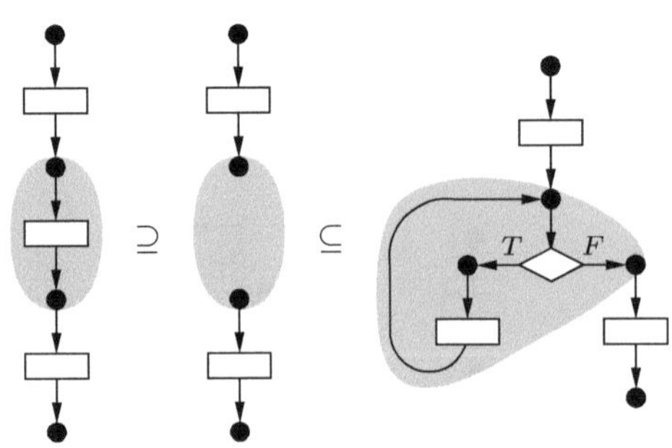

Fig. 9.5. An application of the rule $r_{\mathit{while\text{-}do}}$

A rule application of $r = (L \supseteq K \subseteq R)$ can be depicted by the following diagram where the graph morphisms $d\colon K \to Z$ and $h\colon R \to H$ are given by $d(v) = g(v)$ for all $v \in V_K$, $d(e) = g(e)$ for all $e \in E_K$, $h(v) = d(v)$ if $v \in V_K$, $h(v) = v$ if $v \in V_R - V_K$, $h(e) = d(e)$ if $e \in E_K$, and $h(e) = e$ if $e \in E_R - E_K$.

$$\begin{array}{ccccc} L & \supseteq & K & \subseteq & R \\ \downarrow g & & \downarrow d & & \downarrow h \\ G & \supseteq & Z & \subseteq & H \end{array}$$

It is worth noting that if the subgraph relations in the diagram are interpreted as inclusion morphisms, both squares of the diagram are pushouts in the category of graphs. This is why the presented approach is also called double-pushout approach (cf. [3]). Here the identification condition is significant because the left diagram is not a pushout if g does not obey the identification condition.

9.3.3 Derivation and Application Sequence

The application of a rule r to a graph G is denoted by $G \underset{r}{\Longrightarrow} H$ where H is a graph resulting from an application of r to G. A rule application is called a *direct derivation*, and the iteration of direct derivations $G_0 \underset{r_1}{\Longrightarrow} G_1 \underset{r_2}{\Longrightarrow} \cdots \underset{r_n}{\Longrightarrow} G_n$ $(n \in \mathbb{N})$ is called a *derivation* from G_0 to G_n. As usual, the derivation from G_0 to G_n can also be denoted by $G_0 \overset{n}{\underset{P}{\Longrightarrow}} G_n$ where $\{r_1, \ldots, r_n\} \subseteq P$, or by $G_0 \overset{*}{\underset{P}{\Longrightarrow}} G_n$ if the number of direct derivations is not of interest.

The string $r_1 \cdots r_n$ is called an *application sequence* of the derivation $G_0 \underset{r_1}{\Longrightarrow} G_1 \underset{r_2}{\Longrightarrow} \cdots \underset{r_n}{\Longrightarrow} G_n$.

Example 4 (flow diagrams). Figure 9.6 contains a derivation using the rules from Figure 9.2. Its last direct derivation is the one detailed in Figure 9.5, and the application sequence of the whole derivation is $r_{compound}\, r_{compound}\, r_{while\text{-}do}$.

In the literature one encounters various approaches to graph transformation, among them specific ones, like edge replacement [6] or node replacement [12], and general ones, like the double-pushout approach [3], the single-pushout approach [9], or the PROGRES approach [30].

9.4 Graph Grammars and Graph Transformation Units

Analogously to Chomsky grammars in formal language theory, graph transformation can be used to generate graph languages. A graph grammar consists

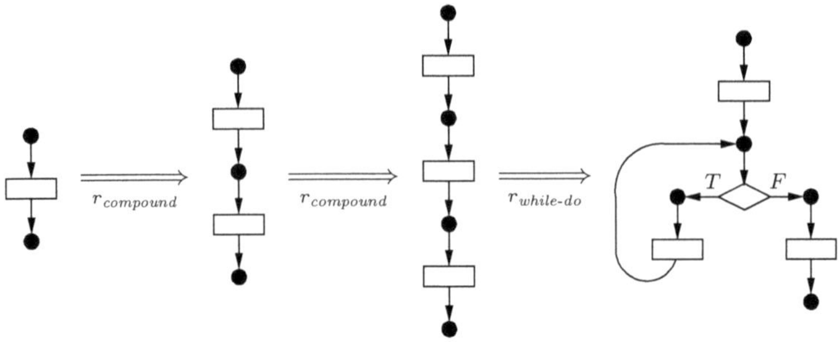

Fig. 9.6. A derivation of a flow diagram

of a set of rules, a start graph, and a terminal expression fixing the set of terminal graphs. Such a terminal expression may consist of a set $\Delta \subseteq \Sigma$ of terminal labels admitting all graphs that are labeled over Δ.

9.4.1 Graph Grammar

A *graph grammar* is a system $GG = (S, P, \Delta)$ where $S \in \mathcal{G}_\Sigma$ is the *initial graph* of GG, P is a finite set of graph transformation rules, and $\Delta \subseteq \Sigma$ is a set of *terminal symbols.* The *generated language* of GG consists of all graphs $G \in \mathcal{G}_\Sigma$ that are labeled over Δ and that are derivable from the initial graph S via successive application of the rules in P, i.e. $L(GG) = \{G \in \mathcal{G}_\Delta \mid S \overset{*}{\underset{P}{\Longrightarrow}} G\}$.

Example 5 (connected graphs). As an example of a graph grammar consider *connected* $= (\bullet, P, \{*\})$ where the start graph consists of a single node and the terminal expression allows all graphs labeled only with $*$. Recall that the symbol $*$ denotes a special label in Σ standing for *unlabeled* and being invisible in displayed graphs. The rules in $P = \{p_1, p_2, p_3\}$ are depicted in Figure 9.7. The rule p_1 adds a node v and an edge e such that v is the target of e, and takes as source of e an already existing node. The rule p_2 is similar, the only difference being that the direction of the new edge e is inverted. The third rule p_3 generates a new edge between two existing nodes. The new edge can also be a loop if the two nodes in the left-hand side of p_3 are identified, i.e. if they are one and the same node in the match of the left-hand side. It can be shown that the generated language of *connected*, $L($*connected*$)$, consists of all non-empty connected unlabeled graphs.

Example 6 (Petri nets). A place/transition system (N, m_0) consists of a Petri net $N = (S, T, F)$ with a set of places S, a set of transitions T, and a flow relation $F \subseteq (S \times T) \cup (T \times S)$, and an initial marking $m_0 \colon S \to \mathbb{N}$. To model the firing of transitions by graph transformation in the introduced way, one may represent a net $N = (S, T, F)$ with a marking $m \colon S \to \mathbb{N}$ by the graph $G(N, m)$ as indicated in Figure 9.8 where all tokens become nodes. Moreover,

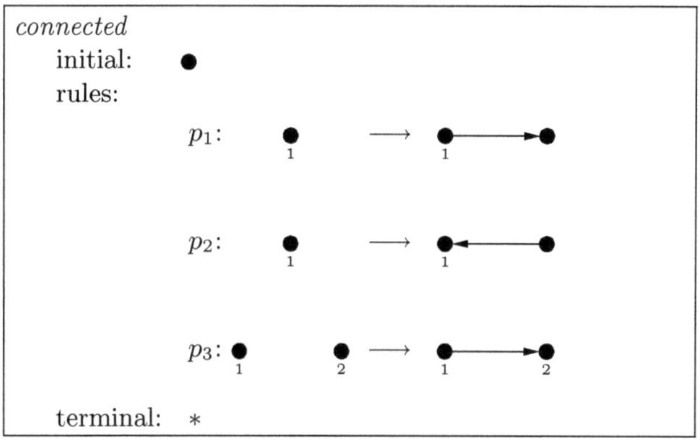

Fig. 9.7. A graph grammar generating connected graphs

all places and transitions of N are labeled with their respective names (i.e., a loop – that is not drawn here – carrying the respective label is attached).

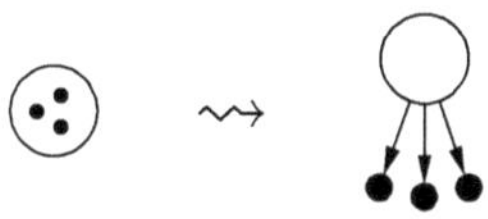

Fig. 9.8. Turning tokens into nodes

Then we can transform a place/transition system (N, m_0) into a graph grammar $GG(N, m_0) = (G(N, m_0), P(N), P \cup T \cup \{*\})$ where $P(N)$ contains a rule $r(t)$ for each transition t as depicted in Figure 9.9. Note that the labels attached to the nodes are needed to make sure that distinct pre- or postplaces cannot be identified in a match.

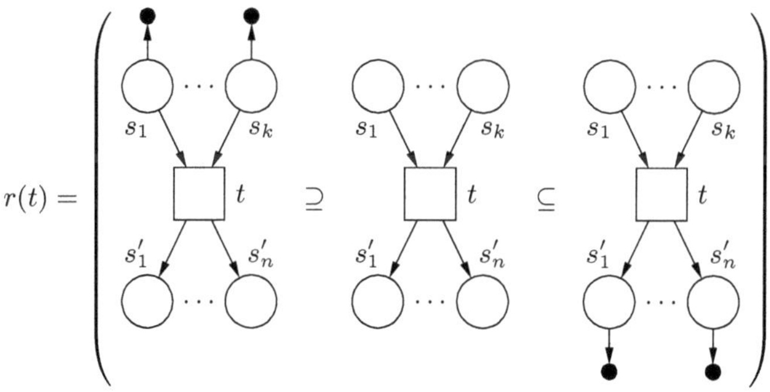

Fig. 9.9. Firing rule for a transition t with ${}^\bullet t = \{s_1, \ldots, s_k\}$ and $t^\bullet = \{s'_1, \ldots, s'_n\}$

It is not difficult to show that a marking m is reachable from m_0 if and only if there is a derivation $G(N, m_0) \overset{*}{\Longrightarrow} G(N, m)$. Thus, $L(GG(N, m_0))$ consists of all net representations $G(N, m)$ where m is a marking reachable from m_0 in N.

One may also model Petri nets with weights assigned to the edges of the flow relation in this way. The left-hand side of the firing rule for a transition t has as many token nodes attached to each preplace s of t as the weight assigned to the edge (s, t) of the flow relation specifies, and analogously for the right-hand side and the postplaces. Then the identification condition ensures that token nodes cannot be identified in a match, so that the application of such a rule works correctly.

As for formal string languages, one does not only want to generate languages, but also to recognize them or to verify certain properties. Moreover, for modeling and specification aspects one wants to have additional features like the possibility to cut down the non-determinism inherent in rule-based graph transformation. This can be achieved with the concept of transformation units (see, e.g., [1, 20, 21]), which generalize graph grammars in the following aspects:

- Transformation units allow a set of initial graphs instead of a single one.
- The class of terminal graphs can be specified in a more general way.
- The derivation process can be controlled.
- Transformation units provide a structuring concept for graph transformation.
- Transformation units do not depend on a specific graph transformation approach.

The first two points are achieved by replacing the initial graph and the terminal alphabet of a graph grammar by a graph class expression specifying sets of initial and terminal graphs. The regulation of rule application is obtained by means of so-called control conditions.

9.4.2 Graph Class Expressions

A *graph class expression* may be any syntactic entity X that specifies a class of graphs $SEM(X) \subseteq \mathcal{G}_\Sigma$. A typical example is the above-mentioned subset $\Delta \subseteq \Sigma$ with $SEM(\Delta) = \mathcal{G}_\Delta \subseteq \mathcal{G}_\Sigma$. Similarly, the expression *all edges* x for $x \in \Sigma$ specifies the class of all graphs G with $l_G(e) = x$ for every $e \in E_G$, i.e. all edges are labeled with x. Another useful type of graph class expressions is given by sets of rules. More precisely, for a set P of rules $SEM(P)$ contains all P-*reduced* graphs, i.e. graphs to which none of the rules in P can be applied. Finally, it is worth noting that a graph grammar GG itself may serve as a graph class expression with $SEM(GG) = L(GG)$.

9.4.3 Control Conditions

A *control condition* may be any syntactic entity that cuts down the derivation process. A typical example is a regular expression over a set of rules (or any other string-language-defining device). Let C be a regular expression specifying the language $L(C)$. Then a derivation with application sequence s is *permitted* by C if $s \in L(C)$. As a special case of this type of control condition, the condition *true* allows every application sequence, i.e. $L(C) = P^*$ where P is the set of underlying graph transformation rules. Another useful control condition is *as long as possible*, which requires that all rules be applied as long as possible. More precisely, let P be the set of underlying rules. Then $SEM(\textit{as long as possible})$ allows all derivations $G \underset{P}{\Longrightarrow} G'$ such that no rule of P is applicable to G'. Hence, this control condition is similar to the graph class expression P introduced above. Also similar to *as long as possible* are *priorities* on rules being partial orders on rules such that if $p_1 > p_2$, p_1 must be applied as long as possible before any application of p_2. More details on control conditions for transformation units can be found in [23].

Now we have collected all components for defining unstructured transformation units.

9.4.4 Transformation Units

A *transformation unit* (without import) is a system $tu = (I, P, C, T)$ where I and T are graph class expressions to specify the *initial* and the *terminal* graphs respectively, P is a set of rules, and C is a control condition.

Such a transformation unit specifies a binary relation $SEM(tu) \subseteq SEM(I) \times SEM(T)$ that contains a pair (G, H) of graphs if and only if there is a derivation $G \overset{*}{\underset{P}{\Longrightarrow}} H$ permitted by C.

Example 7 (Eulerian graphs). As an example consider the transformation unit *Eulerian* shown in Figure 9.10. It takes as initial graphs all those generated by the graph grammar *connected* introduced in Example 5, i.e. all connected unlabeled graphs. The terminal graphs are all graphs whose edges are labeled only with *ok*. The five rules label edges in a specific way, and the control condition requires that the rules $p_2, \ldots, p_5$ can only be applied if p_1 is not applicable. In more detail, the transformation unit *Eulerian* checks whether every node in a connected graph has the same number of incoming and outgoing edges, where loops count as one incoming and one outgoing edge. Hence it checks whether the input graphs are Eulerian. The rule p_1 labels loops with *ok*. The control condition requires that first all loops are labeled. This is necessary because otherwise a rule out of $p_2, \ldots, p_5$ could also be applied to a loop (by identifying the target and the source node of an edge in a match of a left-hand side), which would also admit non-Eulerian graphs as output. The label s in the rules indicates that the edge has already been counted as outgoing edge of its source node. Analogously, t means that it has been counted

Eulerian
initial: *connected*
rules:

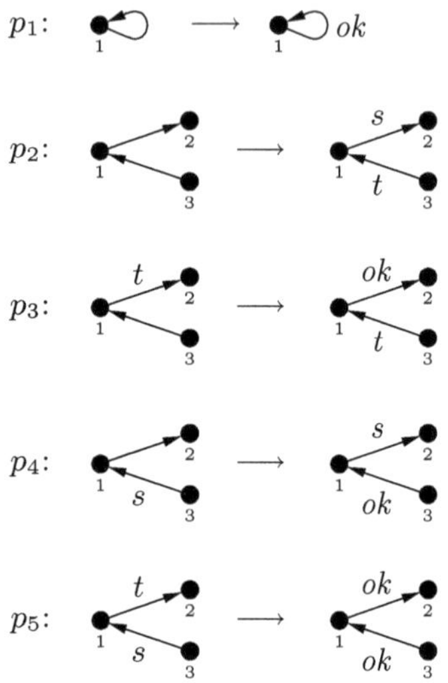

cond: $p_1 > p_x$ for every $x \in \{2, \ldots, 5\}$
terminal: *all edges ok*

Fig. 9.10. A transformation unit that specifies Eulerian graphs

as incoming edge of its target. The label *ok* means that the edge has been counted once for its source and once for its target.

In every rule application there is a node for which exactly one incoming and one outgoing edge are counted. Hence, for each node the difference between the number of uncounted in- and outgoing edges is invariant under a transformation step. Conversely, it can be shown by induction on the number of simple cycles that every Eulerian graph is recognized by the unit *Eulerian*. Hence, the semantics of *Eulerian* consists of all pairs (G, G') where G is connected, unlabeled, and Eulerian, and G' is obtained from G by labeling every edge with *ok*.

It is worth noting that in general transformation units have an import component that allows to use the semantic relations specified by other units as transformation steps [21]. Moreover, transformation units have been generalized to arbitrary m, n-relations on graphs in [17, 18], to distributed graph transformation in [19], to other data structures than graphs in [22], and to parameterized transformation units in [25]. A thorough study of transformation units can also be found in [24].

9.5 Transformation of Chomsky Grammars into Graph Grammars

Intuitively, it may be clear that graph transformation is computationally complete. This claim is made precise in this section by translating Chomsky grammars into graph grammars. The translation is based on the observation that a string $x = a_1 \cdots a_k$ with $a_i \in \Sigma$ for $i = 1, \ldots, k$ can be represented by a so-called string graph $x^\bullet$ that consists of $k+1$ nodes and k edges, where for $i = 1, \ldots, k$ the source of the ith edge is the ith node, the target is the $(i+1)$th node, and the label is a_i (see Figure 9.11). The first and the last nodes are denoted by $b(x^\bullet)$ and $e(x^\bullet)$, respectively.

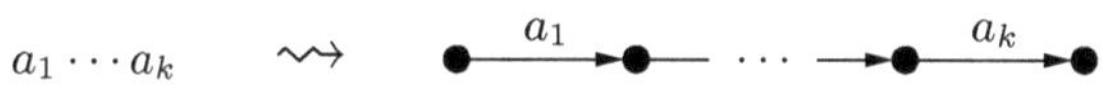

Fig. 9.11. Translating a string into a string graph

Let $CG = (N, T, S, P)$ be a Chomsky grammar. For the sake of convenience, we assume that the right-hand side of every production is not empty, i.e. for all productions $u \to v$ in P we have $v \neq \lambda$. Such a production p is translated into a graph transformation rule r_p as follows. Let $u^\bullet$ and $v^\bullet$ be string graphs associated with u and v, respectively, such that $b(u^\bullet) = b(v^\bullet)$ and $e(u^\bullet) = e(v^\bullet)$. Then let $r_p = (u^\bullet \supseteq be \subseteq v^\bullet)$ with be the graph consisting of the two nodes $b(u^\bullet)$ and $e(u^\bullet)$ be the graph transformation rule associated with p.

Since the edges in a string graph are directed, there exists a match of $u^\bullet$ in a string graph $x^\bullet$ if and only if u is a substring of x. In other words, the rule r_p can be applied to $x^\bullet$ if and only if the production p can be applied to x. The results of the applications correspond, too, so that we have the following theorem.

Theorem 1 (Correct Translation). *Let $CG = (N, T, S, P)$ be a Chomsky grammar with $v \neq \lambda$ for all $u \to v$ in P.*

1. *Let $x, y \in (N \cup T)^*$ and $p \in P$. Then $x \Longrightarrow_p y$ if and only if $x^\bullet \underset{r_p}{\Longrightarrow} y^\bullet$.*
2. *Let $CG^\bullet = (S^\bullet, P^\bullet, T)$ with $P^\bullet = \{r_p \mid p \in P\}$ be the graph grammar associated with CG. Then $L(CG^\bullet) = L(CG)^\bullet = \{x^\bullet \mid x \in L(CG)\}$.*

The reason for excluding productions of the form $u \to \lambda$ from the construction given above is that $\lambda^\bullet$ is a single-node graph with $b(\lambda^\bullet) = e(\lambda^\bullet)$. Since $|u| > 0$, this implies that there is no string graph $u^\bullet$ with $b(u^\bullet) = b(\lambda^\bullet)$ and $e(u^\bullet) = e(\lambda^\bullet)$. We can deal with this problem as follows.

In the case of a Chomsky grammar of type 1 or higher, we may assume w.l.o.g. that there is only the production $S \to \lambda$ with empty right-hand side, and this only if S does not occur in the right-hand side of any other production.

Thus, we may use the graph transformation rule $S^\bullet \supseteq EMPTY \subseteq \lambda^\bullet$ where $EMPTY$ denotes the empty graph.

In the case of a Chomsky grammar of type 0, we may eliminate each production $u \to \lambda$ by replacing it with all productions of the form $ua \to a$ and $au \to a$, where $a \in N \cup T$. If the original grammar generates the empty word, a new axiom S' and productions $S' \to S \mid \lambda$ must be added. Then the construction(s) given above can be used.

As a consequence of Theorem 1, all undecidability results known for Chomsky grammars transfer to graph grammars. In particular, one gets the following results.

Corollary 1 (Undecidability Results). *For graph grammars, the emptiness, finiteness, membership, inclusion, and equivalence problems are undecidable.*

9.6 Parallelism and Concurrency

Parallelism is one of the key concepts of computer science. On the one hand, parallel computing may speed up computational processes such that, for example, data processing problems with exponential time complexity become solvable in polynomial time. On the other hand, parallelism may allow one to model certain applications in a realistic way, like the growing of plants or the transportation of goods. Graph transformation provides a framework in which parallelism can be studied in various respects.

The parallel application of rules is easily introduced into the double-pushout approach because the disjoint unions of rules are rules, which are called parallel rules. Consequently, simultaneous applications of rules are just ordinary direct derivations using parallel rules. Moreover, these parallel derivations have some nice properties with respect to sequentialization and parallelization. Given a direct derivation through a parallel rule, the component rules can be applied in arbitrary order yielding the same result. Conversely, a derivation the steps of which are independent of each other in a certain sense can be composed into a single derivation step. Together, this yields pure concurrency, meaning that independent derivation steps in arbitrary order and the parallel application of the same rules perform the same computation.

9.6.1 Parallel Rules

1. Let $r_i = (L_i \supseteq K_i \subseteq R_i)$ for $i = 1, 2$ be two rules. Then

$$r_1 + r_2 = (L_1 + L_2 \supseteq K_1 + K_2 \subseteq R_1 + R_2)$$

is the *parallel rule* of r_1 and r_2.

2. Let P be a set of rules. Then P_+ denotes the set of parallel rules over P which is recursively defined to be the smallest set with
   (i) $P \subseteq P_+$ and
   (ii) $r_1 + r_2 \in P_+$ for $r_1, r_2 \in P_+$.

The definition of parallel rules makes use of the disjoint union of graphs as defined and discussed in Section 9.2.5. Since the disjoint union of graphs is associative and commutative, the disjoint union of rules is associative and commutative, too. Hence, P_+ can be considered as the free commutative semigroup over P.

Theorem 2 (Sequentialization). *Let $G \underset{r_1+r_2}{\Longrightarrow} H$ be a direct derivation. Then there are two direct derivations $G \underset{r_1}{\Longrightarrow} G_1 \underset{r_2}{\Longrightarrow} H$ for some graph G_1.*

Because $r_1 + r_2 = r_2 + r_1$, Theorem 2 means that there are also two direct derivations $G \underset{r_2}{\Longrightarrow} G_2 \underset{r_1}{\Longrightarrow} H$ for some G_2.

Let $r_i = (L_i \supseteq K_i \subseteq R_i)$ for $i = 1, 2$ be two rules and $incl_i \colon L_i \to L_1 + L_2$ be the inclusions of the left-hand sides into the disjoint union. Let $g \colon L_1 + L_2 \to G$ be the graph morphism underlying the direct derivation $G \underset{r_1+r_2}{\Longrightarrow} H$. Then $g \circ incl_i \colon L_i \to G$ are the graph morphisms inducing the direct derivations $G \underset{r_i}{\Longrightarrow} G_i$. Using the gluing condition of g one can show $g(L_2) \subseteq G_1$ and $g(L_1) \subseteq G_2$. This allows to define graph morphisms $g_2' \colon L_2 \to G_1$ and $g_1' \colon L_1 \to G_2$ which induce the direct derivations $G_1 \underset{r_2}{\Longrightarrow} H_1$ and $G_2 \underset{r_1}{\Longrightarrow} H_2$. Finally, it is not difficult to show that H, H_1 and H_2 are equal up to isomorphism.

Corollary 2. *Let P be a set of rules. Then $\overset{*}{\underset{P}{\Longrightarrow}} = \overset{*}{\underset{P_+}{\Longrightarrow}}$.*

Theorem 2 implies $\underset{P_+}{\Longrightarrow} \subseteq \overset{*}{\underset{P}{\Longrightarrow}}$, and consequently $\overset{*}{\underset{P_+}{\Longrightarrow}} \subseteq \overset{*}{\underset{P}{\Longrightarrow}}$. The converse inclusion follows from $P \subseteq P_+$.

9.6.2 Independence

1. Let $G \underset{r_1}{\Longrightarrow} G_1 \underset{r_2}{\Longrightarrow} H$ be two direct derivations. Let $h_1 \colon R_1 \to G_1$ be the right graph morphism of the first step and $g_2' \colon L_2 \to G_1$ be the (left) graph morphism of the second step. Then the two derivation steps are *sequentially independent* if

$$h_1(R_1) \cap g_2'(L_2) \subseteq h_1(K_1) \cap g_2'(K_2),$$

i.e. the matches overlap in G_1 in gluing parts only.

2. Let $G \underset{r_i}{\Longrightarrow} G_i$ be two direct derivations and $g_i \colon L_i \to G$ their underlying graph morphisms. Then the two derivation steps are *parallel independent* if
$$g_1(L_1) \cap g_2(L_2) \subseteq g_1(K_1) \cap g_2(K_2),$$
i.e. the matches overlap in G in gluing parts only.

Sequential independence is equivalently described by the property
$$h_1(R_1) \cap (g_2'(L_2) \setminus (g_2'(K_2)) = \emptyset = (h_1(R_1) \setminus (h_1(K_1)) \cap g_2'(L_2).$$

In other words, the second step does not remove anything of the right match of the first step and the first step does not add anything of the match of the second step.

Parallel independence is equivalently described by the property
$$g_1(L_1) \cap (g_2(L_2) \setminus g_2(K_2)) = \emptyset = (g_1(L_1) \setminus (g_1(K_1)) \cap (g_2(L_2),$$

meaning that none of the derivation steps removes anything of the match of the other one.

The sequentializations $G \underset{r_1}{\Longrightarrow} G_1 \underset{r_2}{\Longrightarrow} H$ and $G \underset{r_2}{\Longrightarrow} G_2 \underset{r_1}{\Longrightarrow} H$ of a parallel derivation step $G \underset{r_1+r_2}{\Longrightarrow} H$ are sequentially independent, and the two first steps are parallel independent.

Theorem 3 (Parallelization).

1. *Let $G \underset{r_1}{\Longrightarrow} G_1 \underset{r_2}{\Longrightarrow} H$ be two sequentially independent direct derivations. Then there is a direct derivation $G \underset{r_1+r_2}{\Longrightarrow} H$ such that the given derivation is one of its sequentializations.*
2. *Let $G \underset{r_i}{\Longrightarrow} G_i$ for $i = 1, 2$ be two parallel independent direct derivations. Then there is a direct derivation $G \underset{r_1+r_2}{\Longrightarrow} H$ such that the given direct derivations are the first steps of its sequentializations.*

Let $g_1 \colon L_1 \to G$ be the graph morphism of the first step and $g_2' \colon L_2 \to G_1$ be the graph morphism of the second step. Then the sequential independence implies that $g_2'(L_2) \subseteq G$ (up to some renaming of nodes and edges). Therefore the graph morphism $g \colon L_1 + L_2 \to G$ that induces $G \underset{r_1+r_2}{\Longrightarrow} H$ can be defined by $g(x) = g_1(x)$ for x of L_1 and $g(x) = g_2'(x)$ for x of L_2.

Let $g_i \colon L_i \to G$ for $i = 1, 2$ be the graph morphism of the direct derivation $G \underset{r_i}{\Longrightarrow} G_i$. Then the graph morphism $g \colon L_1 + L_2 \to G$ that induces $G \underset{r_1+r_2}{\Longrightarrow} H$ is given by $g(x) = g_i(x)$ for x of L_i. The parallel independence and the application conditions of g_1 and g_2 imply the application conditions of g.

Altogether, the sequentialization and parallelization theorems show that a parallel derivation step, each of two sequentially independent derivations, and two parallel independent direct derivations imply each other.

The situation is illustrated in Figure 9.12, where $||$ indicates independence. This phenomenon is known as *pure concurrency*.

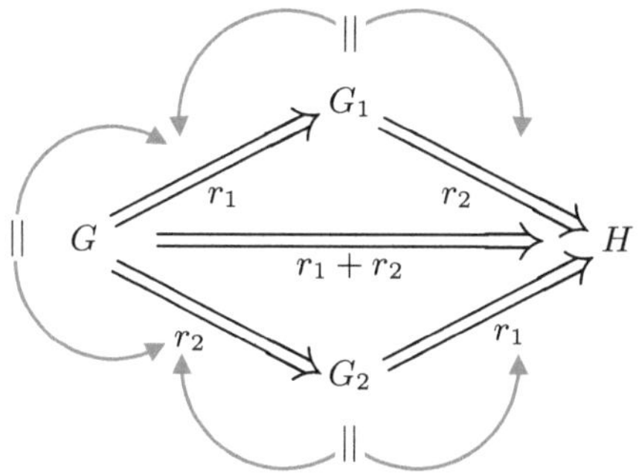

Fig. 9.12. Parallel and sequential independence

Example 8 (Petri nets). In the case of place/transition systems, the only non-gluing items are the token nodes and their incident edges. Hence, the applications of two firing rules for two transitions are independent if and only if they do not share tokens on some common preplace, i.e. if they are concurrent in the sense of net theory. Therefore, the application of the parallel firing rule of the involved transitions – see for an example the parallel rule in Figure 9.13 and its application in Figure 9.14 – corresponds to the parallel firing of these transitions.

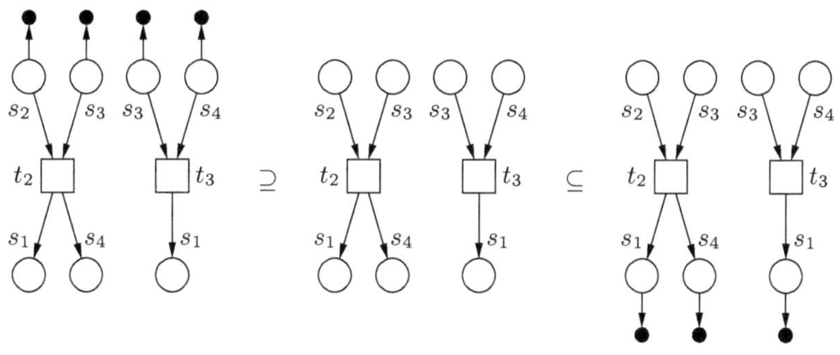

Fig. 9.13. A parallel rule $r(t_2) + r(t_3)$

More about parallelism and concurrency in the line of this section can be found in Corradini et al. [3]. The counterpart for the so-called single-pushout approach is surveyed by Ehrig et al. in [9]. And the third volume of the Handbook of Graph Grammars and Computing by Graph Transformation [10] provides a collection of seven chapters on various graph-transformational approaches to parallelism, concurrency, distribution, and coordination.

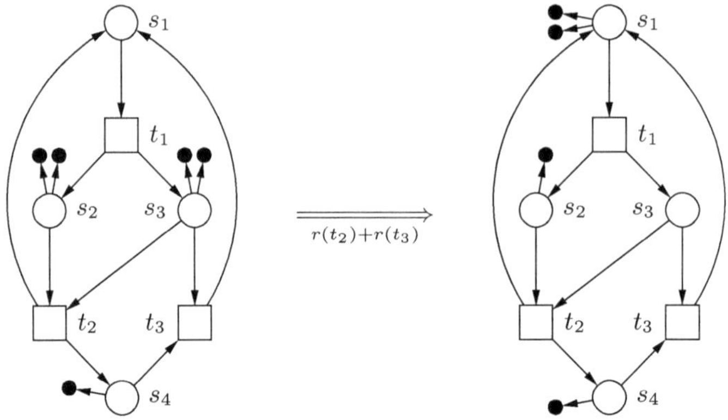

Fig. 9.14. Modeling the parallel firing of transitions t_2 and t_3

9.7 Context-Freeness of Hyperedge Replacement

Graph transformation is a general modeling framework that allows one to specify arbitrary computable relations on graphs. Consequently, any non-trivial semantic property of transformation units and graph grammars is undecidable. If one looks for subclasses with decidable properties (and other nice properties), graph-transformational counterparts of context-freeness are candidates. One of these is hyperedge replacement (see, e.g., [15, 6]), which is usually formulated for hypergraphs the edges of which may be incident to more than two nodes. But hyperedge replacement can also be seen as a special case of graph grammars as introduced in Section 9.4.

To make this precise, we assume some subset $N \subseteq \Sigma$ of *nonterminals* which are typed, i.e. there is an integer $k(A) \in \mathbb{N}$ for each $A \in N$. Moreover we assume that Σ contains the numbers $1, \ldots, max$ for some $max \in \mathbb{N}$ with $k(A) \leq max$ for all $A \in N$. A hyperedge with label $A \in N$ is meant to be an atomic item which is attached to a sequence of nodes $v_1 \cdots v_{k(A)}$. It can be represented by a node with an A-labeled loop and $k(A)$ edges the labels of which are $1, \ldots, k(A)$, respectively, and the targets of which are $v_1, \ldots, v_{k(A)}$, respectively, as depicted in Figure 9.15. Accordingly, we call such a node with its incident edges an A-hyperedge. A graph is said to be N-*proper* if each occurring nonterminal and each number between 1 and max belongs to some hyperedge.

Each $A \in N$ induces a particular N-proper graph $A^\bullet$ with the nodes $\{0, \ldots, k(A)\}$ and a single hyperedge where the A-loop is attached to 0 and $i \in \{1, \ldots, k(A)\}$ is the target of the edge labeled with i for $i = 1, \ldots, k(A)$. Let $[k(A)]$ denote the discrete graph with the nodes $\{1, \ldots, k(A)\}$. Thus a rule of the form $A^\bullet \supseteq [k(A)] \subseteq R$ for some N-proper graph R is a *hyperedge replacement rule*, which can be denoted by $A ::= R$ for short.

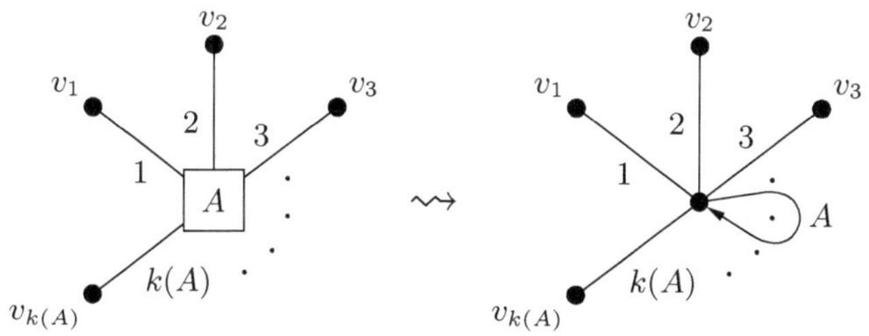

Fig. 9.15. Graph representation of a hyperedge

A *hyperedge replacement grammar* is a system $HRG = (N, T, P, S)$ with $S \in N$, $T \subseteq \Sigma$ with $T \cap N = \emptyset$, and a set of hyperedge replacement rules P. Its generated language $L(HRG)$ is defined as the graph language generated by the graph grammar $(S^\bullet, P, T)$.

Example 9 (flow diagrams). The graph transformation rules $r_{compound}$ and $r_{while\text{-}do}$ in Section 9.3.1 become proper examples of hyperedge replacement rules with hyperedges of type 2 if one replaces every statement (i.e. every rectangle node) with its two incident edges by a node with a box-labeled node and two outgoing numbered edges. This conversion is shown in Figure 9.16.

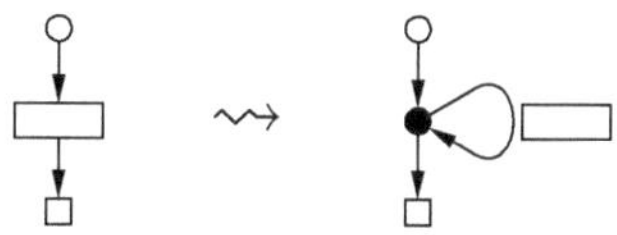

Fig. 9.16. Conversion of statements into the graph representation of a hypergraph

Example 10 (Sierpiński triangles). Another example, this time with hyperedges of type 3, is the hyperedge replacement grammar depicted in Figure 9.17 that generates Sierpiński triangles. A sample derivation is depicted in Figure 9.18, where the second and third step each are replacing three hyperedges in parallel.

Further examples can be found in the literature (see, e.g. [15, 6]).

In this way, hyperedge replacement is just a special case of graph transformation, but with some very nice properties. Some simple observations are the following, giving first indications to the context-freeness of hyperedge replacement.

1. Let $r = (A ::= R)$ be a hyperedge replacement rule and G an N-proper graph with an A-hyperedge y. Then there is a unique graph morphism $g\colon A^\bullet \to G$ mapping $A^\bullet$ to the A-hyperedge y such that the contact condition is satisfied and therefore r is applicable to G.

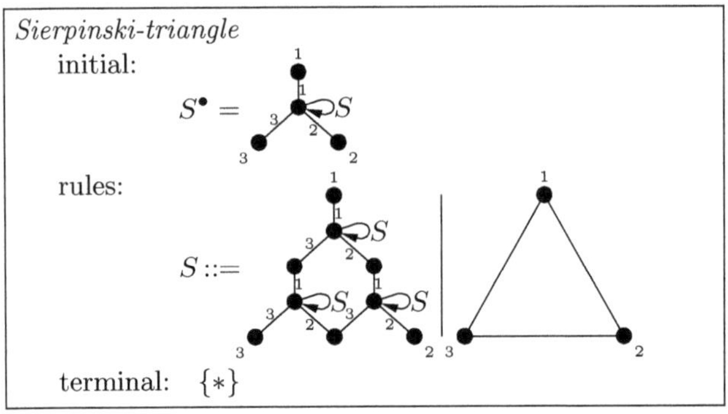

Fig. 9.17. A hyperedge replacement grammar generating Sierpiński triangles

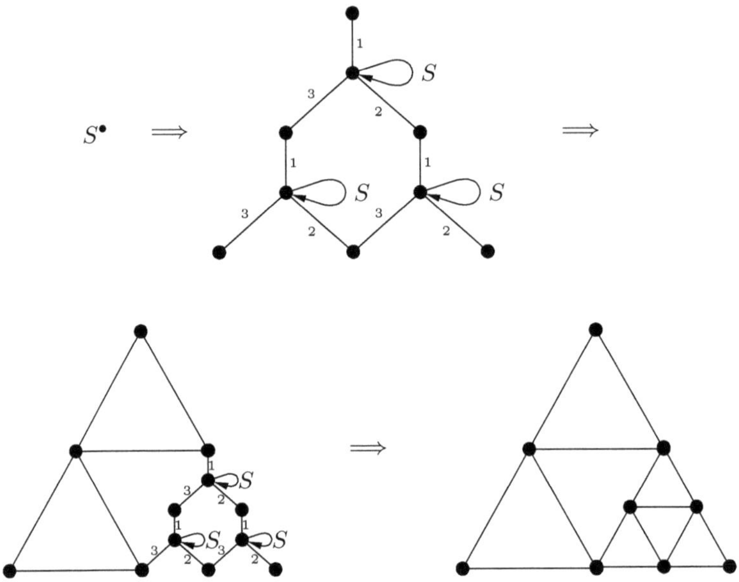

Fig. 9.18. A derivation generating a Sierpiński triangle

2. The directly derived graph H is N-proper and is obtained by removing y, i.e. by removing the node with the A-loop and all other incident edges, and by adding R up to the nodes $1, \ldots, k(A)$ where edges of R incident to $1, \ldots, k(A)$ are redirected to $g(1), \ldots, g(k(A))$, respectively. Due to this construction, H may be denoted by $G[y/R]$.
3. Two direct derivations $G \underset{r_1}{\Longrightarrow} H_1$ and $G \underset{r_2}{\Longrightarrow} H_2$ are parallel independent if and only if they replace distinct hyperedges.
4. A parallel rule $r = \sum_{i \in I} r_i$ of hyperedge replacement rules $r_i = (A ::= R_i)$ for $i \in I$ is applicable to G if and only if there are pairwise distinct A_i-

hyperedges y_i for all $i \in I$. In analogy to the application of a single rule, the resulting graph may be denoted by $G[y_i/R_i \mid i \in I]$.

5. If $I = I_1 \uplus I_2$, then we have in addition

$$G[y_i/R_i \mid i \in I] = (G[y_i/R_i \mid i \in I_1])[y_i/R_i \mid i \in I_2].$$

6. Two successive direct derivations $G \underset{r_1}{\Longrightarrow} G_1 \underset{r_2}{\Longrightarrow} H$ are sequentially independent if and only if the hyperedge replaced by the second step is not created by the first one.

Observation 1 holds because the A-hyperedge of $A^\bullet$ and y have the same structure. The only identifications that are possible concern the targets of numbered edges. But they are gluing nodes such that the identification condition holds. The only node to be removed is the one with the A-loop. But all its incident edges are removed, too, so that the contact condition holds. Observation 2 rephrases the definition of a direct derivation for the special case of a hyperedge replacement rule. Observation 3 holds because the matches of two direct derivations are either equal and then dependent on each other because the non-gluing node is removed by both of them, or they share only target nodes of numbered edges and then they are parallel independent. If a parallel rule is applicable, the identification condition is satisfied in particular. Therefore, as observation 4 states, no two rules can remove the same hyperedge as they would share non-gluing items. Then, observation 5 is a consequence of the sequentialization theorem as $\sum_{i \in I} r_i = \sum_{i \in I_1} r_i + \sum_{i \in I_2} r_i$. Finally, observation 6 holds by definition and the special form of the hyperedge replacement rules. Altogether, the direct derivations through hyperedge replacement rules can be ordered arbitrarily as long as they deal with different hyperedges. This observation leads to the following result.

Theorem 4 (Context-Freeness Lemma). *Let HRG* $= (N, T, P, S)$ *be a hyperedge replacement grammar and let* $A^\bullet \overset{n+1}{\underset{P}{\Longrightarrow}} H$ *be a derivation. Then there are some rule* $A ::= R$ *and a derivation* $A(y)^\bullet \overset{n(y)}{\underset{P}{\Longrightarrow}} H(y)$ *for each hyperedge* y *of* R *with label* $A(y)$ *such that* $H = R[y/H(y) \mid y \in Y_R]$ *and* $\sum_{y \in Y_R} n(y) = n$*, where* Y_R *is the set of hyperedges of* R.

If one varies the start symbol of *HRG* through all nonterminals, one gets a family of hyperedge replacement grammars $(HRG(A))_{A \in N}$ with $HRG(A) = (N, T, P, A)$. The Context-Freeness Lemma relates the graphs derived by this family to each other. Reformulated for the generated languages, hyperedge replacement languages turn out to be fixed points.

Theorem 5 (Fixed-Point Theorem). *Let* $(HRG(A))_{A \in N}$ *with* $HRG(A) = (N, T, P, A)$ *be a family of hyperedge replacement grammars (which share rules* *as well as terminals and nonterminals). Then, for each $A \in N$, the following equality holds:*

$$L(HRG(A)) = \bigcup_{(A ::= R) \in P} \{(R[y/H(y) \mid y \in Y_R] \mid H(y) \in L(HRG(A(y)))\}.$$

The Context-Freeness Lemma and Fixed-Point Theorem characterize generated graphs as covered each by a right-hand side of a rule without the hyperedges and smaller derived graphs. This provides a recursive way to prove and decide properties of the generated languages and their members if the properties are compatible with the composition of generated graphs as subsitution of hyperedges by derived graphs in right-hand sides. Many graph-theoretic properties like connectedness, planarity, Hamiltonicity, k-colorability, and many more are compatible. The explicit decidability results of hyperedge replacement grammars can be found in [15, 6] where also further structured results are surveyed. For other context-free graph-transformational approaches which are mainly based on node replacement, the reader may consult Engelfriet and Courcelle [11, 5].

9.8 Conclusion

In this chapter, we have given an introductory survey on some essentials of graph transformation focussing on theoretical aspects. For further reading, we recommend the three volumes of the Handbook of Graph Grammars and Computing by Graph Transformation [29, 7, 10]. Language-theoretic topics with respect to node and hyperedge replacement are addressed in Chapters 1, 2 and 5 of Volume 1. Collage grammars as a picture-generating device based on hyperedge replacement are the subject of Chapter 11 of Volume 2. Graph transformation as a general computational framework is presented in Chapters 3 and 4 of Volume 1 with respect to the double and single pushout approaches while an alternative approach can be found in Chapter 7. Chapters 3 and 4 discuss aspects of parallelism and concurrency in particular, which are also studied in the whole of Volume 3. Finally, Volume 2 is devoted to potential applications of graph transformation relating it to term rewriting and functional programming (Part 1), to visual and object-oriented languages (Part 2), to software engineering (Part 3), to other engineering disciplines (Part 4), to picture processing (Part 5), to the implementation of graph-transformational specification languages and tools (Part 6), and to structuring and modularization (Part 7). In recent years, main topics of interest have become the syntactic and semantic foundation of visual languages in the broadest sense and of model transformation, as graphs seem to be obvious candidates to represent visual models, diagrams, and all kinds of complex structures in a precise way (see, e.g., [4, 8, 26] and the SeGraVis webpage `www.segravis.org`).

References

1. Marc Andries, Gregor Engels, Annegret Habel, Berthold Hoffmann, Hans-Jörg Kreowski, Sabine Kuske, Detlef Plump, Andy Schürr, and Gabriele Taentzer. Graph transformation for specification and programming. *Science of Computer Programming*, 34(1):1–54, 1999.
2. Kenneth Appel and Wolfgang Haken. *Every Planar Map is Four Colorable*, volume 98 of *Contemporary Mathematics*. Amer. Mathematical Society, 1989.
3. Andrea Corradini, Hartmut Ehrig, Reiko Heckel, Michael Löwe, Ugo Montanari, and Francesca Rossi. Algebraic approaches to graph transformation Part I: Basic concepts and double pushout approach. In Rozenberg [29].
4. Andrea Corradini, Hartmut Ehrig, Hans-Jörg Kreowski, and Grzegorz Rozenberg, editors. *Proc. 1st Int. Conference on Graph Transformation (ICGT 2002)*, volume 2505 of *Lecture Notes in Computer Science*. Springer, 2002.
5. Bruno Courcelle. The expression of graph properties and graph transformations in monadic second-order logic. In Rozenberg [29], pages 313–400.
6. Frank Drewes, Annegret Habel, and Hans-Jörg Kreowski. Hyperedge replacement graph grammars. In Rozenberg [29], pages 95–162.
7. Hartmut Ehrig, Gregor Engels, Hans-Jörg Kreowski, and Grzegorz Rozenberg, editors. *Handbook of Graph Grammars and Computing by Graph Transformation, Vol. 2: Applications, Languages and Tools*. World Scientific, Singapore, 1999.
8. Hartmut Ehrig, Gregor Engels, Francesco Parisi-Presicce, and Grzegorz Rozenberg, editors. *Proc. 2nd Int. Conference on Graph Transformation (ICGT 2004)*, volume 3256 of *Lecture Notes in Computer Science*. Springer, 2004.
9. Hartmut Ehrig, Reiko Heckel, Martin Korff, Michael Löwe, Leila Ribeiro, Annika Wagner, and Andrea Corradini. Algebraic approaches to graph transformation Part II: Single pushout approach and comparison with double pushout approach. In Rozenberg [29], pages 247–312.
10. Hartmut Ehrig, Hans-Jörg Kreowski, Ugo Montanari, and Grzegorz Rozenberg, editors. *Handbook of Graph Grammars and Computing by Graph Transformation, Vol. 3: Concurrency, Parallelism, and Distribution*. World Scientific, Singapore, 1999.
11. Joost Engelfriet. Context-free graph grammars. In Grzegorz Rozenberg and Arto Salomaa, editors, *Handbook of Formal Languages, Volume 3: Beyond Words*, pages 125–213. Springer, 1997.
12. Joost Engelfriet and Grzegorz Rozenberg. Node replacement graph grammars. In Rozenberg [29], pages 1–94.
13. Alan Gibbons. *Algorithmic Graph Theory*. Cambridge University Press, 1985.
14. Claude Girault and Rüdiger Valk. *Petri Nets for Systems Engineering*. Springer, 2003.
15. Annegret Habel. *Hyperedge Replacement: Grammars and Languages*, volume 643 of *Lecture Notes in Computer Science*. Springer, 1992.
16. Frank Harary. *Graph Theory*. Addison Wesley, 1969.
17. Renate Klempien-Hinrichs, Hans-Jörg Kreowski, and Sabine Kuske. Rule-based transformation of graphs and the product type. In Patrick van Bommel, editor, *Transformation of Knowledge, Information, and Data: Theory and Applications*, pages 29–51. Idea Group Publishing, Hershey, Pennsylvania, USA, 2004.
18. Renate Klempien-Hinrichs, Hans-Jörg Kreowski, and Sabine Kuske. Typing of graph transformation units. In Ehrig et al. [8], pages 112–127.

19. Peter Knirsch and Sabine Kuske. Distributed graph transformation units. In Corradini et al. [4], pages 207–222.
20. Hans-Jörg Kreowski and Sabine Kuske. Graph transformation units and modules. In Ehrig et al. [7], pages 607–638.
21. Hans-Jörg Kreowski and Sabine Kuske. Graph transformation units with interleaving semantics. *Formal Aspects of Computing*, 11(6):690–723, 1999.
22. Hans-Jörg Kreowski and Sabine Kuske. Approach-independent structuring concepts for rule-based systems. In Martin Wirsing, Dirk Pattison, and Rolf Hennicker, editors, *Proc. 16th Int. Workshop on Algebraic Development Techniques (WADT 2002)*, volume 2755 of *Lecture Notes in Computer Science*, pages 299–311. Springer, 2003.
23. Sabine Kuske. More about control conditions for transformation units. In Hartmut Ehrig, Gregor Engels, Hans-Jörg Kreowski, and Grzegorz Rozenberg, editors, *Proc. Theory and Application of Graph Transformations*, volume 1764 of *Lecture Notes in Computer Science*, pages 323–337. Springer, 2000.
24. Sabine Kuske. *Transformation Units—A structuring Principle for Graph Transformation Systems.* PhD thesis, University of Bremen, 2000.
25. Sabine Kuske. Parameterized transformation units. In *Proc. GETGRATS Closing Workshop*, volume 51 of *Electronic Notes in Theoretical Computer Science*, 2002.
26. John L. Pfaltz, Manfred Nagl, and Boris Böhlen, editors. *Proc. 2nd Int. Workshop and Symposium on Applications of Graph Transformations with Industrial Relevance (AGTIVE 2003)*, volume 3062 of *Lecture Notes in Computer Science.* Springer, 2004.
27. Wolfgang Reisig. *Petri Nets. An Introduction.* Springer, 1985.
28. Wolfgang Reisig. *Elements of Distributed Algorithms. Modeling and Analysis with Petri Nets.* Springer, 1998.
29. Grzegorz Rozenberg, editor. *Handbook of Graph Grammars and Computing by Graph Transformation, Vol. 1: Foundations.* World Scientific, Singapore, 1997.
30. Andy Schürr. Programmed graph replacement systems. In Rozenberg [29], pages 479–546.

10

Molecular Computation

Mitsunori Ogihara

Department of Computer Science
University of Rochester
Rochester, NY 14627, USA
E-mail: ogihara@cs.rochester.edu

This chapter provides the basic principles of molecular computation and introduces some algorithms.

10.1 The Basics of Molecular Biology

A DNA molecule consists of a *sugar backbone*, a *phosphate group*, and a *nitrogen base*. There are four types of DNA, A (Adenine), C (Cytosine), G (Guanine), and T (Thymine). The four have the same nitrogen base and the same phosphate group. The difference is in the nitrogen base. In water solution the nitrogen base of an Adenine molecule and that of a Thymine molecule can be coupled with two hydrogen bonds. Similarly, the nitrogen base of a Guanine molecule and that of a Cytosine molecule can be coupled with three hydrogen bonds. These are the *DNA base pairs*. We say that Adenine and Thymine are complementary to each other and that Cytosine and Guanine are complementary to each other.

The DNA molecules are polymers. The sugar backbone of a DNA molecule can be connected with the sugar backbone of another DNA molecule. The linkage of two sugar backbones always occurs between the fifth carbon atom of one backbone and the third of the other. Thus, the polymeric structure of DNA is linear. This linear structure is called *single-stranded DNA* (*ssDNA* for short). One free end of the linear structure is always the fifth carbon atom and the other end the third carbon atom. The $5'$-*end* and the $3'$-*end* are the words that refer to these ends.

When *base pairing* occurs between two complementary DNA molecules, the backbones are at the farthest from the two or three hydrogen bonds; also, they are in the opposite directions, in the sense that the $5'$-end of one molecule is on the same side as the $3'$-end of the other. This alignment feature extends to pairwise base pairing of two single-stranded DNA. That is, two single-stranded DNA are paired molecule-wise when one strand is molecule-wise complementary (with respect to base pairs) to the other. For example,

M. Ogihara: *Molecular Computation*, Studies in Computational Intelligence (SCI) **25**, 255–267 (2006)
www.springerlink.com

ACCAAAGGGG is complementary to TGGTTTCCCC. Because of how molecule-wise base pairing occurs, the 5′-ends of the strands are on the opposite side from each other; that is, the 5′-end of the former is at the leftmost A if and only if the 5′-end of the latter is at the rightmost C; equivalently, the 3′-end of the former is at the leftmost A if and only if the 3′-end of the latter is at the rightmost C. To signify that the complementary strands have the 5′-ends opposite from each other the two strands are said to be *complementary anti-parallel.* A *double-stranded DNA* (dsDNA for short) is a pair of complementary anti-parallel strands joined by molecule-wise base pairing.

The factors that govern the formation of double-stranded DNA are the patterns of the ssDNA, the pH, and the temperature. The hydrogen bonds joining complementary base pairs are cut due to vibration of the DNA molecules. The higher the temperature is, the more vigorously the molecules vibrate. Thus, double-stranded DNA are ripped apart into single-stranded DNA when the temperature is raised high enough (*denaturing*). The two strands form the double-stranded DNA again if the temperature is slowly reduced (*annealing* or *hybridization*). The temperature at which denaturing takes places (the *melting temperature*) is a function of the base-length, the number of G-C pairs in the pattern, and the patterns at the ends. Roughly speaking, the longer the strands are, the higher the denaturing temperature is. The speed at which hybridization takes place depends on the temperature, the concentration, and the pH. Below the melting temperature, when two single-stranded DNA freely moving in the solution bump into each other, they quickly decide whether to form double-stranded DNA. If they decide not to, they move away from each other. Also, annealing and denaturing are kinetic reactions, and thus, both of them reach an equilibrium.

In the annealing process, two strands many form partially complementary double-stranded DNA. For example, 5′-AACCTTT-3′ and 5′-CCAAAGG-3′ form double-stranded DNA in which the two-molecule pattern of AA sticks and the molecule pattern of GG stick out from the ends:

$$\begin{array}{llllllllllll} 5'\text{-} & A & A & G & G & T & T & T & & & \text{-}3' \\ 3'\text{-} & & & C & C & A & A & A & G & G & \text{-}5', \end{array}$$

and 5′-ACACTGGTTGG-3′ and 5′-AACCA-3′ form double-stranded DNA with the ACAC and the GG of the former strand sticking out:

$$\begin{array}{llllllllllllll} 5'\text{-} & A & C & A & C & T & G & G & T & T & G & G & \text{-}3' \\ 3'\text{-} & & & & & A & C & C & A & A & & & \text{-}5' \end{array}$$

More than two strands can anneal to form double-stranded DNA. In the latter of the two examples, if 5′-GTGT-3′ is present, then it may anneal with the double-stranded DNA to form:

$$\begin{array}{llllllllllllll} 5'\text{-} & A & C & A & C & T & G & G & T & T & G & G & \text{-}3' \\ 3'\text{-} & T & G & T & G| & A & C & C & A & A & & & \text{-}5'. \end{array}$$

The vertical line immediately to the right of TGTG on the bottom pattern indicates that the two strands on the bottom part are yet to be joined. The

DNA ligase is an enzyme that connects such unconnected two strands in the presence of of strand that is complementary anti-parallel to the concatenated pattern. Such a pattern is called a *template*. A template does not have to cover the whole length of the concatenation of the two, but it is necessary that the template straddle the gap between the two.

The process of DNA hybridization (more precisely, called DNA-DNA hybridization to signify that the two hybridizing strands are DNA sequences) is not perfect. Single-stranded DNA with almost complete anti-parallel complementarity can be mistaken for the one with complete anti-parallel complementarity, in particular, when the double-stranded part is long and the unmatched base pairs (called *mismatches*) are located toward the middle of the part. The affinity between mismatched single-stranded DNA is influenced by many factors, including the temperature, the pH, the type of mismatches, and the length of the double-stranded part. If everything else being equal, a mismatch containing a Cytosine or a Guanine is less stable that those not containing one.

DNA polymerase is an enzyme that extends incomplete double-stranded DNA by repeatedly adding single DNA molecules (*oligonucleotides*) toward the 3′-end. For example, given enough supply of A, C, G, and T, DNA polymerase turns

$$\begin{array}{l} 5'\text{- A C A C} \qquad\qquad\qquad\qquad \text{-}3' \\ 3'\text{- T G T G A C G T A C G T -}5' \end{array}$$

into

$$\begin{array}{l} 5'\text{- A C A C T G C A T G C A -}3' \\ 3'\text{- T G T G A C G T A C G T -}5'. \end{array}$$

This process is called *DNA polymerization*. The temperature at which DNA polymerization occurs is certainly below the melting temperature of the initial partially complete double-stranded DNA.

It is possible to use DNA polymerase to create many copies of DNA sequences of known end patterns. To create copies of a single-stranded DNA w with the 5′-end pattern S and the 3′-end pattern T, we synthesize many copies of S and the pattern complementary anti-parallel to T. We add these patterns and oligonucleotides into the test tube containing w together with DNA polymerase. Then we repeat the cycle of three steps

$$\text{annealing} \to \text{polymerization} \to \text{denaturing.}$$

In the *annealing* step, the short stretches of DNA (S and T's complement) hybridize with the fully long single-stranded DNA. In the *polymerization* step, the short stretches are fully extended; from the original the complementary pattern is produced, and from the complement, the original pattern is produced. In the *denaturing* step, the complement and the original are pulled apart. The amplitude of the original is almost doubled in each round (not quite doubling because hybridization arrives at an equilibrium and because the full complementary pairs may hybridize—in such a hybrid polymerization doesn't occur), so the amplitude increases exponentially in the number of rounds. This technique is called *polymerase chain reaction* (or *PCR*), and the sequence S and the complement of T from which copy strings start with are called *primers*.

DNA molecules are electrically (negatively) charged molecules. When an electronic field is cast on a DNA solution, DNA molecules are attracted to the positive pole. When DNA strands are solved in a gel, the speed at which they move within the solution is determined by their base length; the shorter they are, the more swiftly they move.u This property enables sorting DNA strands according to their base length, according to the distance they travel from the starting point in a gel. This sorting procedure is called *gel electrophoresis*. The usual type of gel requires that the DNA be single-stranded, but there are other types that allow sorting of double-stranded DNA.

Oligonucleotides can be chemically manufactured. Oligonucleotides are linked into any particular sequence pattern using a DNA synthesizer. Natural DNA strands that reside in the living cells can be very long. However, in water solution DNA strands degenerate relatively quickly and long ones tend to break apart. It is also possible to attach various molecules to oligonucleotides and DNA strands to endow with them special properties. Two types come in handy for manipulating DNA molecules. First, one can attach fluorescent molecules to DNA. Given enough quantify of fluorescently labeled DNA it is possible to detect fluorescent light emitted from the molecules. Second one can attach magnetically potential molecules (biotin) to DNA. Biotin has affinity with strepavidin. By letting a solution of DNA slowly flow on immobilized strepavidin molecules, it is possible collect biotin-labeled DNA strands.

10.2 Fundamental Operations in Molecular Computation

Using the aforementioned properties of DNA, one can design various operations on DNA for computation.

- **Synthesis** Synthesize single-stranded DNA of a given sequence pattern.
- **Merger** Mix two test tubes, each containing DNA, into one test tube.
- **Annealing** Let the DNA strands in a test tube for double-stranded DNA.
- **Denaturing** Remove double-stranded DNA into single-stranded DNA.
- **Detection** Using fluorescently labeled DNA, test whether DNA is contained in a test tube.
- **Length** Using gel-electrophoresis separate DNA according to their base-length.
- **Separation** Extract from a test tube single-stranded DNA with a particular pattern. To do this operation with respect to a pattern p, we

synthesize the biotin-labeled complementary anti-parallel of p, pour this into the test tube, perform annealing, and extract those with biotin.

- **Ligating** Connect single-stranded DNA ending with a pattern p and single-stranded DNA starting with a pattern q. To do this operation we synthesize the complementary anti-parallel of $p.q$, pour this into the test tube, perform annealing with DNA ligase, and then denature. To remove the complementary anti-parallel "linker" we either use length-bases separation or use separation by biotin-labeling the linker.
- **Amplification** Amplify the sequences using primers p and q.

In molecular computation, each pattern is represented by multiple copies (it is not possible to create a single copy; even if it is, you cannot detect a single copy of DNA strand). So, the reactors in these operations have to be amply present.

10.3 Adleman's Algorithm

In 1994 Leonard Adleman published the very first molecular-based algorithm [2]. The field of molecular computation was born then. The algorithm of Adleman was to solve the Hamiltonian Path Problem, the program of testing, given a directed graph G and a pair of nodes s and t, whether G contains a directed path from s to t that visits every node of G exactly once.

Let $G = (V, E)$ be an input graph. Let $V = \{1, ..., n\}$ for some integer n and let $s = 1$ and $t = n$. We assume that s has no incoming edges and that t has no outgoing edges.

Let L be an integer parameter. We encode each node as a $2L$-base-long single-stranded DNA. For each i, $1 \leq i \leq n$, let w_i be the pattern for node i. The patterns are designed so that the L bases either at the $5'$-end or at the $3'$-end of any of these sequences are not "close" to any L-base pattern or its complement appearing elsewhere. This is to ensure that mismatches will not occur during the execution of the algorithm. The exact conditions under which such mismatches occur has been the subject of intensive scientific investigations, but no definitive answers have been given. Here we use the following: Two L-base-long sequences are close if their Hamming distance is at most $L/3$ and they match at the first and the last three bases. The value of L can be within some constant factor of $\log n$.

Let us now return to the presentation of the algorithm of Adleman. For each edge $(i, j) \in E$, let λ_{ij} be the single-stranded DNA that is complementary anti-parallel to the pattern constructed by appending the first half of w_j to the second half of w_i. Because of the "close-to-nothing-else" of the terminal L-bases, λ_{ij} hybridizes only with w_i and w_j. Also, when it hybridizes with both, there is no gap between w_i and w_j. If this hybrid is existent together with DNA ligase, the gap between w_i and w_j is eliminated and they become a $4L$-base-long sequence. We call the strands λ_{ij} *linkers*.

To execute the algorithm we need to synthesize many copies of the following: the w_i's, the linkers, the first half of w_i's and the sequence complementary anti-parallel to the second half of w_n.

The first step of the algorithm is to mix all w_i's and the linkers together with DNA ligase and then to let the sequences anneal. This reaction connects the w_i's into paths according to the edges represented by the linkers. If there are enough components, for each "not so long" (recall that long DNA strands *in vitro* get broken) path in G, a sequence of the w_i's corresponding to the path is generated. Note that, since DNA ligase acts both on the "node" strands and on the linkers, the sequence of the linker corresponding to the paths is generated.

The next step is selection. The following two-phase cycle is repeated:

- Use the Length operation to select single-stranded DNA having length $2nL$.
- Use the first half of w_1 and the complementary anti-parallel of the second half of w_n to run the Amplification operation.

Let u be any $2nL$-base-long single-stranded DNA. This u is amplified in the second phase if and only if either it starts with the first half of w_1 and ends with the second half of w_n or it is complementary anti-parallel to such a pattern. Thus, such u is either a length-$2nL$ path from w_1 to w_n or the anti-parallel complement of such a path. The single-stranded DNA remaining in the test tube after the repetition of the two-phase cycle may contain other types of $2nL$-base-long strands, but they will never be amplified. So, the amplitude of such a type will never increase.

The next step is pattern-based selection. Let T_0 be the test tube at the end of the previous step. We do the following for i, $1 \leq i \leq n$: use the Separate operation to select from T_{i-1} to T_i all the strands that having pattern w_i.

Let u be an arbitrary single-stranded DNA in T_n. The strand u has length $2nL$, it starts with w_1, ends with w_n, and contains w_i for each i, $2 \leq i \leq n$. This means that u is a permutation of $w_1, \ldots, w_n$. Since u is a path and the anti-parallel complement of u cannot be in T_n, u represents a Hamiltonian path in G. We now apply the Detect operation to test whether T_n contains a single-stranded DNA to see whether G is Hamiltonian.

The Detect operation finds whether there is a Hamiltonian path at all but does not find a path. The solution T_n may contain more than one paths. If it is possible to remove a single copy of single-stranded DNA out of T_n, then one has only to amplify the single copy and decode it. However, such a single-molecule selection is far beyond the reach of the current biotechnology.

Thus, to find one Hamiltonian path we will use some other method. The technique called *graduated PCR*. The idea is the following: We find for which i, $2 \leq i \leq n-1$, w_i is appearing as the second node of any Hamiltonian path in T_n. We then select one such i. We collect all paths having w_i as the second node. Similarly, we find for which j, $2 \leq j \leq n-1$, w_j is appearing as the third node of any Hamiltonian path whose second node is w_i. We select one such j. By repeating this for the fourth node, the fifth node, and so one, we will be able to find a single Hamiltonian path in G.

The identification of the nodes appearing as the second node is done as follows: Amplify T_n and then split T_n into $n-2$ almost equal parts. Call these $S_2, \ldots, S_{n-1}$. We hope that the Amplification operation creates enough copy of each path so that every path is represented in each one of S_i's. For each i, $2 \leq i \leq n-1$, we run the Amplification operation with the first half of w_i and the anti-parallel complement of the second half of w_n as the primers. Each strand in S_i has w_1 at the beginning, w_n at the end, and w_i somewhere. By the Amplification operation, the suffix of such a strand starting precisely at its (unique) appearance of w_i will be amplified. If there is a path whose second node is w_i, the S_i must contain a strand having length $2(n-1)L$. We split S_i into halves. Apply the Length operation of one half to extract all the $2(n-1)L$-bases-long sequences, and then run the Detect operation on the extracted part. We pick one i for which the Detect operation returned an affirmative answer. Use the remaining half for the next round. In the next round, we will do the same, but the base-length to be used is $2(n-2)L$ instead of $2(n-1)L$.

10.4 The Algorithm for 3SAT by Lipton

The Adleman algorithm generates all the paths in a single Annealing, step so it may also generate paths having length larger than n. To avoid the problem of generating paths unnecessarily long, one can generate paths in multiple steps. A paper by Lipton [41], which came out immediately after the celebrated Adleman paper, addresses this issue. Lipton presented an algorithm for 3SAT that uses gradual extension.

Let φ be an input formula of n variables, $x_1, \ldots, x_n$. We will design $2n$ patterns, each having length $L+2K$: $v_{i,T}$, $v_{i,F}$ $1 \leq i \leq n$. Here $2K < L$. For each i, $1 \leq i \leq n$, the first K-bases of $v_{i,T}$ is equal to those of $v_{i,F}$ and the last K-bases of $v_{i,T}$ is equal to those of $v_{i,F}$. We think of a truth-assignment to φ as a $(L+2K)n$-base-long sequence such that for each i, $1 \leq i \leq n$, its i-th $(L+2K)$-base-long component is one of $v_{i,T}$ and $v_{i,F}$.

The goal of the algorithm is to generate all truth-assignments, collect all satisfying assignments while discarding all non-satisfying assignments, and then test whether any assignment survived.

The computation starts with a test tube containing many copies of $v_{1,T}$ and $v_{1,F}$. For $i = 2, \ldots, n$, we split the test tube into halves, extend one with $v_{i,T}$ and the other with $v_{i,F}$, and then mix the two together. Here the extension is carried out by the $2K$-base-long "linker" between $v_{i-1,\cdot}$ and $v_{i,\cdot}$, which is complementary anti-parallel to the $2k$-base-long sequence whose first K bases are the K $3'$-end bases of $v_{i,\cdot}$ and whose last K bases are the K $5'$-end bases of $v_{i,\cdot}$. After extension, the test tube is denatured and using the Length operation the strands having length $(L+2K)i$ are extracted. Given enough quantity of the strands, we can expect that all truth-assignments will be generated.

In the next phase, for every clause of φ, sequentially one at a time, we discard the assignments that fail to satisfy the clause Let C be a clause with three literals, say x_1, x_2, and $\overline{x_3}$. Let T_0 be the test tube containing all the assignments that have survived the screening process for all the clauses preceding C. Then, we do the following:

- Collect from T_0 all the strands containing $x_{1,T}$.
- Collect from the rest of T_0 all the strands containing $x_{2,T}$.
- Collect from the rest of T_0 all the strands containing $x_{3,F}$.
- Mix the three collections together, call it T_0.

The strands discarded have $x_{1,F}$, $x_{2,F}$, and $x_{3,T}$, so they encode non-satisfying assignments.

10.5 The Field of Molecular Computation

The computation using molecules that Adleman proposed had two advantages. First, DNA molecules potentially have information capacity much higher than that of silicon — in one milliliter of DNA solution one can store 10^{15} bits of information. Second, molecular operations do not generate heat. However, molecular operations are slow and error-prone. The goal of molecular computation is to explore the possibility of molecular operations for computation and to design and analyze molecular algorithms. The Lipton algorithm shows that a simpler method that uses gradual extend and repeated selection can solve 3SAT.

An algorithm of a flavor similar to that of the Lipton algorithm can be designed for the Hamiltonian Path Problem so as to replace the Adleman algorithm. Indeed, molecular algorithms are not unique. For a single problem, it is possible to design different algorithms, using different sets of operation and different encoding principles. That observation leads one to a number of intriguing questions:

- What types of operations can be permitted in molecular computation?
- Other than oligonucleotides, what types of nucleotides can be used?
- What are the feasibility of molecular computation in terms of laboratory experiments?
- How can you efficiently design sequences?
- What problems can be solved efficiently with molecular computation?
- What is the complexity of molecular computational models?

The field of molecular computation has developed since the paper of Adleman. Researchers from various disciplines have worked on various issues in the field, including the above fundamental questions. Following is a short list of representative papers.

1. **Molecular computation models** including the "sticker" model [3, 57], the "surface" model [17], the "hairpin" model [28, 29, 40, 59, 64], the enzymatic models [12, 34, 55], the self-assembly models [65, 62, 63, 66, 45], the circuit-based models [1, 52, 48, 51, 46], the RNA model [18, 23, 43], and the PRAM model [27, 54].
2. **Biological experiments** [2, 14, 9, 22, 24, 25, 26, 31, 32, 39, 33, 49, 61, 60].
3. **DNA sequence design** [21, 20].
4. **Various molecular algorithms** [6, 4, 5, 8, 7, 13, 15, 16, 10, 11, 19, 26, 35, 37, 44, 41, 42, 47, 50, 58]
5. **Connection with theoretical models** [1, 12, 30, 36, 38, 51, 52, 53, 54, 56, 66, 67, 68].

References

1. M. Amos, P. E. Dunne, and A. Gibbons. DNA simulation of boolean circuits. In J. R. Koza, W. Banzhaf, K. Chellapilla, K. D. Deb, D. B. Fogel, M. H. Garzon, D. E. Goldberg, H. Iba, and R. L. Riolo, editors, *Proceedings of 3rd Annual Genetic Programming Conference*, pages 679–683, San Francisco, CA, 1998. Morgan Kaufmann.
2. L. Adleman. Molecular computation of solutions to combinatorial problems. *Science*, 266:1021–1024, 1994.
3. L. Adleman. On constructing a molecular computer. In R. Lipton and E. Baum, editors, *DNA Based Computers*, pages 1–21. The American Mathematical Society DIMACS Series in Discrete Mathematics and Theoretical Computer Science Volume **27**, 1996.
4. M. Amos, A. Gibbons, and D. Hodgson. Error-resistant implementation of DNA computations. In L. Landweber and E. Baum, editors, *DNA Based Computers II*, pages 151–162. The American Mathematical Society DIMACS Series in Discrete Mathematics and Theoretical Computer Science Volume **44**, 1999.
5. M. Arita, M. Hagiya, and A. Suyama. Joining and rotating data with molecules. In *Proceedings of International Conference on Evolutionary Computing*, pages 243–248. IEEE Computer Society Press, Los Alamitos, CA, 1997.
6. L. Adleman, P. Rothemund, S. Roweis, and E. Winfree. On applying molecular computation to the Data Encryption Standard. In L. Landweber and E. Baum, editors, *DNA Based Computers II*, pages 31–44. The American Mathematical Society DIMACS Series in Discrete Mathematics and Theoretical Computer Science Volume **44**, 1999.
7. E. Baum and D. Boneh. Running dynamic programming algorithms on a DNA computer. In L. Landweber and E. Baum, editors, *DNA Based Computers II*, pages 77–80. The American Mathematical Society DIMACS Series in Discrete Mathematics and Theoretical Computer Science Volume **44**, 1999.
8. E. Bach, A. Condon, E. Glaser, and C. Tanguay. DNA models and algorithms for NP-complete problems. In *Proceedings of 11th Conference on Computational Complexity*, pages 290–299. IEEE Computer Society Press, Los Alamitos, CA, 1996.

9. R. S. Braich, N. Chelapov, C. Johnson, P. W. K. Rothemund, and L. Adleman. Solution of a 20-variable 3-SAT problem on a DNA computer. *Science*, 296:499–502, 2002.
10. D. Boneh, C. Dunworth, and R. Lipton. Breaking DES using a molecular computer. In R. Lipton and E. Baum, editors, *DNA Based Computers*, pages 37–65. The American Mathematical Society DIMACS Series in Discrete Mathematics and Theoretical Computer Science Volume **27**, 1996.
11. D. Boneh, C. Dunworth, R. Lipton, and J. Sgall. On the computational power of DNA. *Discrete Applied Mathematics*, 71:79–94, 1996.
12. D. Beaver. Computing with DNA. *Journal of Computational Biology*, 2(1):1–8, 1995.
13. R. Beigel and B. Fu. Molecular computing, bounded nondeterminism, and efficient recursion. *Algorithmica*, 25:222–238, 1999.
14. R. S. Braich, C. Johnson, P. W. K. Rothemund, D. Hwang, N. Chelapov, and L. Adleman. Solution of a satisfiability problem on a gel-based DNA computer. In A. Condon and G. Rozenberg, editors, *DNA Computing, 6th International Workshop on DNA-Based Computers*, pages 27–42. Springer-Verlag *Lecture Notes in Computer Science 2054*, 2000.
15. D. Boneh and R. Lipton. Batching DNA computations. manuscript.
16. D. Boneh and R. Lipton. A divide and conquer approach to sequencing. manuscript.
17. W. Cai, A. Condon, R. Corn, E. Glaser, Z. Fei, T. Frutos, Z. Guo, M. Lagally, Q. Liu, L. Smith, and A. Thiel. The power of surface-based DNA computation. In *Proceedings of 1st International Conference on Computational Molecular Biology*, pages 67–74. ACM Press, 1997.
18. A. Cukras, D. Faulhammer, R. Lipton, and L. Landweber. Chess game: a model for RNA-based computation. In *Preliminary Proceedings of 4th DIMACS Workshop on DNA Based Computers*, pages 27–37, 1998.
19. S. Díaz, J. L. Esteban, and M. Ogihara. A DNA-based random walk method for solving k-SAT. In A. Condon, editor, *Proceedings of 6th Workshop on DNA-based Computers*, pages 209–219. Springer-Verlag *Lecture Notes in Computer Science 2054*, 2000.
20. R. Deaton, M. Garzon, R. Murphy, J. Rose, D. Franceschetti, and Jr. S. Stevens. Genetic search of reliable encodings for DNA-based computation. In *Proceedings of 1st Annual Genetic Programming Conference*, San Francisco, CA, 1996. Morgan Kaufmann.
21. R. Deaton, R. Murphy, M. Garzon, D. Franceschetti, and Jr. S. Stevens. Good encodings for DNA-based solutions to combinatorial problems. In L. Landweber and E. Baum, editors, *DNA Based Computers II*, pages 247–258. The American Mathematical Society DIMACS Series in Discrete Mathematics and Theoretical Computer Science Volume **44**, 1999.
22. A. D. Ellington, M. P. Robertson, K. D. James, and J. Colin Cox. Strategies for DNA computing. In H. Rubin and D. H. Wood, editors, *DNA Based Computers III*, pages 173–184, 1997.
23. D. Faulhammer, A. Cukras, R. Lipton, and L. Landweber. Molecular computation: DNA solutions to chess problems. *Proceedings of the National Academy of Science*, 97:1385–1389, 2000.
24. D. Faulhammer, R. Lipton, and L. Landweber. Counting DNA: estimating the complexity of a test tube of DNA. In *Preliminary Proceedings of 4th DIMACS Workshop on DNA Based Computers*, pages 249–250, 1998.

25. A. G. Frutos, Q. Liu, A. J. Thiel, A. M. W. Sanner, A. E. Condon, L. M. Smith, and R. M. Corn. Demonstration of a word design strategy for DNA computation on surfaces. *Nucleic Acids Research*, 25:4748–4757, 1997.
26. F. Guarnieri, M. Fliss, and C. Bancroft. Making DNA add. *Science*, 273:220–223, 1995.
27. A. Gehani and J. Reif. Microflow bio-molecular computation. *Biosystems*, 52(1–3):197–216, 1999.
28. M. Hagiya. Towards autonomous molecular computers. In J. R. Koza, W. Banzhaf, K. Chellapilla, K. D. Deb, D. B. Fogel, M. H. Garzon, D. E. Goldberg, H. Iba, and R. L. Riolo, editors, *Proceedings of 3rd Annual Genetic Programming Conference*, pages 691–699, San Francisco, CA, 1998. Morgan Kaufmann.
29. M. Hagiya, M. Arita, D. Kiga, K. Sakamoto, and S. Yokoyama. Towards parallel evaluation and learning of Boolean μ-formulas with molecules. In H. Rubin and D. H. Wood, editors, *DNA Based Computers III*, pages 57–72, 1997.
30. T. Head. Formal language theory and DNA: an analysis of the generative capacity of specific recombinant behaviors. *Bulletin of Mathematics Biology*, 49(6):737–759, 1987.
31. A. J. Hartemink and D. K. Gifford. Theormodynamic simulatin of deoxyoligonucleotide hybridization for DNA computation. In *DNA Based Computers III*, pages 25–37, 1997.
32. A. Hartemink, D. Gifford, and J. Khodor. Automated constraint-based nucleotide sequence selection for DNA computation. In *Preliminary Proceedings of 4th DIMACS Workshop on DNA Based Computers*, pages 287–297, 1998.
33. P. Kaplan, G. Cecchi, and A. Libchaber. DNA-based molecular computation: template-template interactions in PCR. In L. Landweber and E. Baum, editors, *DNA Based Computers II*, pages 97–104. The American Mathematical Society DIMACS Series in Discrete Mathematics and Theoretical Computer Science Volume **44**, 1999.
34. J. Khodor and D. Gifford. Design and implementation of computational systems based on programmed mutagenesis. In *Preliminary Proceedings of 4th DIMACS Workshop on DNA Based Computers*, pages 101–108, 1998.
35. R. Karp, C. Kenyon, and O. Waarts. Error-resilient DNA computation. In *Proceedings of 7th ACM-SIAM Symposium on Discrete Algorithms*, pages 458–467. ACM Press/SIAM, 1996.
36. L. Kari and L. Landweber. The evolution of DNA computing: nature's solution to a computational problem. In J. R. Koza, W. Banzhaf, K. Chellapilla, K. D. Deb, D. B. Fogel, M. H. Garzon, D. E. Goldberg, H. Iba, and R. L. Riolo, editors, *Proceedings of 3rd Annual Genetic Programming Conference*, pages 700–708, San Francisco, CA, 1998. Morgan Kaufmann.
37. S. Kurtz, S. Mahaney, J. Royer, and J. Simon. Biological computing. In L. Hemaspaandra and A. Selman, editors, *Complexity Theory Retrospective II*, pages 179–195. Springer-Verlag, New York, NY, 1997.
38. L. Kari, Gh. Paŭn, G. Rozenberg, A. Salomaa, and S. Yu. DNA computing, sticker sustems and universality. *Acta Informatica*, 35:401–420, 1998.
39. P. D. Kaplan, D. S. Thaler, and A. Libchaber. Parallel overlap assembly of paths through a directed graph. In *Preliminary Proceedings of 3rd DIMACS Workshop on DNA Based Computers*, pages 127–141, 1997.
40. S. Kobayashi, T. Yokomori, G. Sanpei, and K. Mizobuchi. DNA implementation of simple Horn clause comptuation. In *Proceedings of International Conference*

on Evolutionary Computing, pages 213–217. IEEE Computer Society Press, Los Alamitos, CA, 1997.
41. R. Lipton. DNA solutions of hard computational problems. *Science*, 268:542–545, 1995.
42. R. Lipton. A memory based attack on cryptosystems with application to DNA computing. In *Preliminary Proceedings of 4th DIMACS Workshop on DNA Based Computers*, pages 267–272, 1998.
43. L. F. Landweber, T.-C. Kuo, and E. A. Curtis. Evolution and assembly of an extremely scrambled gene. *Proceedings of the National Academy of Science*, 97:3298–3303, 2000.
44. L. Landweber and R. Lipton. DNA^2DNA computations: a potential 'killer app'? In *Preliminary Proceedings of 3rd DIMACS Workshop on DNA Based Computers*, pages 59–68, 1997.
45. Q. Liu, L. Wang, A. G. Frutos, R. M. Corn, and L. M. Smith. DNA computing on surfaces. *Nature*, 403:175–178, 2000. January, 13.
46. G. G. Owensen, M. Amos, D. A. Hodgson, and A. Gibbons. Dna-based logic. *Soft Computing*, 5(2):102–105, 2001.
47. M. Ogihara. Breadth first search 3SAT algorithms for DNA computers. Technical Report TR-629, Department of Computer Science, University of Rochester, Rochester, NY, July 1996.
48. M. Ogihara. Relating the minimum model for DNA computation and Boolean circuits. In W. Banzhaf, J. Daida, A. E. Eiben, M. H. Garzon, V. Honavar, M. Jakiela, and R. E. Smith, editors, *Genetic and Evolutionary Computation Conference*, pages 1817–1822. Morgan Kaufmann Publishers, San Francisco, CA, 1999.
49. Q. Ouyang, P. D. Kaplana, S. Liu, and A. Libchaber. DNA solution of the maximal clique problem. *Science*, 278:446–449, 1997.
50. M. Ogihara and A. Ray. DNA-based parallel computation by counting. In H. Rubin and D. H. Wood, editors, *DNA Based Computers III*, pages 255–264, 1997.
51. M. Ogihara and A. Ray. Simulating boolean circuits on DNA computers. In *Proceedings of 1st International Conference on Computational Molecular Biology*, pages 326–331. ACM Press, 1997.
52. M. Ogihara and A. Ray. The minimum DNA model and its computational power. In *Unconventional Models of Computation*, pages 309–322. Springer, Singapore, 1998.
53. G. Păun. The splicing as an operation on formal languages. In *Proceedings of International Symposium on Intelligence in Neural and Biological Systems*, pages 176–180. IEEE Computer Society Press, Los Alamitos, CA, 1995.
54. J. Reif. Parallel molecular computation. In *Proceedings of 7th ACM Symposium on Parallel Algorithms and Architecture*, pages 213–223. ACM Press, 1995.
55. P. Rothemund. A DNA and restriction enzyme implementation of Turing machines. In R. Lipton and E. Baum, editors, *DNA Based Computers*, pages 75–119. The American Mathematical Society DIMACS Series in Discrete Mathematics and Theoretical Computer Science Volume **27**, 1996.
56. D. Rooß and K. Wagner. On the power of DNA computing. *Information and Computation*, 131:95–109, 1996.
57. S. Roweis, E. Winfree, R. Burgoyne, N. Chelapov, M. Goodman, P. Rothemund, and L. Adleman. A sticker based model for DNA computation. In L. Landweber

and E. Baum, editors, *DNA Based Computers II*, pages 1–30. The American Mathematical Society DIMACS Series in Discrete Mathematics and Theoretical Computer Science Volume **44**, 1999.
58. Y. Sakakibara. Solving computational learning problems of boolean formulae on DNA computers. In A. Condon and G. Rozenberg, editors, *DNA Computing, 6th International Workshop on DNA-Based Computers*, pages 220–230. Springer-Verlag *Lecture Notes in Computer Science 2054*, 2000.
59. J. Sakamoto, H. Gouzu, K. Komiya, D. Kiga, S. Yokoyama, T. Yokomori, and M. Hagiya. Molecular computation by DNA hairpin formation. *Science*, 288:1223–1226, 2000.
60. N. Seeman, F. Liu, C. Mao, X. Yang, L. Wenzler, and E. Winfree. DNA nanotechnology: control of 1-D and 2-D arrays and the construction of a nanomechanical device. In *Preliminary Proceedings of 4th DIMACS Workshop on DNA Based Computers*, pages 241–242, 1998.
61. N. Seeman, H. Wang, B. Liu, J. Qi, X. Li, X. Yang, F. Liu, W. Sun, Z. Shen, R. Sha, C. Mao, Y. Wang, S. Zhang, T. Fu, S. Du, J. Mueller, Y. Zhang, and J. Chen. The perils of polynucleotides: the experimental gap between the design and assembly of unusual DNA structures. In L. Landweber and E. Baum, editors, *DNA Based Computers II*, pages 215–234. The American Mathematical Society DIMACS Series in Discrete Mathematics and Theoretical Computer Science Volume **44**, 1999.
62. E. Winfree. Complexity of restricted and unrestricted models of molecular computation. In R. Lipton and E. Baum, editors, *DNA Based Computers*, pages 187–198. The American Mathematical Society DIMACS Series in Discrete Mathematics and Theoretical Computer Science Volume **27**, 1996.
63. E. Winfree. On the computational power of DNA annealing and ligation. In R. Lipton and E. Baum, editors, *DNA Based Computers*, pages 199–221. The American Mathematical Society DIMACS Series in Discrete Mathematics and Theoretical Computer Science Volume **27**, 1996.
64. E. Winfree. Whiplash PCR for $O(1)$ computing. In *Preliminary Proceedings of 4th DIMACS Workshop on DNA Based Computers*, pages 175–188, 1998.
65. E. Winfree, F. Liu, L. A. Wenzler, and N. C. Seeman. Design and self-assembly of two-dimensional DNA crystals. *Nature*, 394:539–544, 1998.
66. E. Winfree, X. Yang, and N. Seeman. Universal comptuation via self-assembly of DNA: some theory and experiments. In L. Landweber and E. Baum, editors, *DNA Based Computers II*, pages 191–214. The American Mathematical Society DIMACS Series in Discrete Mathematics and Theoretical Computer Science Volume **44**, 1999.
67. T. Yokomori and S. Kobayashi. DNA evolutionary linguistics and RNA structure modeling: a computational approach. In *Proceedings of International Symposium on Intelligence in Neural and Biological Systems*, pages 38–45. IEEE Computer Society Press, Los Alamitos, CA, 1995.
68. T. Yokomori, S. Kobayashi, and C. Ferretti. On the power of circular splicing systems and DNA computability. In *Proceedings of International Conference on Evolutionary Computing*, pages 219–224. IEEE Computer Society Press, Los Alamitos, CA, 1997.

11

Restarting Automata

Friedrich Otto

Fachbereich Mathematik/Informatik
Universität Kassel
D-34109 Kassel
E-mail: `otto@theory.informatik.uni-kassel.de`
`http://www.theory.informatik.uni-kassel.de`

Summary. The restarting automaton was introduced by Jančar et al. in 1995 in order to model the so-called 'analysis by reduction,' which is a technique used in linguistics to analyse sentences of natural languages that have a free word order. By now there are many different models of restarting automata, and their investigation has proved very fruitful in that they offer an opportunity to study the influence of various kinds of resources on their expressive power. Here we introduce and discuss the main variants of these automata, and present the main results that have been obtained until now. In particular, we investigate the relationship of the language classes that are defined through the various types of restarting automata to the classes of the Chomsky hierarchy, and we address some open problems.

11.1 Introduction

Analysis by reduction is a technique used in linguistics to analyse sentences of natural languages that have a free word order [48, 49]. This analysis consists of a stepwise simplification of a sentence in such a way that the syntactical correctness or incorrectness of the sentence is not affected. After a finite number of steps either a correct simple sentence is obtained, or an error is detected. In the former case the given sentence is accepted as being syntactically correct; if, however, all possible sequences of simplifications yield errors, then the given sentence is not syntactically correct. In this way it is also possible to determine dependencies between various parts of the given sentence, and to disambiguate between certain morphological ambiguities contained in the sentence. To illustrate this, we consider the following (admittedly rather artificial) example:

They mean that the means she means are very mean.

To analyse this sentence we read it from left to right until we discover a phrase that can be simplified. For example we can delete the word that or the word very, thus obtaining the following sentences:

F. Otto: *Restarting Automata*, Studies in Computational Intelligence (SCI) **25**, 269–303 (2006)
`www.springerlink.com`

(1) They mean the means she means are very mean.

(2) They mean that the means she means are mean.

As both these simplifications are correct, we can conclude that the phrases that and very are not dependent on each other, that is, they are *independent.* From (1) as well as from (2) we can obtain the following sentence in the next step:

(3) They mean the means she means are mean.

Next we simplify (3) in two different ways by deleting the phrase the means and the phrase They mean, respectively:

(4) They mean she means are mean.

(5) The means she means are mean.

As (5) is correct, while (4) is not, we conclude that the phrase They mean *depends* on the phrase the means. Continuing with (5) we obtain a simple sentence:

(6) The means are mean.

This simple sentence is easily recognised as being syntactically correct, and therewith our analysis finishes with success. Thus, we conclude that the original sentence is syntactically correct. In addition we have obtained some information on dependencies and independencies between certain parts of the sentence. Further, we can also use this analysis to determine that the first occurrence of the word mean in the given sentence is a verb, while the second occurrence of this word is an adjective, in this way solving some disambiguities occurring in the given sentence.

A generative device capable of simulating the analysis by reduction is the *sorted dependency 1-contextual grammar with regular selectors* by Gramatovici and Martín-Vide [12], which combines the *contextual grammars* of Marcus (see, e.g., [9, 28]) with dependency grammars (see, e.g., [8]).

As illustrated by Jančar et al. (cf., e.g., [21]) the restarting automaton was invented to model the analysis by reduction. In fact, many aspects of the work on restarting automata are motivated by the basic tasks of computational linguistics. The notions developed in the study of restarting automata give a rich taxonomy of constraints for various models of analysers and parsers. Already several programs are being used in Czech and German (corpus) linguistics that are based on the idea of restarting automata (cf., e.g., [40, 44]).

As defined in [17] the *restarting automaton* is a nondeterministic machine model that processes strings which are stored in a list (or a 'rubber' tape) with end markers. It has a finite control, and it has a read/write window with a finite look-ahead working on the list of symbols. As such it can be seen as a modification of the *list automaton* presented in [6] and the *forgetting au-*

tomaton [15, 16]. However, the restarting automaton can only perform two kinds of operations: *move-right transitions*, which shift the read/write window one position to the right, thereby changing the actual state, and combined *delete/restart transitions*, which delete some symbols from the read/write window, place this window over the left end of the list, and put the automaton back into its initial state. Hence, after performing a delete/restart transition a restarting automaton has no way to remember that it has already performed some steps of a computation. Further, by each application of a delete/restart transition the list is shortened. In this aspect the restarting automaton is similar to the *contraction automaton* [47]. It follows that restarting automata are linearly space-bounded.

Subsequently Jančar et al. extended their model in various ways. Instead of simply deleting some symbols from the actual content of the read/write window during a delete/restart transition, a restarting automaton *with rewriting* has combined *rewrite/restart* transitions that replace the content of the read/write window by a shorter string [18]. Further, the use of auxiliary (that is, non-input) symbols was added to the model in [20], which yields the so-called RWW-*automaton*. Also in [20] the restart transition was separated from the rewrite transition so that, after performing a rewrite step, the automaton can still read the remaining part of the tape before performing a restart transition. This gives the so-called RRWW-*automaton*, which is required to execute exactly one rewrite transition between any two restart transitions. In addition, various notions of *monotonicity* have been discussed for the various types of restarting automata. It turned out that monotone RWW- and RRWW-automata accept exactly the context-free languages, and that all the various types of monotone deterministic RWW- and RRWW-automata accept exactly the deterministic context-free languages. Finally, *move-left transitions* were added to the restarting automaton, which gave the so-called *two-way restarting automaton* (RLWW-automaton) [43]. This automaton can first scan its tape completely, and then move left to apply a rewrite transition to some factor of the tape content.

Here we give an introduction to the restarting automaton and its many variants. We compare the expressive power of the various types of restarting automata to each other and to the classes of the Chomsky hierarchy. Of particular interest will be the class *GCS* of *growing context-sensitive languages*, which was considered in detail by Dahlhaus and Warmuth, who proved that the membership problem for each growing context-sensitive language is solvable in polynomial time [7]. A detailed discussion of this class, which has many nice closure properties, can be found in [4] (see also [29]). We will see that even the most restricted type of restarting automaton is still sufficiently powerful to accept all deterministic context-free languages. On the other hand, the unrestricted RLWW-automaton is more expressive than the class *GCS*. However, it is still open whether there is any context-sensitive language that cannot be accepted by any RLWW-automaton.

This chapter is structured as follows. In the next section the basic variants of the restarting automaton are defined. The definitions given here differ slightly from the ones found in the literature (cf., e.g., [21, 39]), but they are easily seen to be equivalent to them. In Section 11.3 monotone restarting automata are considered, and the relationship between the various language classes thus obtained and the (deterministic) context-free languages is described. Then we turn to various types of deterministic restarting automata (Section 11.4), and we discuss their relationship to the Church-Rosser languages [30, 33]. In Section 11.5 weakly monotone restarting automata are introduced, and it is shown that they characterise the growing context-sensitive languages. Then in Section 11.6 the language classes defined by the unrestricted types of nondeterministic restarting automata are compared to each other and to the classes of the Chomsky hierarchy, and in Section 11.7 we discuss in short the notion of *left-monotonicity* for restarting automata. This notion, which at first glance appears to be symmetric to the notion of monotonicity, shows a completely different behaviour for the deterministic types of restarting automata. In the final section further variants of the restarting automaton are mentioned, and some open problems are presented.

Due to the introductory nature of this presentation many aspects of the restarting automaton are only touched upon. In addition we give almost no proofs, but for all results presented the appropriate references are cited. Furthermore, we illustrate the concepts presented through detailed examples.

Throughout this chapter all alphabets considered are finite. For a string w and a non-negative integer n, w^n is defined inductively by $w^0 := \lambda$ and $w^{n+1} := w^n w$, and w^R denotes the *mirror image* of w. For a language L, we use the notation L^R to denote the language $L^R := \{ w^R \mid w \in L \}$. Finally, for an automaton M, the language accepted by M is denoted as $L(M)$, and for a class A of automata, $\mathcal{L}(\mathsf{A})$ denotes the class of languages that are accepted by automata from that class.

11.2 Definitions and Examples

In defining the restarting automaton and its main variants we will not follow the historical development outlined above, but we will first present the most general model, the RLWW-automaton, and then describe the other variants as restrictions thereof.

As indicated above a restarting automaton is a nondeterministic machine model that has a finite control and a read/write window that works on a list of symbols delimited by end markers (see Figure 11.1).

Formally, a *two-way restarting automaton*, RLWW-automaton for short, is a one-tape machine that is described by an 8-tuple $M = (Q, \Sigma, \Gamma, ¢, \$, q_0, k, \delta)$, where Q is a finite set of states, Σ is a finite input alphabet, Γ is a finite tape alphabet containing Σ, $¢, \$ \notin \Gamma$ are symbols that serve as markers for the left and right border of the work space, respectively, $q_0 \in Q$ is the initial state,

$k \in \mathbf{N}_+$ is the size of the *read/write window*, and

$$\delta : Q \times \mathcal{PC}^{(k)} \to 2^{((Q\times(\{\mathsf{MVR},\mathsf{MVL}\}\cup\mathcal{PC}^{\leq(k-1)}))\cup\{\mathsf{Restart},\mathsf{Accept}\})}$$

is the *transition relation*. Here $\mathcal{PC}^{(k)}$ is the set of *possible contents* of the read/write window of M, where

$$\mathcal{PC}^{(i)} := (\text{¢} \cdot \Gamma^{i-1}) \cup \Gamma^{i} \cup (\Gamma^{\leq i-1} \cdot \$) \cup (\text{¢} \cdot \Gamma^{\leq i-2} \cdot \$) \quad (i \geq 0),$$

and

$$\Gamma^{\leq n} := \bigcup_{i=0}^{n} \Gamma^{i} \quad \text{and} \quad \mathcal{PC}^{\leq(k-1)} := \bigcup_{i=0}^{k-1} \mathcal{PC}^{(i)}.$$

Fig. 11.1. Schematic representation of a restarting automaton

The transition relation describes five different types of transition steps:

(1) A *move-right step* is of the form $(q', \mathsf{MVR}) \in \delta(q, u)$, where $q, q' \in Q$ and $u \in \mathcal{PC}^{(k)}$, $u \neq \$$. If M is in state q and sees the string u in its read/write window, then this move-right step causes M to shift the read/write window one position to the right and to enter state q'. However, if the content u of the read/write window is only the symbol \$, then no shift to the right is possible.

(2) A *move-left step* is of the form $(q', \mathsf{MVL}) \in \delta(q, u)$, where $q, q' \in Q$ and $u \in \mathcal{PC}^{(k)}$ does not begin with the symbol ¢. It causes M to shift the read/write window one position to the left and to enter state q'. This, however, is only possible if the window is not already at the left end of the tape.

(3) A *rewrite step* is of the form $(q', v) \in \delta(q, u)$, where $q, q' \in Q$, $u \in \mathcal{PC}^{(k)}$, $u \neq \$$, and $v \in \mathcal{PC}^{\leq(k-1)}$ such that $|v| < |u|$. It causes M to replace the content u of the read/write window by the string v, thereby shortening the tape, and to enter state q'. Further, the read/write window is placed immediately to the right of the string v. However, some additional restrictions apply in that the border markers ¢ and \$ must not disappear from the tape nor that new occurrences of these markers are created. Further,

the read/write window must not move across the right border marker \$, that is, if the string u ends in \$, then so does the string v, and after performing the rewrite operation, the read/write window is placed on the \$-symbol.

(4) A *restart step* is of the form Restart $\in \delta(q, u)$, where $q \in Q$ and $u \in \mathcal{PC}^{(k)}$. It causes M to place its read/write window over the left end of the tape, so that the first symbol it sees is the left border marker ¢, and to reenter the initial state q_0.

(5) An *accept step* is of the form Accept $\in \delta(q, u)$, where $q \in Q$ and $u \in \mathcal{PC}^{(k)}$. It causes M to halt and accept.

If $\delta(q, u) = \emptyset$ for some $q \in Q$ and $u \in \mathcal{PC}^{(k)}$, then M necessarily halts, and we say that M *rejects* in this situation. Further, the letters in $\Gamma - \Sigma$ are called *auxiliary symbols*.

A *configuration* of M is a string $\alpha q \beta$, where $q \in Q$, and either $\alpha = \lambda$ and $\beta \in \{¢\} \cdot \Gamma^* \cdot \{\$\}$ or $\alpha \in \{¢\} \cdot \Gamma^*$ and $\beta \in \Gamma^* \cdot \{\$\}$; here $q \in Q$ represents the current state, $\alpha\beta$ is the current content of the tape, and it is understood that the read/write window contains the first k symbols of β or all of β when $|\beta| \leq k$. A *restarting configuration* is of the form $q_0 ¢ w \$$, where $w \in \Gamma^*$; if $w \in \Sigma^*$, then $q_0 ¢ w \$$ is an *initial configuration*. Thus, initial configurations are a particular type of restarting configurations. Further, we use Accept to denote the *accepting configurations*, which are those configurations that M reaches by executing an Accept instruction. A configuration of the form $\alpha q \beta$ such that $\delta(q, \beta_1) = \emptyset$, where β_1 is the current content of the read/write window, is a *rejecting configuration*. A *halting configuration* is either an accepting or a rejecting configuration.

In general, the automaton M is *nondeterministic*, that is, there can be two or more instructions with the same left-hand side (q, u), and thus, there can be more than one computation for an input word. If this is not the case, the automaton is *deterministic*. We use the prefix det- to denote deterministic classes of restarting automata.

We observe that any finite computation of a two-way restarting automaton M consists of certain phases. A phase, called a *cycle*, starts in a restarting configuration, the head moves along the tape performing MVR, MVL, and Rewrite operations until a Restart operation is performed and thus a new restarting configuration is reached. If no further Restart operation is performed, any finite computation necessarily finishes in a halting configuration – such a phase is called a *tail*. We require that M performs *exactly one* Rewrite operation during any cycle – thus each new phase starts on a shorter word than the previous one. During a tail at most one Rewrite operation may be executed. By $\vdash_M^c$ we denote the execution of a complete cycle, and $\vdash_M^{c^*}$ is the reflexive transitive closure of this relation. It can be seen as the *rewrite relation* that is realised by M on the set of restarting configurations.

An input $w \in \Sigma^*$ is accepted by M, if there exists a computation of M which starts with the initial configuration $q_0 ¢ w \$$, and which finally ends with executing an Accept instruction.

Given an input of length n, an RLWW-automaton can execute at most n cycles. Thus, we have the following result, where NP and P denote the well-known complexity classes, and DCS denotes the class of *deterministic context-sensitive languages*, that is, the space complexity class $DSPACE(n)$.

Proposition 1. (a) $\mathcal{L}(\mathsf{RLWW}) \subseteq NP \cap CS$.
(b) $\mathcal{L}(\text{det-}\mathsf{RLWW}) \subseteq P \cap DCS$.

Before we continue with more definitions, we present a simple example.

Example 1. Let $M := (Q, \Sigma, \Sigma, ¢, \$, q_0, 3, \delta)$ be the deterministic RLWW-automaton that is defined by taking $Q := \{q_0, q_c, q_d, q_r\}$, $\Sigma := \{a, b, c, d\}$, and δ as given by the following table:

(1) $\delta(q_0, x) = (q_0, \mathsf{MVR})$ for all $x \in \{aaa, aab, abb, abc, bbb, bbc, bbd\}$,
(2) $\delta(q_0, ¢c\$) =$ Accept,
(3) $\delta(q_0, ¢d\$) =$ Accept,
(4) $\delta(q_0, ¢ab) = (q_0, \mathsf{MVR})$,
(5) $\delta(q_0, ¢aa) = (q_0, \mathsf{MVR})$,
(6) $\delta(q_0, bc\$) = (q_c, \mathsf{MVL})$,
(7) $\delta(q_0, bd\$) = (q_d, \mathsf{MVL})$,
(8) $\delta(q_r, -) =$ Restart,
(9) $\delta(q_c, abc) = (q_r, c)$,
(10) $\delta(q_c, bbc) = (q_c, \mathsf{MVL})$,
(11) $\delta(q_c, bbb) = (q_c, \mathsf{MVL})$,
(12) $\delta(q_c, abb) = (q_r, b)$,
(13) $\delta(q_d, bbd) = (q_d, \mathsf{MVL})$,
(14) $\delta(q_d, bbb) = (q_d, \mathsf{MVL})$,
(15) $\delta(q_d, abb) = (q_r, \lambda)$.

Here q_r is a *restart state*, that is, in state q_r, M excutes a Restart operation, independently of the current content of the read/write window. Obviously, M accepts c and d immediately. So let $w \in \Sigma^+ - \{c, d\}$. Starting from the initial configuration $q_0 ¢ w \$$, M will get stuck (and therewith reject) while scanning w from left to right unless w is of the form $w = a^m b^n c$ or $w = a^m b^n d$ for some positive integers m and n. In these cases the configuration $¢ a^m b^{n-1} q_0 bc\$$ or $¢ a^m b^{n-1} q_0 bd\$$ is reached by performing the transition steps (4) or (5) and (1) repeatedly, and then either the state q_c (6) or the state q_d is entered (7). Now M performs MVL-steps until the read/write window gets back to the boundary between the syllables a^m and b^n. If the actual state is q_c, that is, if w ends in c, then a factor ab is deleted by (9) or (12); if the actual state is q_d, that is, if w ends in d, then a factor abb is deleted (15). In both cases M enters the state q_r and restarts (8). Thus, we see that M accepts the following language

$$L_1 := \{ a^n b^n c \mid n \geq 0 \} \cup \{ a^n b^{2n} d \mid n \geq 0 \},$$

which is a well-known example of a context-free language that is not deterministic context-free.

Next we restate some basic facts about computations of restarting automata. These facts will be used repeatedly in the following, mostly without mentioning them explicitly.

Proposition 2. (Error Preserving Property).
Let $M = (Q, \Sigma, \Gamma, ¢, \$, q_0, k, \delta)$ *be an* RLWW-*automaton, and let* u, v *be words over its* input *alphabet* Σ. *If* $q_0 ¢ u \$ \vdash_M^{c^*} q_0 ¢ v \$$ *holds and* $u \notin L(M)$, *then* $v \notin L(M)$, *either.*

Proposition 3. (Correctness Preserving Property).
Let $M = (Q, \Sigma, \Gamma, ¢, \$, q_0, k, \delta)$ *be an* RLWW-*automaton, and let* u, v *be words over its* input *alphabet* Σ. *If* $q_0 ¢ u \$ \vdash_M^{c^*} q_0 ¢ v \$$ *is an initial segment of an* accepting *computation of* M, *then* $v \in L(M)$.

Also the following fact from [43] will be used repeatedly.

Proposition 4. (Pumping Lemma).
For any RLWW-*automaton* $M = (Q, \Sigma, \Gamma, ¢, \$, q_0, k, \delta)$, *there exists a constant* p *such that the following holds. Assume that* $q_0 ¢ uvw \$ \vdash_M^{c} q_0 ¢ uv'w \$$, *where* $u = u_1 u_2 u_3$ *and* $|u_2| = p$. *Then there exists a factorisation* $u_2 = z_1 z_2 z_3$ *such that* z_2 *is non-empty, and*

$$q_0 ¢ u_1 z_1 (z_2)^i z_3 u_3 v w \$ \vdash_M^{c} q_0 ¢ u_1 z_1 (z_2)^i z_3 u_3 v' w \$$$

holds for all $i \geq 0$, *that is,* z_2 *is a 'pumping factor' in the above cycle. Similarly, such a pumping factor can be found in any factor of length* p *of* w. *Such a pumping factor can also be found in any factor of length* p *of a word accepted in a tail computation.*

Each cycle of each computation of an RLWW-automaton M consists of three phases: first M scans its tape performing MVR- and MVL-instructions, then it executes a Rewrite step, and finally it scans its tape again performing MVR- and MVL-steps. Hence, in the first and the last phase of each cycle M behaves like a nondeterministic two-way finite-state acceptor (2NFA). In analogy to the proof that the language accepted by a 2NFA is regular (see, e.g, [14]), the following result can be established. Here a restarting automaton is called an RRWW-*automaton* if it does not use any MVL-transitions. Thus, in each cycle an RRWW-automaton can scan its tape only once from left to right.

Theorem 1. [43] *Let* $M_L = (Q_L, \Sigma, \Gamma, ¢, \$, q_0, k, \delta_L)$ *be an* RLWW-*automaton. Then there exists an* RRWW-*automaton* $M_R = (Q_R, \Sigma, \Gamma, ¢, \$, q_0, k, \delta_R)$ *such that, for all* $u, v \in \Gamma^*$,

$$q_0 ¢ u \$ \vdash_{M_L}^{c} q_0 ¢ v \$ \quad \textit{if and only if} \quad q_0 ¢ u \$ \vdash_{M_R}^{c} q_0 ¢ v \$,$$

and the languages $L(M_L)$ *and* $L(M_R)$ *coincide.*

Thus, as far as nondeterministic restarting automata are concerned, the MVL-instruction is not needed. However, this does not hold for deterministic restarting automata, as we will see later. For RRWW-automata we have the following normalisation result, which is easily proved by using standard techniques from automata theory.

Lemma 1. *Each* RRWW*-automaton M is equivalent to an* RRWW*-automaton M' that satisfies the following additional restriction:*

($*$) *M' makes an accept or a restart step only when it sees the right border marker \$ in its read/write window.*

This lemma means that in each cycle of each computation and also during the tail of each accepting computation the read/write window moves all the way to the right before a restart is made, respectively, before the machine halts and accepts.

Based on this fact each cycle (and also the tail) of a computation of an RRWW-automaton consists of three phases. Accordingly, the transition relation of an RRWW-automaton can be described through a sequence of so-called *meta-instructions* [38] of the form $(E_1, u \to v, E_2)$, where E_1 and E_2 are regular languages, called the *regular constraints* of this instruction, and u and v are strings such that $|u| > |v|$. The rule $u \to v$ stands for a Rewrite step of the RRWW-automaton M considered. On trying to execute this meta-instruction M will get stuck (and so reject) starting from the configuration $q_0 \cent w \$$, if w does not admit a factorisation of the form $w = w_1 u w_2$ such that $\cent w_1 \in E_1$ and $w_2 \$ \in E_2$. On the other hand, if w does have a factorisation of this form, then one such factorisation is chosen nondeterministically, and $q_0 \cent w \$$ is transformed into $q_0 \cent w_1 v w_2 \$$. In order to describe the tails of accepting computations we use meta-instructions of the form $(\cent \cdot E \cdot \$, \mathsf{Accept})$, where the strings from the regular language E are accepted by M in tail computations. We illustrate this concept by describing an RRWW-automaton for the language L_1 of Example 1.

Example 2. Let M_1 be the RRWW-automaton with input alphabet $\{a, b, c, d\}$ and without auxiliary symbols that is described by the following sequence of meta-instructions:

$$(1)\ (\cent \cdot a^*,\ ab \to \lambda,\ b^* \cdot c\,\$), \qquad (3)\ (\cent c\,\$,\ \mathsf{Accept}),$$
$$(2)\ (\cent \cdot a^*,\ abb \to \lambda,\ b^* \cdot d\,\$), \qquad (4)\ (\cent d\,\$,\ \mathsf{Accept}).$$

It is easily seen that $L(M_1) = L_1$ holds.

This way of describing an RRWW-automaton corresponds to the characterisation of the class $\mathcal{L}(\mathsf{RRWW})$ by certain infinite prefix-rewriting systems as given in [39], Corollary 6.4.

Finally, we introduce some restricted types of restarting automata. A restarting automaton is called an RWW-*automaton* if it makes a restart immediately after performing a rewrite operation. In particular, this means that it cannot perform a rewrite step during the tail of a computation.

A cycle of a computation of an RWW-automaton M consists of two phases only. Accordingly, the transition relation of an RWW-automaton can be described by a finite sequence of *meta-instructions* of the form $(E, u \to v)$, where E is a regular language, and u and v are strings such that $|u| > |v|$, and meta-instructions of the form $(¢ \cdot E \cdot \$, \mathsf{Accept})$ for describing tail computations. This description corresponds to the characterisation of the class $\mathcal{L}(\mathsf{RWW})$ by certain infinite prefix-rewriting systems as given in [39], Corollary 6.2.

Example 3. The following sequence of meta-instructions defines an RWW-automaton M_2 with input alphabet $\Sigma := \{a, b, c, d\}$ and tape alphabet $\Gamma := \Sigma \cup \{C, D\}$ for the language L_1:

$$
\begin{array}{llll}
(1)\ (¢ \cdot a^*, & ab \to C), & (5)\ (¢Cc\$, & \mathsf{Accept}),\\
(2)\ (¢ \cdot a^*, & abb \to D), & (6)\ (¢Dd\$, & \mathsf{Accept}),\\
(3)\ (¢ \cdot a^*, & aCb \to C), & (7)\ (¢c\$, & \mathsf{Accept}),\\
(4)\ (¢ \cdot a^*, & aDbb \to D), & (8)\ (¢d\$, & \mathsf{Accept}).
\end{array}
$$

An RLWW-automaton is called an RLW-*automaton* if its tape alphabet Γ coincides with its input alphabet Σ, that is, if no auxiliary symbols are available. It is an RL-*automaton* if it is an RLW-automaton for which the right-hand side v of each Rewrite step $(q', v) \in \delta(q, u)$ is a scattered subword of the left-hand side u. Analogously, we obtain RRW- and RR-automata from the RRWW-automaton and RW- and R-automata from the RWW-automaton, respectively. The restarting automaton described in Example 1 is a deterministic RL-automaton, and the automaton M_1 of Example 2 is an RR-automaton. Further, for RLW- and RL-automata a result analogous to Theorem 1 holds.

Figure 11.2 summarises the obvious inclusions between the language classes defined by the various types of restarting automata, the inclusions of Proposition 1, and the equalities of Theorem 1. In the following we are concerned with two main topics:

(a) Which of the inclusions in Figure 11.2 are proper?
(b) Where do the language classes defined by the various types of restarting automata lie in relation to other well-known language classes?

To answer these questions we will derive additional characterisations for some of these classes, and we will present some closure properties. The next result explains the role that the auxiliary symbols play in restarting automata. It is a slight generalisation of a result that has been presented in [39] for RRWW- and for RWW-automata.

Theorem 2. *A language L is accepted by a (deterministic)* RLWW-*automaton if and only if there exist a (deterministic)* RLW-*automaton M_1 and a regular language E such that $L = L(M_1) \cap E$.*

This characterisation yields the following closure property.

Corollary 1. *The language classes* $\mathcal{L}$(RLWW), $\mathcal{L}$(RRWW), *and* $\mathcal{L}$(RWW) *and their deterministic counterparts are closed under the operation of intersection with regular languages.*

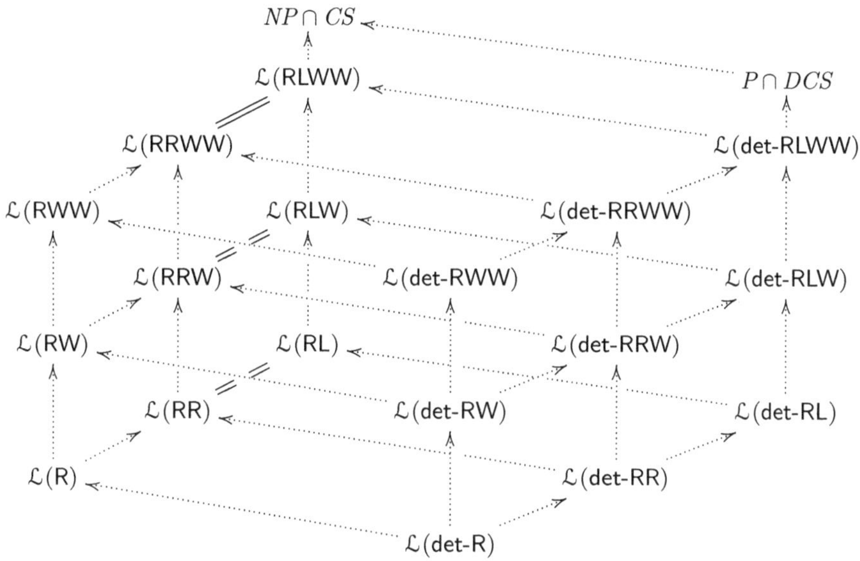

Fig. 11.2. Inclusions between the language classes defined by the basic types of restarting automata. The equalities, expressed by =, follow from Theorem 1, while $\cdots >$ denotes a (not necessarily proper) inclusion.

Let $M = (Q, \Sigma, \Gamma, ¢, \$, q_0, k, \delta)$ be an RLWW-automaton. The corresponding RLW-automaton M_1 and the regular language E of Theorem 2 are obtained by taking $M_1 := (Q, \Gamma, \Gamma, ¢, \$, q_0, k, \delta)$ and $E := \Sigma^*$, that is, M_1 is obtained from M by simply declaring all auxiliary letters of M to input symbols. The identity mapping on Σ^* is obviously a reduction from the language $L(M)$ to the language $L(M_1)$. Thus, we have the following consequence.

Corollary 2. *The language class* $\mathcal{L}$(RLWW) *is reducible in linear time and constant space to the language class* $\mathcal{L}$(RLW).

An analogous result holds for the classes $\mathcal{L}$(RRWW) and $\mathcal{L}$(RWW) and the corresponding deterministic classes.

Next we turn to a reduction that replaces arbitrary rewrite operations by delete operations, that is, we reduce the class $\mathcal{L}$(RLW) to the class $\mathcal{L}$(RL).

Let $\Gamma_1 = \{a_1, \ldots, a_m\}$, let $k \in \mathbf{N}_+$, and let $\Gamma_2 := \{0, 1, c, d\}$. We define an encoding $\varphi_{k,m} : \Gamma_1 \to \Gamma_2^+$ as follows:

$$a_i \mapsto c1^{m+1-i}0^i(cd1^{m+1}0^{m+1})^k \quad (1 \le i \le m).$$

Then, for all $1 \le i \le m$,

$$|\varphi_{k,m}(a_i)| = (m+1) \cdot (2k+1) + 2k + 1 = (m+2) \cdot (2k+1).$$

The encoding $\varphi_{k,m}$ is naturally extended to strings by taking

$$\varphi_{k,m}(x_1 x_2 \ldots x_n) := \varphi_{k,m}(x_1) \ldots \varphi_{k,m}(x_n) \text{ for all } x_1, \ldots, x_n \in \Gamma_1,\ n \geq 0.$$

Observe that $\varphi_{k,m} : \Gamma_1^* \to \Gamma_2^*$ is indeed an encoding, that is, it is an injective mapping. It has the following important property.

Lemma 2. [26] *For all $u \in \Gamma_1^k$ and $v \in \Gamma_1^*$, if $|v| < k$, then $\varphi_{k,m}(v)$ is a scattered subword of $\varphi_{k,m}(u)$.*

Based on this kind of encoding we obtain our second reduction.

Theorem 3. *If a language L is accepted by a (deterministic)* RLW*-automaton M with tape alphabet Γ_1 of cardinality m and read/write window of size k, then there exists a (deterministic)* RL*-automaton M' which accepts the language $\varphi_{k,m}(L) \subseteq \Gamma_2^*$.*

Proof. Let $M = (Q, \Gamma_1, \Gamma_1, ¢, \$, q_0, k, \delta)$ be an RLW-automaton for L with $\text{card}(\Gamma_1) = m$. Let r denote the number $r := (m+2) \cdot (2k+1)$. Then we can construct an RL-automaton M' with tape alphabet Γ_2 and read/write window of size $k \cdot r$ as follows. Whenever the distance of the read/write window of M' from the left end of its tape is a multiple of r, then M' checks whether the read/write window contains a string that is the image $\varphi_{k,m}(u)$ for some $u \in \Gamma_1^*$. In the negative it rejects immediately, in the affirmative it performs an action that simulates the actual transition of M. It follows that M' accepts the language $\varphi_{k,m}(L)$, and that M' is deterministic, if M is. □

Obviously each encoding of the form $\varphi_{k,m}$ can be computed in linear time and constant space. Thus, we have the following result.

Corollary 3. *The language class $\mathcal{L}$(RLW) is reducible in linear time and constant space to the language class $\mathcal{L}$(RL).*

Again, an analogous result holds for the classes $\mathcal{L}$(RRW) and $\mathcal{L}$(RW) and the corresponding deterministic classes.

We close this section with an example. Let $L_{\text{copy}} := \{ w\#w \mid w \in \{a,b\}^* \}$. This language is obviously context-sensitive, but it is not growing context-sensitive [5, 27].

Example 4. Let $M := (Q, \Sigma, \Gamma, ¢, \$, q_0, 3, \delta)$ be the RRWW-automaton on $\Sigma := \{a, b, \#\}$ and $\Gamma := \Sigma \cup \{ [c,d] \mid c,d \in \{a,b\} \}$ that is given by the following sequence of meta-instructions, where $c, d \in \{a,b\}$:

(1) $(¢ \cdot \{a,b\}^*, cd\# \to [c,d]\#, \{a,b\}^* \cdot \$)$,
(2) $(¢ \cdot \{a,b\}^* \cdot [c,d]\# \cdot \{a,b\}^*, cd\$ \to [c,d]\$, \lambda)$,
(3) $(¢ \cdot \{a,b\}^*, [c,d]\# \to \#, \{a,b\}^* \cdot [c,d]\$)$,
(4) $(¢ \cdot \{a,b\}^* \cdot \# \cdot \{a,b\}^*, [c,d]\$ \to \$, \lambda)$,
(5) $(¢c\#c\$, \text{Accept})$,
(6) $(¢\#\$, \text{Accept})$.

Then it is easily verified that $L(M) = L_{\text{copy}}$ holds.

This example shows that the language class $\mathcal{L}(\mathsf{RRWW})$ contains languages that are not growing context-sensitive. But we can do better than that.

Example 5. Let $M' := (Q, \Sigma, \Gamma, \text{¢}, \$, q_0, 4, \delta)$ be the RWW-automaton with input alphabet $\Sigma := \{a, b, \#\}$ and tape alphabet $\Gamma := \Sigma \cup \{\#'\} \cup \Gamma'$, where

$$\Gamma' := \{\, [c, d, S] \mid c, d \in \{a, b\}, S \in \{0, 1\}\,\},$$

that is given through the following sequence of meta-instructions. Here $c, d, e, f \in \{a, b\}$, $S, S' \in \{0, 1\}$, and $\bar{S} := 1 - S$:

(1) $(\text{¢}, cd \to [c, d, 0])$,
(2) $(\text{¢} \cdot \Gamma'^* \cdot [e, f, S], cd \to [c, d, \bar{S}])$,
(3) $(\text{¢}[c, d, 0] \cdot \{a, b\}^* \cdot \#, cd \to [c, d, 0])$,
(4) $(\text{¢} \cdot \Gamma'^* \cdot [c, d, S] \cdot \{a, b\}^* \cdot \# \cdot \Gamma'^* \cdot [e, f, \bar{S}], cd \to [c, d, S])$,
(5) $(\text{¢} \cdot \Gamma'^*, [c, d, S] \cdot X \cdot [e, f, S'] \to \#')$ for all $X \in \{\#, \#'\}$,
(6) $(\text{¢} \cdot \Gamma'^*, [c, d, S] \cdot g \cdot X \cdot [e, f, S'] \to g\#')$ for all $g \in \{a, b\}, X \in \{\#, \#'\}$,
(7) $(\text{¢}X\$, \mathsf{Accept})$ for all $X \in \{\#, \#'\}$,
(8) $(\text{¢}gXg\$, \mathsf{Accept})$ for all $g \in \{a, b\}, X \in \{\#, \#'\}$.

Given the string $abaab\#abaab$ as input, M' can execute the following computation:

$$\begin{aligned} q_0\text{¢}\underline{ab}aab\#abaab\$ &\vdash^c_{(1)} q_0\text{¢}[a, b, 0]aab\#\underline{ab}aab\$ \\ &\vdash^c_{(3)} q_0\text{¢}[a, b, 0]\underline{aa}b\#[a, b, 0]aab\$ \\ &\vdash^c_{(2)} q_0\text{¢}[a, b, 0][a, a, 1]b\#[a, b, 0]\underline{aa}b\$ \\ &\vdash^c_{(4)} q_0\text{¢}[a, b, 0]\underline{[a, a, 1]b\#[a, b, 0]}[a, a, 1]b\$ \\ &\vdash^c_{(6)} q_0\text{¢}\underline{[a, b, 0]b\#'[a, a, 1]}b\$ \\ &\vdash^c_{(6)} q_0\text{¢}b\#'b\$ \\ &\vdash_{(8)} \mathsf{Accept}. \end{aligned}$$

We see that the meta-instructions (1) and (2) compress the first $\{a, b\}$-syllable $a_1a_2a_3a_4a_5a_6\ldots$ into the form $[a_1, a_2, 0][a_3, a_4, 1][a_5, a_6, 0]\ldots$. The meta-instructions (3) and (4) compress the second $\{a, b\}$-syllable in the same way. The alternating values in the third components of the Γ'-symbols guarantee that between any two applications of meta-instructions (3) or (4) at least one application of meta-instruction (2) takes place. It follows that the meta-instructions (1) to (4) can compress the first two $\{a, b\}$-syllables completely only if the second syllable is a scattered subword of the first syllable. Now the meta-instructions (5) and (6) delete the compressed syllables in a synchronous manner, and M' accepts if and only if the two syllables have the same length. Together these observations imply that M' accepts an input w if and only if w has the form $w = u\#u$ for some string $u \in \{a, b\}^*$, that is, we see that $L(M') = L_{\text{copy}}$.

By the corresponding analogue to Theorem 2, there exists an RW-automaton M'_1 and a regular language E such that $L(M'_1) \cap E = L(M') = L_{\text{copy}}$.

As $L_{\text{copy}} \notin GCS$, and as the class GCS is closed under the operation of intersection with regular languages, it follows that $L(M_1') \notin GCS$. Finally, as the tape alphabet of M_1' has cardinality 12, it follows from Theorem 3 that $\varphi_{4,12}(L(M_1'))$ is accepted by some R-automaton. As the class GCS is closed under inverse morphisms, this implies that the language $\varphi_{4,12}(L(M_1'))$ is not growing context-sensitive, either. Thus, we obtain the following non-inclusion result.

Corollary 4. $\mathcal{L}(\mathsf{R}) \nsubseteq GCS$.

Hence, already the R-automaton has a fairly large expressive power.

11.3 Monotone Restarting Automata

In [18] the notion of monotonicity is introduced for restarting automata. Here we consider a slightly more general notion, which is taken from [22]. Let M be an RLWW-automaton. Each computation of M can be described by a sequence of cycles $C_1, C_2 \ldots, C_n$, where C_n is the last cycle, which is followed by the tail of the computation. Each cycle C_i of this computation contains a unique configuration of the form $¢xquy\$$ such that q is a state and $(q', v) \in \delta(q, u)$ is the Rewrite step that is applied during this cycle. By $D_r(C_i)$ we denote the *right distance* $|uy\$|$ of this cycle, and $D_l(C_i) := |¢x|$ is the *left distance* of this cycle. The sequence of cycles $C_1, C_2, \ldots, C_n$ is called *monotone* if $D_r(C_1) \geq D_r(C_2) \geq \cdots \geq D_r(C_n)$ holds. A computation of M is called *monotone* if the corresponding sequence of cycles is monotone. Observe that the tail of the computation is not taken into account here. Finally, the RLWW-automaton M is called *monotone* if each of its computations that starts from an initial configuration is monotone.

Here we want to compare the expressive power of the various types of monotone restarting automata to each other. We use the prefix mon- to denote the classes of monotone restarting automata. The first result characterises the expressive power of monotone restarting automata with auxiliary symbols.

Theorem 4. [21, 43]

$$\mathcal{L}(\text{mon-RLWW}) = \mathcal{L}(\text{mon-RRWW}) = \mathcal{L}(\text{mon-RWW}) = CF.$$

Proof. From a context-free grammar G in Chomsky normal form we can construct a monotone RWW-automaton M such that, given an input word w, M tries to construct a rightmost derivation of w in G in reverse order. M will succeed if and only if such a derivation exists, that is, if and only if $w \in L(G)$. Hence, $L(M) = L(G)$, which shows that each context-free language is accepted by a monotone RWW-automaton.

Conversely, assume that the language L is accepted by a monotone RLWW-automaton. Because of Theorem 1 there is a monotone RRWW-automaton

M' that also accepts the language L. Let w be an input string, and let $C_1, C_2, \ldots, C_m$ be the sequence of cycles that corresponds to a computation of M' that starts from the initial configuration $q_0¢w\$$. For each cycle the tape content can be divided into three parts: a prefix w_1, the part u within the read/write window during the execution of the actual Rewrite step, and the remaining suffix w_2. As M' is monotone, the positions where the Rewrite steps are performed move from left to right across the tape. Accordingly, the prefixes w_1 can be stored in a pushdown store, and the suffices w_2 can be interpreted as the remaining part of the input. Thus, M' can be simulated by a pushdown automaton (see [21] for the details which are technically quite involved, as M' can already scan the suffix w_2 completely in each cycle), that is, $L(M)$ is context-free. □

If the monotone RRWW-automaton M is deterministic, then the above construction can be used to show that the language $L(M)$ is deterministic context-free. On the other hand, based on the characterisation of the deterministic context-free languages by LR(0)-grammars (see, e.g., [14]) it can be shown that already monotone det-R-automata accept all deterministic context-free languages, which yields the following characterisation.

Theorem 5. [18, 19, 20, 21]

For all types $\mathsf{X} \in \{\mathsf{R}, \mathsf{RR}, \mathsf{RW}, \mathsf{RRW}, \mathsf{RWW}, \mathsf{RRWW}\}, \mathcal{L}(\text{det-mon-}\mathsf{X}) = DCF$.

As the deterministic RL-automaton for the language L_1 that is given in Example 1 is monotone, the first inclusion in the following sequence is proper:

$$DCF = \mathcal{L}(\text{det-mon-RR}) \subset \mathcal{L}(\text{det-mon-RL}) \subseteq \mathcal{L}(\text{det-mon-RLWW}).$$

To investigate the classes of nondeterministic monotone restarting automata without auxiliary symbols, we consider the example language L_1 from the previous section and the following languages:

$$\begin{aligned} L_2 &:= \{\, a^n b^n \mid n \geq 0 \,\} \cup \{\, a^n b^m \mid m > 2n \geq 0 \,\}, \\ L_3 &:= \{f, ee\} \cdot \{\, a^n b^n \mid n \geq 0 \,\} \cup \{g, ee\} \cdot \{\, a^n b^m \mid m > 2n \geq 0 \,\}, \\ L_4 &:= \{\, a^n b^m \mid 0 \leq n \leq m \leq 2n \,\}. \end{aligned}$$

Lemma 3. $L_1 \in \mathcal{L}(\text{mon-RR}) - \mathcal{L}(\text{RW})$.

Proof. As L_1 is accepted by a monotone deterministic RL-automaton (see above), there exists a monotone (nondeterministic) RR-automaton for this language by Theorem 1.

Now assume that L_1 were accepted by some RW-automaton M_1 with read/write window of size k. For each input of the form $a^n b^{2n} d$, M_1 has an accepting computation. If $n > k$ is sufficiently large, then the corresponding computation must contain Restart steps, that is, it consists of a sequence of cycles followed by a tail computation. During the first cycle M_1 must delete a factor of the form $a^r b^{2r}$ because of the Correctness Preserving Property (Proposition 3). Here $3r \leq k < n$. As M_1 restarts immediately after performing a Rewrite step, it can apply the same Rewrite operation to the input $a^n b^{n+r} c$, which yields the cycle $q_0 ¢ a^n b^{n+r} c\$ \vdash^c_{M_1} q_0 ¢ a^{n-r} b^{n-r} c\$$. Now $a^{n-r} b^{n-r} c \in L_1$, while $a^n b^{n+r} c \notin L_1$, contradicting the Error Preserving Property (Proposition 2). Hence, L_1 is not accepted by any RW-automaton. □

Using the same kind of reasoning it can be shown that the context-free language L_2 is not in $\mathcal{L}$(RRW), and that $L_3 \in \mathcal{L}$(mon-RW) $-$ $\mathcal{L}$(RR) and $L_4 \in \mathcal{L}$(mon-R)$-\mathcal{L}$(det-RLW) hold. Together with the above results these facts yield the relations displayed in Figure 11.3, where $\mathcal{L}_1 \xrightarrow[L]{} \mathcal{L}_2$ denotes the proper inclusion of $\mathcal{L}_1$ in $\mathcal{L}_2$, with the language L separating the two language classes.

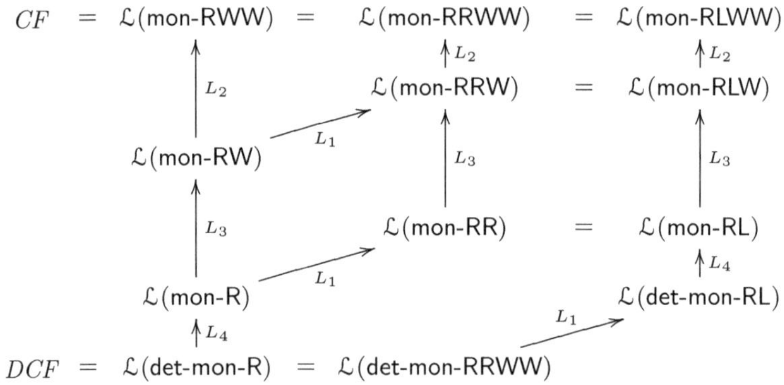

Fig. 11.3. The taxonomy of monotone restarting automata

Finally we turn to the problem of deciding whether a given restarting automaton is monotone. The solution rests on the following two technical results, the first of which is a straightforward generalisation of the result that each monotone RRWW-automaton can be simulated by a pushdown automaton.

Lemma 4. *Given an* RRWW-*automaton* M *with input alphabet* Σ, *a pushdown automaton* P_M *can be constructed such that* $L(P_M) = \{ w \in \Sigma^* \mid$ *There exists an accepting monotone computation for* M *on input* $w \}$.

Let M be an RRWW-automaton with tape alphabet Γ. We say that a word $w \in \Gamma^*$ *causes a non-monotone computation* of M, if there exist two cycles

$$C_1 : q_0 ¢ w\$ \vdash^c_M q_0 ¢ v\$ \ \text{and} \ C_2 : q_0 ¢ v\$ \vdash^c_M q_0 ¢ z\$$$

such that $D_r(C_1) < D_r(C_2)$. By $L_{\text{nm}}(M)$ we denote the language consisting of all words $w \in \Gamma^*$ that cause a non-monotone computation of M.

Lemma 5. *Given an* RRWW-*automaton M, a finite-state acceptor can be constructed for the language $L_{\mathrm{nm}}(M)$.*

Proof. The finite-state acceptor simulates two successive cycles C_1 and C_2 of M in parallel. It accepts if and only if the cycles simulated satisfy the condition that $D_r(C_1) < D_r(C_2)$. □

It follows in particular that the language $L_{\mathrm{nm}}(M)$ is regular. From Lemma 4 we obtain a pushdown automaton P'_M for M such that, for each input word $w \in \Sigma^*$ of M and each restarting configuration $q_0 ¢ z \$$ that M can reach from the initial configuration $q_0 ¢ w \$$ through a monotone sequence of cycles, P'_M has an accepting computation that ends with its pushdown store containing the word z. Let $SC_{P'_M}(\Sigma^*)$ denote the language that consists of those words z for which there exists an accepting computation of P'_M on some input $w \in \Sigma^*$ that ends with the pushdown store containing z. Then M is *not* monotone, if and only if the language $SC_{P'_M}(\Sigma^*)$ contains a word that causes a non-monotone computation of M, that is, if and only if $SC_{P'_M}(\Sigma^*) \cap L_{\mathrm{nm}}(M)$ is non-empty. According to [13] the language $SC_{P'_M}(\Sigma^*)$ is regular, and a finite-state acceptor for it can be constructed effectively from P'_M. Thus, we have the following decidability result.

Theorem 6. [22] *It is decidable whether an* RRWW-*automaton is monotone.*

By Theorem 1 we can associate with each RLWW-automaton M an RRWW-automaton M_R such that, in each computation, M_R executes the same sequence of cycles as M. Hence, M is monotone if and only if M_R is. Thus, Theorem 6 extends to RLWW-automata.

11.4 Deterministic Restarting Automata

As we will see, there is a close correspondence between certain types of deterministic restarting automata and the class of Church-Rosser languages. Therefore, we shortly restate the definition of Church-Rosser languages and their main properties.

Let Σ be an alphabet. A *string-rewriting system* R on Σ is a subset of $\Sigma^* \times \Sigma^*$. It induces several binary relations on Σ^*, the most basic of which is the *single-step reduction relation*

$$\rightarrow_R := \{ (u\ell v, urv) \mid u, v \in \Sigma^*, (\ell \rightarrow r) \in R \}.$$

Its reflexive and transitive closure is the *reduction relation* $\rightarrow_R^*$ induced by R. If $u \rightarrow_R^* v$, then u is an *ancestor* of v, and v is a *descendant* of u. If there is no $v \in \Sigma^*$ such that $u \rightarrow_R v$ holds, then the string u is called *irreducible* (mod R). By IRR(R) we denote the set of all irreducible strings. If R is finite, then IRR(R) is obviously a regular language. The string-rewriting system R is called

- *length-reducing* if $|\ell| > |r|$ holds for each rule $(\ell \to r) \in R$,
- *confluent* if, for all $u, v, w \in \Sigma^*$, $u \to_R^* v$ and $u \to_R^* w$ imply that v and w have a common descendant.

If a string-rewriting system R is length-reducing, then each reduction sequence starting with a string of length n has itself at most length n. If, in addition, R is confluent, then each string $w \in \Sigma^*$ has a unique irreducible descendant $\hat{w} \in \mathrm{IRR}(R)$, which can be computed from w in linear time (see, e.g., [3]). This observation was one of the main reasons to introduce the Church-Rosser languages in [30, 33].

Definition 1. *A language $L \subseteq \Sigma^*$ is a* Church-Rosser language *if there exist an alphabet $\Gamma \supset \Sigma$, a finite, length-reducing, confluent string-rewriting system R on Γ, two strings $t_1, t_2 \in (\Gamma - \Sigma)^* \cap \mathrm{IRR}(R)$, and a symbol $Y \in (\Gamma - \Sigma) \cap \mathrm{IRR}(R)$ such that, for all $w \in \Sigma^*$, $t_1 w t_2 \to_R^* Y$ if and only if $w \in L$.*

By *CR* we denote the class of Church-Rosser languages. It is known that the inclusions $DCF \subset CR \subset GCS \subset CS$ are proper, and that the classes *CR* and *CF* are incomparable under set inclusion [5, 30, 33]. The Church-Rosser languages have been characterised by certain types of two-pushdown automata.

Definition 2. *A* two-pushdown automaton (TPDA) *with* pushdown windows *of size k is a nondeterministic automaton $P = (Q, \Sigma, \Gamma, \delta, k, q_0, \perp, t_1, t_2, F)$, where Q is a finite set of states, Σ is a finite input alphabet, Γ is a finite tape alphabet containing Σ, $q_0 \in Q$ is the initial state, $\perp \in \Gamma - \Sigma$ is the bottom marker of the pushdown stores, $t_1, t_2 \in (\Gamma - \Sigma)^*$ is the preassigned content of the first and second pushdown store, respectively, $F \subseteq Q$ is the set of final (or halting) states, and δ is the transition relation. To each triple (q, u, v), where $q \in Q$ is a state, $u \in \Gamma^k \cup \{\perp\} \cdot \Gamma^{<k}$ is the content of the topmost part of the first pushdown store, and $v \in \Gamma^k \cup \Gamma^{<k} \cdot \{\perp\}$ is the content of the topmost part of the second pushdown store, it associates a finite set of triples from $Q \times \Gamma^* \times \Gamma^*$. The automaton P is a* deterministic two-pushdown automaton (DTPDA)*, if δ is a (partial) function.*

A configuration of a (D)TPDA is described by uqv, where $q \in Q$ is the actual state, $u \in \Gamma^*$ is the content of the first pushdown store with the first symbol of u at the bottom and the last symbol of u at the top, and $v \in \Gamma^*$ is the content of the second pushdown store with the last symbol of v at the bottom and the first symbol of v at the top. For an input string $w \in \Sigma^*$, the corresponding initial configuration is $\perp t_1 q_0 w t_2 \perp$. The computation relation of a (D)TPDA P is denoted by $\vdash_P^*$. The (D)TPDA P accepts with empty pushdown stores, that is,

$$L(P) := \{\, w \in \Sigma^* \mid \perp t_1 q_0 w t_2 \perp \ \vdash_P^* q \text{ for some } q \in F \,\}$$

is the *language accepted by P*. Observe that the input is provided to a TPDA as a part of the initial content of its second pushdown store, and that in order to accept a TPDA is required to empty its pushdown stores. Thus, it is forced to consume its input completely.

A (D)TPDA is called *shrinking* if there exists a *weight function* $\varphi : Q \cup \Gamma \to \mathbf{N}_+$ such that, for all transitions $(p, u', v') \in \delta(q, u, v)$, $\varphi(u'pv') < \varphi(uqv)$. Here φ is extended to a morphism $\varphi : (Q \cup \Gamma)^* \to \mathbf{N}$ in the obvious way. A (D)TPDA is called *length-reducing* if $|u'v'| < |uv|$ holds for all transitions $(p, u', v') \in \delta(q, u, v)$. Obviously, the length-reducing TPDA is a special case of the shrinking TPDA.

From the definitions it is easily seen that a language is growing context-sensitive, if it is accepted by a length-reducing TPDA, and that it is Church-Rosser, if it is accepted by a length-reducing DTPDA. Actually, the following characterisations have been established in [5, 34, 35].

Theorem 7.
(a) *A language is Church-Rosser, if and only if it is accepted by a shrinking* DTPDA, *if and only if it is accepted by a length-reducing* DTPDA.
(b) *A language is growing context-sensitive, if and only if it is accepted by a shrinking* TPDA, *if and only if it is accepted by a length-reducing* TPDA.

A (deterministic) TPDA P that is length-reducing can be simulated by a (deterministic) RWW-automaton M [37]. A configuration $\bot uqv \bot$ of P, where q is a state, is encoded by the tape content $¢\hat{u}qv\$$ of M, where $\hat{u}$ is a copy of u that consists of marked symbols. M simply moves its read/write window to the right until it encounters the left-hand side of a transition of P, which it then simulates by applying a rewrite step. We must, however, ensure that during the first cycle of a computation, which starts with the tape content $¢w\$$ for some input word w, M not only simulates the first step of P on input w, but that it also inserts the encoding of the corresponding state of P into its tape content.

Conversely, it can be shown that each deterministic RRWW-automaton M can be simulated by a shrinking DTPDA P [39]. The key observation is the following. If M rewrites $¢xuy\$$ into $¢xvy\$$, then in the next cycle M must read all of x and at least the first symbol of $vy\$$ before it can execute another Rewrite step. However, while reading the prefix x, M will undergo the same state transitions as in the previous cycle. Thus, instead of reading x again, P simply stores the information on the states of M together with the symbols of x on its first pushdown store. Further, after performing a Rewrite step $¢xpuy\$ \vdash ¢xvqy\$$, where p and q are the corresponding states, M will continue with MVR-steps until it reaches the right delimiter \$, and then it either restarts, halts and accepts, or halts without accepting. Hence, with the string y, we can associate two subsets $Q_+(y)$ and $Q_{\mathrm{rs}}(y)$ of the state set Q of M as follows:

A state $q \in Q$ is contained in $Q_+(y)$ ($Q_{\mathrm{rs}}(y)$) if, starting from the configuration $¢qy\$$, M makes only MVR-steps until it scans the \$-symbol, and halts and accepts (respectively, restarts) then.

The DTPDA P stores the string y on its second pushdown. In addition, for each suffix z of y, its stores encodings of the sets $Q_+(z)$ and $Q_{\mathrm{rs}}(z)$ under the first symbol of z. Using this information it does not need to simulate the MVR-steps of M step-by-step, but it can immediately simulate a restart or halt and accept (or reject). As P starts with the input on its second pushdown, it needs a preprocessing stage to determine and to store the sets $Q_+(z)$ and $Q_{\mathrm{rs}}(z)$, but this is easily realised by a shrinking DTPDA. This yields the following results.

Theorem 8. [37, 39] (a) $CR = \mathcal{L}(\text{det-RWW}) = \mathcal{L}(\text{det-RRWW})$.
(b) $GCS \subseteq \mathcal{L}(\text{RWW}) \subseteq \mathcal{L}(\text{RRWW})$.

Finally we turn to the relationships between the language classes that are defined by the various types of deterministic restarting automata. It was pointed out in Section 11.3 that the language

$$L_2 = \{\, a^n b^n \mid n \geq 0 \,\} \cup \{\, a^n b^m \mid m > 2n \geq 0 \,\}$$

is not accepted by any RRW-automaton. However, it is accepted by a deterministic RWW-automaton because of the theorem above and the following result.

Lemma 6. *The language L_2 is a Church-Rosser language.*

Proof. It suffices to describe a shrinking DTPDA P for the language L_2. This DTPDA proceeds as follows, given an input of the form $a^n b^m$:

(1) By repeatedly dividing the numbers of occurrences of a's and b's by two, the binary representations $\mathsf{bin}(n)$ and $\mathsf{bin}(m)$ of the exponents n and m are determined. The string $\mathsf{bin}(n)$ is collected on the bottom of the first pushdown, while the string $\mathsf{bin}(m)$ is collected on the bottom of the second pushdown. Thus, after a finite number of steps P has transformed its input into the combined stack content $\mathsf{bin}(n)^R\#\mathsf{bin}(m)$. Here $\#$ is a special symbol marking the border between the two binary representations, and $\mathsf{bin}(n)^R$ denotes the mirror image of the string $\mathsf{bin}(n)$.

(2) Next P checks whether $n = m$ or whether $m > 2n$ holds. Let $k := |\mathsf{bin}(n)|$ and $l := |\mathsf{bin}(m)|$. Obviously $n = m$ if and only if $\mathsf{bin}(n) = \mathsf{bin}(m)$. Further, $m > 2n$ if and only if

– $l > k + 1$, or
– $l = k + 1$ and (i) $\mathsf{bin}(m) = \mathsf{bin}(n)1$ or
(ii) $\mathsf{bin}(m) = \mathsf{bin}(r)d_0$ for some $r > n$ and any digit d_0.

Comparing $\mathsf{bin}(n)$ and $\mathsf{bin}(m)$ digit by digit, P can check these conditions. Further, it is easily seen that this computation can be realised in a weight-reducing manner. □

For establishing further separation results, we consider the following example languages:

$$
\begin{aligned}
L_5 &:= \{\, (ab)^{2^n} \mid n \geq 0 \,\}, \\
L_6 &:= L_{6,1} \cup L_{6,2} \cup L_{6,3}, \text{ where} \\
&\quad L_{6,1} := \{\, (ab)^{2^n-i}c(ab)^i \mid n \geq 0, 0 \leq i \leq 2^n \,\}, \\
&\quad L_{6,2} := \{\, (ab)^{2^n-2i}(abb)^i \mid n \geq 1, 0 \leq i \leq 2^{n-1} \,\}, \\
&\quad L_{6,3} := \{\, (abb)^{2^n-i}(ab)^i \mid n \geq 0, 0 \leq i \leq 2^n \,\}, \\
L_7 &:= L_{7,1} \cup L_{7,2}, \text{ where} \\
&\quad L_{7,1} := \{\, a^{2^n-2i}ca^i \mid n \geq 1, 0 \leq i < 2^{n-1} \,\}, \\
&\quad L_{7,2} := \{\, a^i da^{2^n-2i} \mid n \geq 1, 0 \leq i < 2^{n-1} \,\}.
\end{aligned}
$$

It can be shown that $L_6 \in \mathfrak{L}(\textsf{det-RR}) - \mathfrak{L}(\textsf{RW})$, and $L_7 \in \mathfrak{L}(\textsf{det-RW}) - \mathfrak{L}(\textsf{RL})$. Further, based on L_5 a non-context-free language L_5' is described in [18] that belongs to $\mathfrak{L}(\textsf{det-R})$. In addition, based on the RRWW-automaton for the language L_{copy} given in Example 4, a deterministic RLWW-automaton can be obtained for this language. Hence, by Theorem 2 there exists a language $L'_{\text{copy}} \in \mathfrak{L}(\textsf{det-RLW})$ and a regular language E such that $L_{\text{copy}} = L'_{\text{copy}} \cap E$, and by Theorem 3 there exists a morphism $\varphi_{k,m}$ such that the language $\tilde{L}_{\text{copy}} := \varphi_{k,m}(L'_{\text{copy}}) \in \mathfrak{L}(\textsf{det-RL})$. As L_{copy} is not growing context-sensitive, and as the class *CR* is closed under the operation of intersection with regular languages and under inverse morphisms, it follows that none of these languages is Church-Rosser. The same reasoning shows that the language $\tilde{L}_{\text{copy}}$ is not context-free. Together with the above results these facts yield the proper inclusions displayed in Figure 11.4.

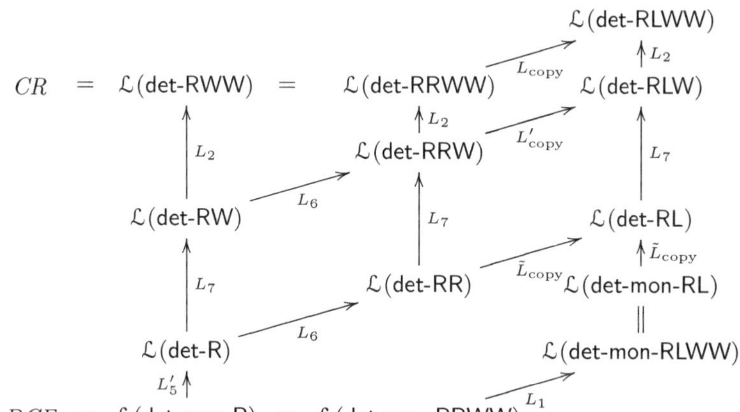

Fig. 11.4. The taxonomy of deterministic restarting automata

As the language L_5' is not context-free, and as the classes *CR* and *CF* are incomparable under inclusion, we obtain the following consequence.

Corollary 5. *The language classes* $\mathcal{L}$(det-R) *and CF are incomparable with respect to inclusion.*

As $\mathcal{L}$(det-mon-RL) $\subset$ $\mathcal{L}$(mon-RLWW) $=$ *CF*, it follows in particular that $\mathcal{L}$(det-R) is not contained in $\mathcal{L}$(det-mon-RL). On the other hand, the language L_1 is not accepted by any deterministic RRW-automaton [39], implying that $\mathcal{L}$(det-mon-RL) is not contained in $\mathcal{L}$(det-RRW). However, the following additional results hold.

Theorem 9. $\mathcal{L}$(det-mon-RLWW) $=$ $\mathcal{L}$(det-mon-RL) $\subset$ *CR*.

Proof. A monotone deterministic RL-automaton M can be simulated by a deterministic TPDA that is shrinking. The simulation is essentially the same as the one used to derive Theorem 8. Here the monotonicity is only used to guarantee that the right distance does not increase by an arbitrarily large amount from one cycle to the next. Hence, $\mathcal{L}$(det-mon-RL) is contained in *CR*. The language L_7 shows that this inclusion is proper.

In [25] it is shown that each language that is accepted by a monotone deterministic RLWW-automaton can also be accepted by a monotone deterministic RL-automaton, which means that monotone deterministic RL-automata don't need auxiliary symbols. □

In order to derive still another separation result, we consider the linear language $L_{\text{pal}} := \{ ww^R \mid w \in \{a,b\}^* \}$. In [23] it is shown that L_{pal} is not a Church-Rosser language, but as it is a context-free language, it is accepted by some monotone RWW-automaton M_{pal} (Theorem 4).

Now let M'_{pal} be the monotone RW-automaton that is obtained from M_{pal} by declaring all symbols to input symbols, and let $L'_{\text{pal}} := L(M'_{\text{pal}})$. Finally, let $\tilde{L}_{\text{pal}} := \varphi_{k,m}(L'_{\text{pal}})$, where $\varphi_{k,m}$ is the encoding of Theorem 3 that corresponds to M'_{pal}. Then $\tilde{L}_{\text{pal}}$ is accepted by a monotone R-automaton. However, as the class *CR* is closed under the operation of intersection with regular languages and inverse morphisms, $\tilde{L}_{\text{pal}}$ is not Church-Rosser, either. As the classes *CR* and *CF* are incomparable [5], this yields the following incomparablily result.

Corollary 6. *The language classes* $\mathcal{L}$(mon-R) *and CR are incomparable with respect to inclusion.*

It is further shown in [36, 37] that $\mathcal{L}$(det-R) forms a *quotient basis* for the recursively enumerable languages. Actually, this result is shown for the class of *confluent internal contextual languages*, which is a restricted version of the *internal contextual languages*, for which a corresponding result is proved in [10]. It follows that $\mathcal{L}$(det-R) is already quite an expressive class of languages.

11.5 Weakly Monotone Restarting Automata

As seen in Theorem 8, the language class *GCS* is contained in $\mathcal{L}$(RWW). On the other hand, by Corollary 4 there exist R-automata that accept languages which are not growing context-sensitive. Hence, we have the following result.

Corollary 7. *The class GCS is properly contained in* $\mathcal{L}$(RWW).

In addition, it has been shown that $\mathcal{L}$(RWW) contains some *NP*-complete languages [24]. By combining this result with the reductions of Corollaries 2 and 3 we obtain the following consequence, where LOG($\mathcal{L}$) denotes the closure of a language class $\mathcal{L}$ under log-space reductions.

Theorem 10.
$\mathcal{L}(R)$ contains NP-complete languages. In particular, LOG($\mathcal{L}$(R)) $= NP$.

In order to obtain a characterisation of the class *GCS* in terms of restarting automata, a weaker notion of monotonicity has been introduced in [24]. Let M be an RLWW-automaton, and let $c \geq 0$ be an integer. We say that a sequence of cycles $C_1, C_2, \ldots, C_n$ of M is *weakly c-monotone* if $D_r(C_{i+1}) \leq D_r(C_i) + c$ holds for all $i = 1, \ldots, n-1$. A computation of M is called *weakly c-monotone* if the corresponding sequence of cycles is weakly c-monotone. Observe that the tail of the computation is not taken into account. Finally, the RLWW-automaton M is called *weakly c-monotone* if, for each restarting configuration $q_0 ¢ w\$$ of M, each computation of M starting with $q_0 ¢ w\$$ is weakly c-monotone. Note that here we do not only consider computations that start with an initial configuration, but that we explicitly consider all computations that start with an arbitrary restarting configuration. The RLWW-automaton M is called *weakly monotone* if it is weakly c-monotone for some integer constant $c \geq 0$. The prefix wmon- is used to denote the classes of weakly monotone restarting automata.

Let M be a deterministic RRWW-automaton with read/write window of size k, and let $C_1, C_2, \ldots, C_n$ be the sequence of cycles of a computation of M. If C_i contains the Rewrite step $¢xquy\$ \to ¢xvq'y\$$, where q, q' are states, then $D_r(C_i) = |uy\$| = |y| + k + 1$. In the next cycle, C_{i+1}, M cannot perform a Rewrite step before it sees at least the first symbol of $vy\$$ in its read/write window, as M is deterministic. Thus, $D_r(C_{i+1}) \leq |vy\$| + k - 1 = D_r(C_i) + |v| - 1$. By taking the constant

$$c := \max(\{\, |v| - 1 \mid u \to v \text{ is a Rewrite step of } M \,\} \cup \{0\}),$$

we see that M is necessarily weakly c-monotone. Thus, each deterministic RRWW-automata is necessarily weakly monotone. This, however, is not true for deterministic RLWW-automata in general. For example, the deterministic RLWW-automaton for the language L_{copy} mentioned in Section 11.4 is not weakly monotone. Further, we will see that for the various types of nondeterministic restarting automata, weak monotonicity makes a big difference. However, it has been observed in [32] that an RLWW-automaton $M = (Q, \Sigma, \Gamma, ¢, \$, q_0, k, \delta)$ is weakly monotone if and only if it is weakly c-monotone for some constant $c < |Q|^2 \cdot |\Gamma|^k + 2k$. Also the following result has been established in [32].

Theorem 11. *It is decidable whether a given* RLWW*-automaton M is weakly monotone. In the affirmative the smallest integer c can be determined for which M is weakly c-monotone.*

For each language $L \in GCS$, there exists a length-reducing TPDA P such that $L = L(P)$ (Theorem 7). This TPDA can be simulated by an RWW-automaton M that encodes a configuration $\bot uqv \bot$ of P by the tape content $¢\hat{u}qv\$$ (see the discussion following Theorem 7). The automaton M simply moves its read/write window from left to right across its tape until it discovers the left-hand side of a transition of P, which it then simulates by a Rewrite step. As this Rewrite step includes the unique state symbol of P contained on the tape, we see that M is weakly monotone.

On the other hand, the simulation of a deterministic RRWW-automaton by a shrinking DTPDA carries over to nondeterministic RRWW-automata that are weakly monotone. Just notice the following two facts, where M denotes a weakly monotone RRWW-automaton:

(i) While performing MVR-steps M behaves like a finite-state acceptor. Hence, the first part of each cycle can be simulated deterministically. Nondeterminism comes in as soon as a Rewrite step is enabled.
(ii) As pointed out in Section 11.4, with each string $w \in \Gamma^*$, two subsets $Q_+(w)$ and $Q_{\mathrm{rs}}(w)$ can be associated. For the deterministic case these sets are necessarily disjoint for each string w. If M is nondeterministic, then this is not true anymore. However, if $Q_+(w)$ is nonempty, then M will accept, and so the simulation simply accepts, and if $Q_+(w)$ is empty, but $Q_{\mathrm{rs}}(w)$ is nonempty, then a Restart step must be simulated.

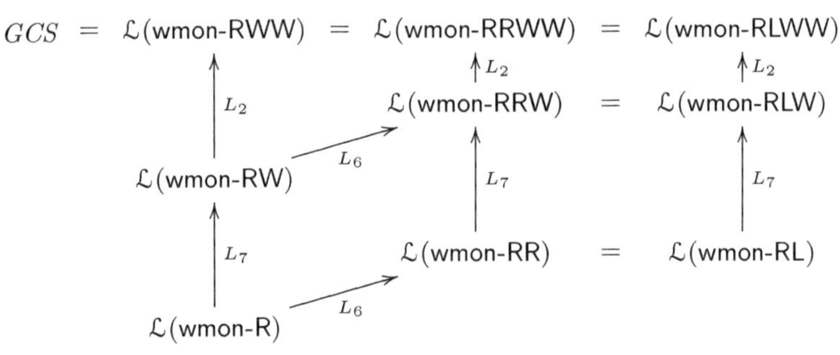

Fig. 11.5. The taxonomy of weakly monotone restarting automata

Because of Theorem 7 (and Theorem 1), this yields the following characterisation.

Theorem 12.

$$GCS = \mathcal{L}(\textsf{wmon-RWW}) = \mathcal{L}(\textsf{wmon-RRWW}) = \mathcal{L}(\textsf{wmon-RLWW}).$$

For each type X ∈ {R, RR, RW, RRW, RWW, RRWW}, the classes $\mathcal{L}$(det-X) and $\mathcal{L}$(mon-X) are obviously contained in the class $\mathcal{L}$(wmon-X). Corollaries 5 and 6 in combination with Theorem 4, Theorem 8(a), and Theorem 12 imply that all these inclusions are proper, and that no class of the form $\mathcal{L}$(wmon-X) is contained in any class of the form $\mathcal{L}$(det-Y) or $\mathcal{L}$(mon-Y), where Y ∈ {R, RR, RW, RRW, RWW, RRWW}. On the other hand, the language class $\mathcal{L}$(det-RL) is not contained in *GCS* because of the language $\tilde{L}_{\text{copy}}$ (see Section 11.4). Figure 11.5 describes the relationships between the various classes of weakly monotone restarting automata.

11.6 On General RWW- and RRWW-Automata

Next we discuss the classes of languages that are characterised by the various types of unrestricted nondeterministic restarting automata.

The languages L_6 and L_7 show that among the classes $\mathcal{L}$(R), $\mathcal{L}$(RR), $\mathcal{L}$(RW), and $\mathcal{L}$(RRW) only the trivial inclusions hold, and that they are proper. The language L_2 is context-free and Church-Rosser (Lemma 6), but not in $\mathcal{L}$(RRW), and the language L'_5 (see Section 11.4) is not context-free, but it belongs to $\mathcal{L}$(R). This shows that *CF* is incomparable to all the above classes, and that the inclusion of $\mathcal{L}$(R(R)W) in $\mathcal{L}$(R(R)WW) is proper. Thus, for the classes of languages characterised by the various types of nondeterministic restarting automata we have the taxonomy presented in Figure 11.6.

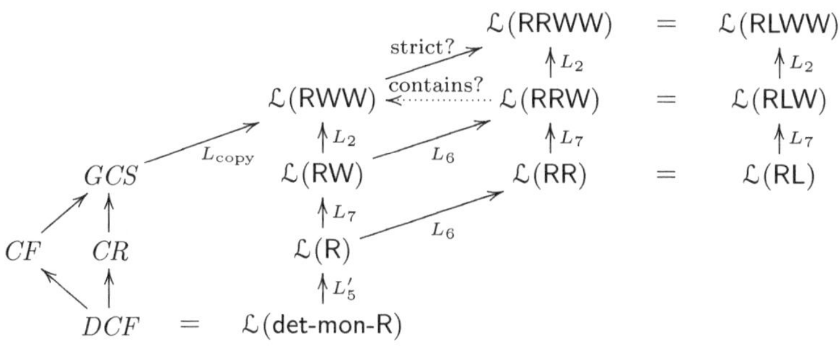

Fig. 11.6. The taxonomy of nondeterministic restarting automata

In addition, we conclude from Theorem 2 that the language classes $\mathcal{L}$(RW) and $\mathcal{L}$(RRW) are not closed under the operation of intersection with regular languages. Further, from Corollary 4 we know that the class $\mathcal{L}$(R) is not even contained in *GCS*. This yields the following consequence.

Corollary 8. $\mathcal{L}$(R) *is incomparable to the language classes CF, CR, and GCS with respect to inclusion.*

It is still open whether the inclusion $\mathcal{L}(\mathsf{RWW}) \subseteq \mathcal{L}(\mathsf{RRWW})$ is strict, and whether $\mathcal{L}(\mathsf{RRW})$ is contained in $\mathcal{L}(\mathsf{RWW})$. If $\mathcal{L}(\mathsf{RRW}) \subseteq \mathcal{L}(\mathsf{RWW})$, then by Theorem 2 we have $\mathcal{L}(\mathsf{RWW}) = \mathcal{L}(\mathsf{RRWW})$, and if $\mathcal{L}(\mathsf{RRW}) \not\subseteq \mathcal{L}(\mathsf{RWW})$, then the inclusion $\mathcal{L}(\mathsf{RWW}) \subseteq \mathcal{L}(\mathsf{RRWW})$ is obviously strict, that is, these two open problems are closely related. Also it is open whether $\mathcal{L}(\mathsf{RR}) \subset \mathcal{L}(\mathsf{RWW})$ holds. Further, the exact relationship of the language classes $\mathcal{L}(\mathsf{det\text{-}RL(W)(W)})$ to the various nondeterministic classes is still open. In particular, it is currently not even known whether or not the class *CF* is contained in $\mathcal{L}(\mathsf{det\text{-}RLWW})$.

Finally we turn to the question of the expressive power of the RWW-, RRWW-, and RLWW-automata. As seen above these automata are more powerful than the class *GCS*, but they can only accept certain context-sensitive languages (Proposition 1). Here we describe in short some other language classes between *GCS* and *CS* and restate a number of open problems.

Let $G = (N, T, S, P)$ be a grammar. By $\hat{L}(G)$ we denote the set of sentential forms that are derivable from the start symbol S in G. For each word $w \in \hat{L}(G)$, let $t_G(w)$ denote the length of a shortest derivation of w from S in G. The *time function* T_G of G is then the following partial function $T_G : \mathbf{N} \to \mathbf{N}$ [2]:

$$T_G(n) := \begin{cases} \max\{\, t_G(w) \mid w \in \hat{L}(G) \cap (N \cup T)^n \,\}, & \text{if } \hat{L}(G) \cap (N \cup T)^n \neq \emptyset, \\ \text{undefined}, & \text{otherwise.} \end{cases}$$

By CS_{lin} we denote the class of all *linear-time bounded context-sensitive languages*. A language L belongs to this class, if there exists a monotone grammar $G = (N, T, S, P)$ that generates L and a constant $c \geq 1$ such that, for all $n \in \mathbf{N}$, if $T_G(n)$ is defined, then $T_G(n) \leq c \cdot n$ [1, 2]. Finally by CS_{quad} we denote the class of all *quadratical-time bounded context-sensitive languages*. A language L belongs to this class, if there exists a monotone grammar $G = (N, T, S, P)$ that generates L and a constant $c \geq 1$ such that, for all $n \in \mathbf{N}$, if $T_G(n)$ is defined, then $T_G(n) \leq c \cdot n^2$.

Proposition 5. ([1] Theorem 2.3.3) $CS_{quad} = \mathsf{NTIMESPACE}_1(n^2, n)$.

Here $\mathsf{NTIMESPACE}_1(n^2, n)$ denotes the class of languages that are accepted by one-tape nondeterministic Turing machines in quadratic time with linear space. It is easily seen that an RLWW-automaton can be simulated by such a Turing machine, that is, $\mathcal{L}(\mathsf{RLWW}) \subseteq CS_{quad}$. Finally by Q we denote the class of *quasi-realtime languages*. This is the class of languages that are accepted by nondeterministic multitape Turing machines in realtime, that is, $Q = \mathsf{NTIME}(\mathrm{n})$.

Proposition 6. [1, 4] $GCS \subset CS_{lin} \subset Q = \mathsf{NTIME}(\mathrm{lin})$.

For proving that *GCS* is only a proper subset of CS_{lin}, the language

$$L_{\mathrm{COPY(n^2)}} = \{\, ww\#^{4|w|^2 - 2|w|} \mid w \in \{a, b\}^* \,\}$$

is considered in [4]. This language is a member of CS_{lin} ([4] Satz VIII.4), but it is not growing context-sensitive ([4] Satz VII.4 and Satz VIII.5).

A nondeterministic multitape Turing machine M that runs in realtime uses linear space only. Hence, it can be simulated by a nondeterministic one-tape Turing machine that is linear space-bounded and quadratically time-bounded. On the other hand, M can be simulated by a deterministic (multi- or one-tape) Turing machine that runs in exponential time, but that only uses linear space. Thus, we obtain the following inclusions, where $DCS = \mathsf{DSPACE}(n)$. It is an open problem which (if any) of these inclusions are proper.

Proposition 7. *$Q \subseteq CS_{quad} \subseteq CS$ and $Q \subseteq DCS$.*

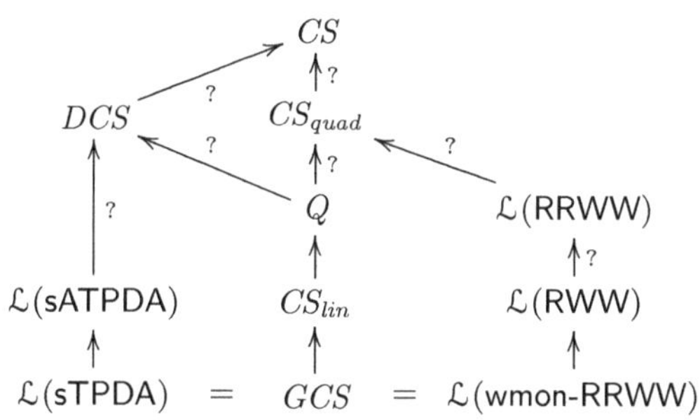

Fig. 11.7. An unmarked arrow indicates that the inclusion is proper, while a question mark indicates that it is an open problem whether the corresponding inclusion is proper. For those classes that are not connected via directed paths in the diagram it is open whether any inclusions hold.

Finally, we turn to a generalisation of the TPDA considered in [42]. An *alternating* TPDA, ATPDA for short, is a TPDA P for which a subset U of the set of states Q is designated as a set of *universal states*. The states in the difference set $Q - U$ are accordingly called *existential states*. An ATPDA is called *shrinking*, if the underlying TPDA is shrinking. By sATPDA we denote the class of all shrinking ATPDAs.

As each sTPDA is a sATPDA without universal states, we see that $GCS \subseteq \mathcal{L}(\mathsf{sATPDA})$ holds. On the other hand, the Gladkij language

$$L_{\mathrm{GL}} := \{\, w\#w^R\#w \mid w \in \{a,b\}^* \,\},$$

which is not growing context-sensitive [1, 4, 11], is accepted by some sATPDA, implying that this inclusion is proper. Actually, the following results on the language class $\mathcal{L}(\mathsf{sATPDA})$ are known.

Proposition 8. [42]

(a) $GCS \subset \mathcal{L}(\mathsf{sATPDA}) \subseteq DCS$.
(b) $\mathsf{LOG}(\mathcal{L}(\mathsf{sATPDA})) = PSPACE$.

Figure 11.7 summarises the language classes considered here and the corresponding inclusion results.

11.7 Left-Monotone Restarting Automata

The notion of monotonicity considered in Section 11.3 is based on the requirement that, within a computation of a restarting automaton, the distance between the place where a Rewrite step is performed and the *right* end of the tape must not increase from one cycle to the next. Here we study the seemingly symmetric notion of *left-monotonicity.*

We say that a *sequence of cycles* $Sq = (C_1, C_2, \ldots, C_n)$ of an RLWW-automaton M is *left-monotone* if $D_l(C_1) \geq D_l(C_2) \geq \cdots \geq D_l(C_n)$. A computation of M is *left-monotone*, if the corresponding sequence of cycles is left-monotone. As before the tail of the computation does not play any role here. Finally, the RLWW-automaton M is called *left-monotone* if each of its computations that starts from an initial configuration is left-monotone. The prefix left-mon- is used to denote the various classes of left-monotone restarting automata.

First we study the influence that left-monotonicity has on the various types of deterministic restarting automata. We begin with a simple example.

Example 6. Let $M := (Q, \Sigma, \Gamma, ¢, \$, q_0, 4, \delta)$ be the deterministic RWW-automaton that is defined by $\Sigma := \{a, b, c, d\}$ and $\Gamma := \Sigma \cup \{C, D\}$, where δ is given through the following meta-instructions:

$$
\begin{array}{llll}
(1)\ (¢ \cdot a^+ \cdot b^*, & bbc\$ \to C\$), & (6)\ (¢ \cdot a^+ \cdot b^*, & bbd\$ \to D\$),\\
(2)\ (¢ \cdot a^+ \cdot b^*, & bbC \to CC), & (7)\ (¢ \cdot a^+ \cdot b^*, & bbD \to DD),\\
(3)\ (¢ \cdot a^+, & abC \to C), & (8)\ (¢ \cdot a^+, & aDD \to D),\\
(4)\ (¢ \cdot a^+, & aaCC \to C), & (9)\ (¢ \cdot aD \cdot \$, & \mathsf{Accept}),\\
(5)\ (¢ \cdot aaC \cdot \$, & \mathsf{Accept}), & (10)\ (¢ \cdot x \cdot \$, & \mathsf{Accept})\ \text{for all } x \in \{c, d, abc\}.
\end{array}
$$

It is easily seen that M accepts the language L_1 of Example 1, and that M is left-monotone.

Observe that the deterministic RL-automaton presented in Example 1 is not only monotone, but that it is also left-monotone, showing that $L_1 \in \mathcal{L}(\text{det-left-mon-RL})$ holds. As L_1 is not deterministic context-free, the above example shows further that $\mathcal{L}(\text{det-mon-RWW})$ and $\mathcal{L}(\text{det-left-mon-RWW})$ do not coincide (see Theorem 5). Further, we have the following result, which represents another difference between the behaviour of the monotone deterministic classes and the left-monotone deterministic classes.

Theorem 13. [25, 41]

(a) $\mathcal{L}(\text{det-left-mon-RWW}) = \mathcal{L}(\text{det-left-mon-RLWW})$.

(b) $\mathcal{L}(\text{det-left-mon-RL}) = \mathcal{L}(\text{det-left-mon-RLWW})$.

It follows that the class $\mathcal{L}(\text{det-left-mon-RWW})$ contains the reversal of every deterministic context-free language. In fact, it also contains some additional languages, as it contains the reversal of every language from the class $\mathcal{L}(\text{det-mon-RLWW})$, which properly contains the class *DCF*.

For separating the class $\mathcal{L}(\text{det-left-mon-RWW})$ from those classes of languages that are accepted by the more restricted types of left-monotone deterministic restarting automata, we introduce some additional example languages:

$$\begin{aligned}
L_8 &:= L_{8,1} \cup L_{8,2} \cup L_{8,3} \cup L_{8,4}, \text{ where}\\
&\quad L_{8,1} := \{\, a^m a^n (bc)^n f a^m \mid m,n > 0 \,\},\\
&\quad L_{8,2} := \{\, a^m a^n (bc)^i b^j a^m \mid m,n,j > 0,\ i \geq 0,\ n = i + j \,\},\\
&\quad L_{8,3} := \{\, a^n (bc)^i c^j \mid n,j > 0,\ i \geq 0,\ n = 2(i+j)\},\\
&\quad L_{8,4} := \{\, a^m (bc)^n f a^k f \mid m,n,k > 0 \,\},\\
L_9 &:= \{\, a^m b^m a^n b^n \mid m,n > 0 \,\},\\
L_{10} &:= \{\, a_5^n a_4^n a_3^m a_2^m a_1^p \mid n,m,p > 0 \,\} \cup \{\, a_5^n a_4^l a_3^m a_2^p a_1^p \mid n \neq l, n,l,m,p > 0 \,\}.
\end{aligned}$$

The language L_8 is accepted by a left-monotone deterministic RR-automaton, but it cannot be accepted by any deterministic RW-automaton. On the other hand, the language L_9 is accepted by a left-monotone deterministic RWW-automaton, but not by any left-monotone deterministic RRW-automaton [46]. Based on the reduction underlying Corollary 2 we obtain a language L_9' from L_9 such that $L_9' \in \mathcal{L}(\text{det-left-mon-RW}) - \mathcal{L}(\text{det-left-mon-RR})$. Finally, as observed in Section 11.4, L_1 is not accepted by any deterministic RRW-automaton. Thus, we have the taxonomy described by Figure 11.8.

On the other hand, it is shown in [41] that $\mathcal{L}(\text{det-left-mon-RWW})$ does not include the deterministic context-free language L_{10}, implying the following non-inclusion result.

Corollary 9. $DCF \not\subseteq \mathcal{L}(\text{det-left-mon-RLWW})$.

We see that the classes of languages specified through the various types of left-monotone deterministic restarting automata show a completely different structure than those classes that are defined through the various types of monotone deterministic restarting automata. By the way, from Theorems 8 and 13 we obtain the inclusion $\mathcal{L}(\text{det-left-mon-RLWW}) \subset CR$, which is proper, because $\mathcal{L}(\text{det-left-mon-RLWW})$ contains only context-free languages, as we will see below.

Now we turn to the nondeterministic types of left-monotone restarting automata. In symmetry to the proof of Theorem 4 a left-monotone RWW-automaton M can be constructed from a context-free grammar G in Chomsky

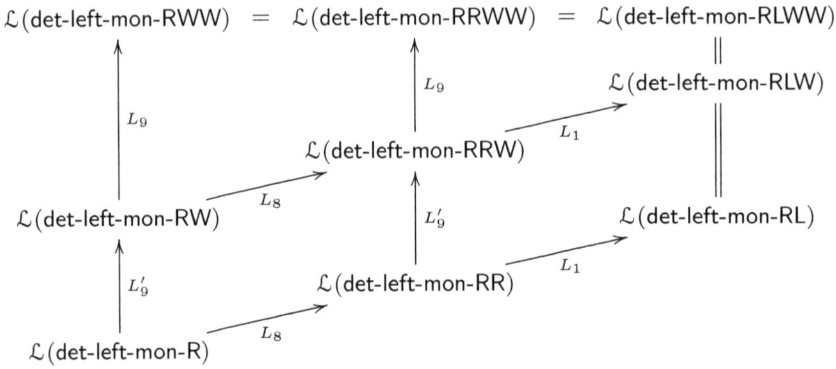

Fig. 11.8. The taxonomy of left-monotone deterministic restarting automata

normal form such that, given an input word w, M tries to construct a leftmost derivation of w in G in reverse order. Conversely, assume that L is accepted by a left-monotone RLWW-automaton. Then its reversal L^R is accepted by an RLWW-automaton that is monotone. Hence, by Theorem 4, L^R is context-free, which in turn implies that L is context-free. Thus, we have the following analogue to the aforementioned theorem.

Theorem 14. [46]

$$\mathcal{L}(\text{left-mon-RLWW}) = \mathcal{L}(\text{left-mon-RRWW}) = \mathcal{L}(\text{left-mon-RWW}) = CF.$$

Consider the language

$$L_3' := \{\, a^n \cdot f \cdot b^n, a^n \cdot ee \cdot b^n \mid n \geq 0 \,\} \cup \{\, a^n \cdot g \cdot b^m, a^n \cdot ee \cdot b^m \mid m > 2n \geq 0 \,\},$$

which is a slight variant of the language L_3. Then it can be shown that L_3' is accepted by a left-monotone RW-automaton, but just as L_3, it is not accepted by any RR-automaton. Further, the language L_1 is not accepted by any RW-automaton, although it is easily seen to be accepted by a left-monotone RR-automaton. Finally, we noticed before that the context-free language L_2 is not accepted by any RRW-automaton. Thus, we have the taxonomy given in Figure 11.9.

It follows in particular that the class $\mathcal{L}$(det-left-mon-RLWW) is contained in the intersection of the language classes CF and CR. Obviously, this is a proper inclusion, as according to Corollary 9, $\mathcal{L}$(det-left-mon-RLWW) does not even contain all deterministic context-free languages. However, the relationships of the language classes defined by the various types of left-monotone (deterministic) restarting automata to each other and to the language classes defined by the other types of restarting automata have not yet been clarified completely.

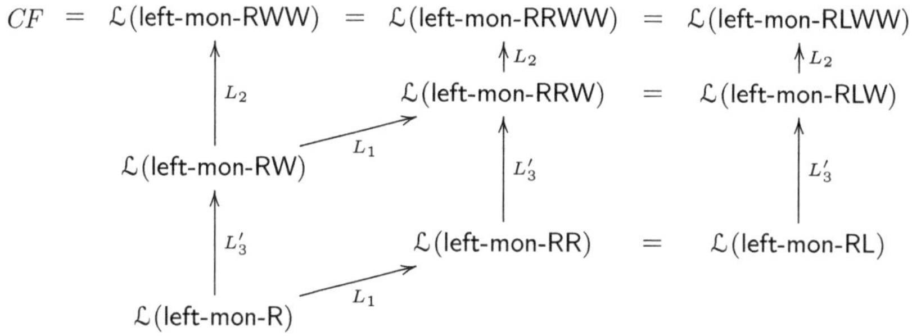

Fig. 11.9. The taxonomy of left-monotone nondeterministic restarting automata

11.8 Further Developments and Open Problems

The variants of the restarting automaton considered up to here were obtained through the following types of restrictions:

- deterministic versus nondeterministic automata,
- automata with or without auxiliary symbols,
- monotone, weakly monotone, left-monotone or non-monotone automata.

Below we discuss in short some results that are obtained by either looking at more fine-grained versions of the above restrictions, or by different variations of the underlying basic model.

11.8.1 Look-ahead Hierarchies

One of the parameters describing a restarting automaton is the size of its read/write window. The influence of this parameter on the expressive power of the various types of restarting automata without auxiliary symbols has been studied in [31]. The results obtained show that with the size of the read/write window also the expressive power of these types of restarting automata increases. It remains open, however, whether corresponding results also hold for the various types of restarting automata with auxiliary symbols.

11.8.2 The Degree of Weak Monotonicity

Another parameter that allows a quantitative analysis is the degree of weak monotonicity. As defined in Section 11.5 a restarting automaton is weakly monotone if it is weakly c-monotone for some integer $c \geq 0$. To each language L that is accepted by some weakly monotone restarting automaton of a certain type X, we can associate the smallest integer $c(L)$ such that L is accepted by a weakly $c(L)$-monotone restarting automaton of type X. This integer can be interpreted as the *degree of weak monotonicity* of the language L with respect to the restarting automaton of type X.

It is shown in [32] that the degree of weak monotonicity yields infinite hierarchies for deterministic as well as for nondeterministic R-, RR-, RW-, and RRW-automata. Again, it is still open whether corresponding results hold for the various types of restarting automata with auxiliary symbols.

11.8.3 Degrees of Non-Monotonicity

Another generalisation of monotonicity is discussed in [45]. Let $j \in \mathbf{N}_+$ be a constant, and let M be a restarting automaton. A sequence of cycles $C_1, C_2, \ldots, C_n$ of M is called *j-monotone* if there is a partition of this sequence into at most j (scattered) subsequences such that each of these subsequences is monotone. Obviously, $C_1, C_2, \ldots, C_n$ is *not* j-monotone if and only if there exist indices $1 \leq i_1 < i_2 < \cdots < i_{j+1} \leq n$ such that $D_r(C_{i_1}) < D_r(C_{i_2}) < \cdots < D_r(C_{i_{j+1}})$ holds. A computation of M is *j-monotone* if the corresponding sequence of cycles is j-monotone, and M is called *j-monotone* if all its computations that start from an initial configuration are j-monotone. It is shown in [45] that by increasing the value of the parameter j, the expressive power of RR- and RRW-automata is increased, and the same is true for R- and RW-automata [46]. On the other hand, it is shown in [26] that already 2-monotone R-automata accept *NP*-complete languages. Also an analogous generalisation of the notion of left-monotonicity is studied in [46]. However, in all these cases it remains open whether infinite hierarchies can be obtained for the various types of restarting automata with auxiliary symbols.

11.8.4 Generalisations of Restarting Automata

Here we mention two further generalisations of the basic model of the restarting automaton.

The first one deals with the number of rewrite steps that a restarting automaton may perform within a single cycle. Motivated by the 'analysis by reduction' each model of restarting automaton considered so far performs exactly one rewrite step in each cycle of each computation. However, as a restarting automaton can be seen as a special type of linear-bounded automaton, it is only natural to also consider variants that perform up to c rewrite steps in each cycle for some constant $c > 1$. How will an increase in the number c influence the expressive power of the various models of restarting automata? A further generalisation might allow $f(n)$ rewrite steps per cycle, where f is a sublinear function and n denotes the length of the input.

Length-reducing TPDAs are as powerful as shrinking TPDAs (Theorem 7). As the Rewrite transitions of a restarting automaton are length-reducing, one might also consider restarting automata for which the Rewrite transitions are only required to be weight-reducing with respect to some weight function. For the deterministic case it can be shown that these automata still characterise the class *CR* (in the presence of auxiliary symbols), but it is currently open whether or not this variant increases the expressive power of the nondeterministic types of restarting automata.

Acknowledgement. The author wants to thank Tomasz Jurdziński, Hartmut Messerschmidt, František Mráz, Martin Plátek, and Heiko Stamer for many fruitful discussions regarding restarting automata and related topics.

References

1. R.V. Book. *Grammars with Time Functions*. PhD dissertation, Harvard University, Cambridge, Massachusetts, 1969.
2. R.V. Book. Time-bounded grammars and their languages. *J. Computer System Sciences* 5 (1971) 397–429.
3. R.V. Book and F. Otto. *String-Rewriting Systems*. Springer, New York, 1993.
4. G. Buntrock. *Wachsende kontext-sensitive Sprachen*. Habilitationsschrift, Fakultät für Mathematik und Informatik, Universität Würzburg, 1996.
5. G. Buntrock and F. Otto. Growing context-sensitive languages and Church-Rosser languages. *Information and Computation* 141 (1998) 1–36.
6. M.P. Chytil, M. Plátek, and J. Vogel. A note on the Chomsky hierarchy. *Bulletin of the EATCS* 27 (1985) 23–30.
7. E. Dahlhaus and M. Warmuth. Membership for growing context-sensitive grammars is polynomial. *J. Computer System Sciences* 33 (1986) 456–472.
8. A. Dikovsky and L. Modina. Dependencies on the other side of the curtain. *Traitement Automatique des Langues* 41 (2000) 79–111.
9. A. Ehrenfeucht, G. Păun, and G. Rozenberg. Contextual grammars and formal languages. In: G. Rozenberg and A. Salomaa (eds.), *Handbook of Formal Languages*, Vol. 2. Springer, Berlin, 1997, 237–293.
10. A. Ehrenfeucht, G. Păun, and G. Rozenberg. On representing recursively enumerable languages by internal contextual languages. *Theoretical Computer Science* 205 (1998) 61–83.
11. A.W. Gladkij. On the complexity of derivations for context-sensitive grammars. *Algebra i Logika* 3 (1964) 29–44 (In Russian).
12. R. Gramatovici and C. Martín-Vide. 1-contextual grammars with sorted dependencies. In: F. Spoto, G. Scollo, and A. Nijholt (eds.), *3rd AMAST Workshop on Language Processing, Proc.*, Universiteit Twente, Enschede, 2003, 99-109.
13. S. Greibach. A note on pushdown store automata and regular systems. *Proc. American Math. Soc.* 18 (1967) 263–268.
14. J.E. Hopcroft and J.D. Ullman. *Introduction to Automata Theory, Languages, and Computation*. Addison-Wesley, Reading, M.A., 1979.
15. P. Jančar, F. Mráz, and M. Plátek. A taxonomy of forgetting automata. In: A.M. Borzyszkowski and S. Sokolowski (eds.), *MFCS'93, Proc., Lecture Notes in Computer Science 711*, Springer, Berlin, 1993, 527–536.
16. P. Jančar, F. Mráz, and M. Plátek. Forgetting automata and context-free languages. *Acta Informatica* 33 (1996) 409–420.
17. P. Jančar, F. Mráz, M. Plátek, and J. Vogel. Restarting automata. In: H. Reichel (ed.), *FCT'95, Proc., Lecture Notes in Computer Science 965*, Springer, Berlin, 1995, 283–292.
18. P. Jančar, F. Mráz, M. Plátek, and J. Vogel. On restarting automata with rewriting. In: G. Păun and A. Salomaa (eds.), *New Trends in Formal Languages, Lecture Notes in Computer Science 1218*, Springer, Berlin, 1997, 119–136.

19. P. Jančar, F. Mráz, M. Plátek, and J. Vogel. Monotonic rewriting automata with a restart operation. In: F. Plášil and K.G. Jeffery (eds.), *SOFSEM'97, Proc.*, *Lecture Notes in Computer Science 1338*, Springer, Berlin, 1997, 505–512.
20. P. Jančar, F. Mráz, M. Plátek, and J. Vogel. Different types of monotonicity for restarting automata. In: V. Arvind and R. Ramanujam (eds.), *FSTTCS'98, Proc.*, *Lecture Notes in Computer Science 1530*, Springer, Berlin, 1998, 343–354.
21. P. Jančar, F. Mráz, M. Plátek, and J. Vogel. On monotonic automata with a restart operation. *J. Automata, Languages and Combinatorics* 4 (1999) 287–311.
22. P. Jančar, F. Mráz, M. Plátek, and J. Vogel. Monotonicity of restarting automata. *J. Automata, Languages and Combinatorics*, to appear.
23. T. Jurdziński and K. Loryś. Church-Rosser languages vs. UCFL. In: P. Widmayer, F. Triguero, R. Morales, M. Hennessy, S. Eidenbenz, and R. Conejo (eds.), *ICALP'02, Proc.*, *Lecture Notes in Computer Science 2380*, Springer, Berlin, 2002, 147–158.
24. T. Jurdziński, K. Loryś, G. Niemann, and F. Otto. Some results on RWW- and RRWW-automata and their relationship to the class of growing context-sensitive languages. *Mathematische Schriften Kassel*, no. 14/01, Fachbereich Mathematik/Informatik, Universität Kassel, 2001. To appear in *J. Automata, Languages and Combinatorics* in revised form.
25. T. Jurdziński, F. Mráz, F. Otto, and M. Plátek. Deterministic two-way restarting automata don't need auxiliary symbols if they are (right-, left-, or right-left-) monotone. *Mathematische Schriften Kassel*, no. 7/04, Fachbereich Mathematik/Informatik, Universität Kassel, 2004.
26. T. Jurdziński, F. Otto, F. Mráz, and M. Plátek. On the complexity of 2-monotone restarting automata. *Mathematische Schriften Kassel*, no. 4/04, Fachbereich Mathematik/Informatik, Universität Kassel, 2004.
27. C. Lautemann. One pushdown and a small tape. In: K.W. Wagner (ed.), *Dirk Siefkes zum 50. Geburtstag*, Technische Universität Berlin and Universität Augsburg, 1988, 42–47.
28. S. Marcus. Contextual grammars and natural languages. In: G. Rozenberg and A. Salomaa (eds.), *Handbook of Formal Languages*, Vol. 2, Springer, Berlin, 1997, 215–235.
29. R. McNaughton. An insertion into the Chomsky hierarchy? In: J. Karhumäki, H. Maurer, G. Păun, and G. Rozenberg (eds.), *Jewels are Forever, Contributions on Theoretical Computer Science in Honor of Arto Salomaa*, Springer, Berlin, 1999, 204–212.
30. R. McNaughton, P. Narendran, and F. Otto. Church-Rosser Thue systems and formal languages. *J. Assoc. Comput. Mach.* 35 (1988) 324–344.
31. F. Mráz. Lookahead hierarchies of restarting automata. *J. Automata, Languages and Combinatorics* 6 (2001) 493–506.
32. F. Mráz and F. Otto. Hierarchies of weakly monotone restarting automata. *Mathematische Schriften Kassel*, no. 8/03, Fachbereich Mathematik/Informatik, Universität Kassel, 2003. To appear in *RAIRO - Theoretical Informatics and Applications* in revised form.
33. P. Narendran. *Church-Rosser and Related Thue Systems.* PhD thesis, Rensselaer Polytechnic Institute, Troy, New York, 1984.
34. G. Niemann. *Church-Rosser Languages and Related Classes.* Doctoral Dissertation, Universität Kassel, 2002.

35. G. Niemann and F. Otto. The Church-Rosser languages are the deterministic variants of the growing context-sensitive languages. In: M. Nivat (ed.), *FoSSaCS'98, Proc.*, *Lecture Notes in Computer Science 1378*, Springer, Berlin, 1998, 243–257
36. G. Niemann and F. Otto. Confluent internal contextual languages. In: C. Martin-Víde and G. Păun (eds.), *Recent Topics in Mathematical and Computational Linguistics*, The Publishing House of the Romanian Academy, Bucharest, 2000, 234–244.
37. G. Niemann and F. Otto. Restarting automata, Church-Rosser languages, and representations of r.e. languages. In: G. Rozenberg and W. Thomas (eds.), *DLT'99, Proc.*, World Scientific, Singapore, 2000, 103–114.
38. G. Niemann and F. Otto. On the power of RRWW-automata. In: M. Ito, G. Păun, and S. Yu (eds.), *Words, Semigroups, and Transductions. Essays in Honour of Gabriel Thierrin, On the Occasion of His 80th Birthday*, World Scientific, Singapore, 2001, 341–355.
39. G. Niemann and F. Otto. Further results on restarting automata. In: M. Ito and T. Imaoka (eds.), *Words, Languages and Combinatorics III, Proc.*, World Scientific, Singapore, 2003, 353–369.
40. K. Oliva, P. Květoň, and R. Ondruška. The computational complexity of rule-based part-of-speech tagging. In: V. Matousek and P. Mautner (eds.), *TSD 2003, Proc.*, *Lecture Notes in Computer Science 2807*, Springer, Berlin, 2003, 82–89.
41. F. Otto and T. Jurdzinski. On left-monotone restarting automata. *Mathematische Schriften Kassel*, no. 17/03, Fachbereich Mathematik/Informatik, Universität Kassel, 2003.
42. F. Otto and E. Moriya. Shrinking alternating two-pushdown automata. *IEICE Transactions on Information and Systems* E87-D (2004) 959–966.
43. M. Plátek. Two-way restarting automata and j-monotonicity. In: L. Pacholski and P. Ružička (eds.), *SOFSEM'01, Proc.*, *Lecture Notes in Computer Science 2234*, Springer, Berlin, 2001, 316–325.
44. M. Plátek, M. Lopatková, and K. Oliva. Restarting automata: motivations and applications. In: M. Holzer (ed.), *Workshop 'Petrinetze' and 13. Theorietag 'Formale Sprachen und Automaten', Proc.*, Institut für Informatik, Technische Universität München, 2003, 90-96.
45. M. Plátek and F. Mráz. Degrees of (non)monotonicity of RRW-automata. In: J. Dassow and D. Wotschke (eds.), *Preproceedings of the 3rd Workshop on Descriptional Complexity of Automata, Grammars and Related Structures,* Report No. 16, Fakultät für Informatik, Universität Magdeburg, 2001, 159–165.
46. M. Plátek, F. Otto, F. Mráz, and T. Jurdzinski. Restarting automata and variants of j-monotonicity. *Mathematische Schriften Kassel*, no. 9/03, Fachbereich Mathematik/Informatik, Universität Kassel, 2003.
47. S.H. von Solms. The characterisation by automata of certain classes of languages in the context sensitive area. *Information and Control* 27 (1975) 262–271.
48. M. Straňáková. Selected types of pg-ambiguity. *The Prague Bulletin of Mathematical Linguistics* 72 (1999) 29–57.
49. M. Straňáková. Selected types of pg-ambiguity: Processing based on analysis by reduction. In: P. Sojka, I. Kopeček, and K. Pala (eds.), *Text, Speech and Dialogue, 3rd Int. Workshop, TSD 2000, Proc.*, *Lecture Notes in Computer Science 1902*, Springer, Berlin, 2000, 139–144.

12

Computable Lower Bounds for Busy Beaver Turing Machines

Holger Petersen

Institut für formale Methoden der Informatik
University of Stuttgart
Universitätsstr. 38, D-70569, Germany
E-mail: petersen@informatik.uni-stuttgart.de

Summary. A relation between a lower bound due to Green on the score of Busy Beavers and Ackermann's function is established. Further, an improved construction outlined by Green is analyzed. A consequence of the analysis is that Green's lower bounds are nonprimitive recursive.

12.1 Introduction

Consider the class of deterministic Turing machines with a binary worktape alphabet operating on a single tape (no separate input tape) with the property that they eventually halt when started on a tape containing initially all 0's. A *Busy Beaver* is a machine of this kind such that no other machine with the same number of states leaves more 1's on its tape upon halting. The function $\Sigma(n)$ defined by Rado [13] is the score in terms of 1's on the tape after halting of a Busy Beaver with n states (excluding the halt state). In spite of its seemingly simple definition as the maximum of a finite set of natural numbers, Rado proved that $\Sigma(n)$ is not computable.

In the paper [7] on Rado's function $\Sigma(n)$, Green established several computable lower bounds on $\Sigma(n)$. Lower bounds on $\Sigma(n)$ are a delicate subject, since Rado proved that $\Sigma(n)$ grows faster than any computable function. So one might wonder, why it should be interesting to exhibit a concrete lower bound, while *every* computable function could serve this purpose. The problem is: Rado's result states that *eventually* $\Sigma(n)$ exceeds our favorite computable function, but figuring out for which n this holds might be difficult for a particular computable function, and is in general undecidable, since otherwise $\Sigma(n)$ would be computable. In contrast, Green's lower bounds hold for *every* n to which the particular construction is applicable.

Despite the fact that Σ is non-computable in general, several researchers have established values $\Sigma(n)$ for small n. Rado found $\Sigma(1) = 1$ and $\Sigma(2) = 4$

H. Petersen: *Computable Lower Bounds for Busy Beaver Turing Machines*, Studies in Computational Intelligence (SCI) **25**, 305–319 (2006)
www.springerlink.com

[13, 14], Lin and Rado showed $\Sigma(3) = 6$ [9], and for $n = 4$, after preliminary reports about computational efforts to determine $\Sigma(4)$ [1, 15, 2], Brady established in [3] that $\Sigma(4) = 13$. For $n \geq 5$ only lower bounds are known. Restricting the search space of possible Busy Beaver candidates by sophisticated methods, Marxen and Buntrock found that $\Sigma(5) \geq 4098$ [11]. One of the machines establishing $\Sigma(5) \geq 4098$ stops after carrying out 47,176,870 steps. In 2001 the same authors established $\Sigma(6) \geq 1.29 \cdot 10^{865}$ [10]. For large n the strongest lower bounds are still due to Green. This general construction will be described in Section 12.4.

It has been pointed out by Chaitin [6] and Brady [4] that the Goldbach Conjecture and similar problems would be settled if $\Sigma(n)$ could be determined for a sufficiently large n. This clearly outlines a limit to the range for which we can hope to compute $\Sigma(n)$, but for a small value such as $n = 5$, Michel was able to relate the behavior of previous and current Busy Beaver candidates to iterated functions similar to the famous Collatz Problem [12]. This might indicate that the computation of $\Sigma(5)$ is still feasible.

In this note I will investigate one of the functions defined by Green more closely and establish that it grows 'approximately' as fast as Ackermann's function. By 'approximately' I mean that both upper and lower bounds can be obtained by composing Ackermann's function and primitive recursive functions.

Green improved his construction and gave a separate analysis of the stronger lower bounds. In contrast, I will show how to express the improved bounds in terms of the weaker. In particular this gives a proof of Brady's remark [4, p. 240] that Green's lower bounds appear to be nonprimitive recursive.

Another construction due to Green, which results in weaker lower bounds, has been briefly discussed by Buro [5, pp. 12–14]. He did however not relate it to Ackermann's function.

12.2 Preliminaries

There are several definitions of Ackermann's function in the literature. We will adopt the one given by Hopcroft and Ullman [8, p. 175]:

$$\begin{aligned} A(0,y) &= 1, \\ A(1,0) &= 2, \\ A(x+2,0) &= x+4, \\ A(x+1,y+1) &= A(A(x,y+1),y). \end{aligned}$$

for any $x, y \geq 0$.

Green's bound $G_n(m)$ is also a function of two parameters, namely number of states n (where n is required to be odd) and the length m of a string of ones written on the tape. It can be defined recursively by:

$$\begin{aligned} G_1(m) &= m+1, \\ G_{2n+3}(0) &= 1+G_{2n+1}(G_{2n+1}(0)), \\ G_{2n+3}(m+1) &= 1+G_{2n+1}(G_{2n+3}(m)-1). \end{aligned}$$

The function $\Sigma(n)$ is defined as the maximum number of ones left on the initially empty (all 0's) tape by a halting deterministic Turing machine having n states. Here the final state is not counted. The Turing machines considered have a binary alphabet $\{0,1\}$ (where 0 is the blank symbol) and work on a one-dimensional and bi-infinite tape. In addition to writing a symbol they move their head in every step either to the left or to the right. Turing machines of this kind can be represented by a table of the form:

		input symbol 0	input symbol 1
current state	1	$w_1^0 m_1^0 s_1^0$	$w_1^1 m_1^1 s_1^1$
	2	$w_2^0 m_2^0 s_2^0$	$w_2^1 m_2^1 s_2^1$
	$\vdots$	$\vdots$	$\vdots$
	n	$w_n^0 m_n^0 s_n^0$	$w_n^1 m_n^1 s_n^1$

where $w_k^i \in \{0,1\}$ indicates the symbol written by M after reading i in state k, $m_k^i \in \{L,R\}$ is the direction of the head movement, and $s_k^i \in \{1,\ldots,n+1\}$ the new state M enters. State 1 is the initial state and state $n+1$ is the halt state.

I will have to discuss computations of Turing machines, which requires some notation for configurations. Let us assume that a machine is in state s and reads symbol x, with u and v finite segments of the tape to the left and right of the current head position. Then the configuration will be denoted

$$^{\omega}0u[sx]v0^{\omega},$$

where $^{\omega}0$ and 0^{ω} stand for semi-infinite sequences of 0's or, equivalently, unlimited supplies of empty tape cells (this is almost the notation used by Michel [12]). If the head never leaves one of the finite segments u or v specified in an initial configuration, then the bordering infinite sequence will be omitted. The instructions of the machine table are applied in the obvious way to generate a successor-configuration c_2 from a given c_1, notation: $c_1 \vdash c_2$. If some finite number of steps leads from c_1 to c_2, we write $c_1 \vdash^* c_2$.

Let me briefly outline the inductive construction of Green's Class G machines that realize the functions defined above. Realizing an integer valued function in this context means that there is a machine doing the following. Starting on an otherwise empty tape containing an input string of m consecutive 1's to the left of its head, the machine eventually halts while moving its head onto the cell immediately to the right of its initial position. This is the first time the cell is visited, and the machine is required to leave a string of 1's immediately to the left of its head denoting in unary the value of the function being computed for input m.

A machine G_1 computing $G_1(m)$ could take the form

	0	1
1	$1R2$	$1R1$

and a machine G_3 for $G_3(m)$ is:

	0	1
1	$1L2$	$1R4$
2	$0L3$	$0L2$
3	$1R3$	$1R4$

For $n \geq 2$ the machine G_{2n+1} computing $G_{2n+1}(m)$ can be defined by the following rules.

- Take the transition table of G_{2n-1}.
- Renumber all states by adding 2, i.e., transform each line of the form

	0	1
k	$w_k^0 m_k^0 s_k^0$	$w_k^1 m_k^1 s_k^1$

into:

	0	1
$k+2$	$w_k^0 m_k^0 (s_k^0+2)$	$w_k^1 m_k^1 (s_k^1+2)$

- Add as the first two states:

	0	1
1	$1L2$	$1R(2n+2)$
2	$0L3$	$0L2$

- Replace the entry $1R(2n+2)$ for state 5 on input 1 by $1R3$.

As an example we carry out the construction of G_5:

	0	1
1	$1L2$	$1R6$
2	$0L3$	$0L2$
3	$1L4$	$1R6$
4	$0L5$	$0L4$
5	$1R5$	$1R3$

A G_{2n+1} machine first marks the initial position of its head with a 1, then transforms the input string of m 1's into 0's, moves an additional cell to the left, and carries out $m+2$ iterations of the submachine G_{2n-1}. A comparison shows that the definition of the Class G machines matches the recursion defining the function $G_{2n+1}(m)$.

Note that, since each machine G_{2n+1} has $2n+1$ states, it is among the machines considered in the definition of $\Sigma(2n+1)$. The first lower bound on Rado's function therefore takes the form

$$G_{2n+1}(0) \leq \Sigma(2n+1).$$

12.3 The Growth of Green's Function

Let us start our discussion by proving some properties of $A(x,y)$ (which are of course well-known).

Lemma 1. *For all* $x \geq 0$

$$A(x,0) = \begin{cases} x+1 & \textit{if } x \leq 1, \\ x+2 & \textit{if } x \geq 2. \end{cases}$$

Lemma 2. *For all* $x \geq 0$

$$A(x,1) = \begin{cases} 1 & \textit{if } x = 0, \\ 2x & \textit{if } x \geq 1. \end{cases}$$

Proof. By definition $A(0,1) = 1$,

$$\begin{aligned} A(1,1) &= A(A(0,1),0) \\ &= A(1,0) \\ &= 2, \end{aligned}$$

and

$$\begin{aligned} A(x+2,1) &= A(A(x+1,1),0) \\ &= A(2x+2,0) \\ &= 2(x+2) \end{aligned}$$

by Lemma 1 and induction. □

Lemma 3. *For all $x \geq 0$*

$$A(x,2) = 2^x.$$

Proof. By definition $A(0,2) = 1$, and

$$\begin{aligned} A(x+1,2) &= A(A(x,2),1) \\ &= 2^{x+1} \end{aligned}$$

by Lemma 2 (where the second case applies since $2^x \geq 1$) and induction. □

Lemma 4. *For all $x \geq 0$*

$$A(x,3) = 2^{2^{\cdot^{\cdot^{\cdot^{2}}}}} \Big\} x.$$

Proof. By definition $A(0,3) = 1$, and

$$\begin{aligned} A(x+1,3) &= A(A(x,3),2) \\ &= 2^{A(x,3)} \\ &= 2^{2^{\cdot^{\cdot^{\cdot^{2}}}}} \Big\} x+1 \end{aligned}$$

by Lemma 3 and induction. □

For some fixed values of n, Green has given concrete formulas for $G_n(m)$ as a function of m. We repeat them here.

Lemma 5. *For all $m \geq 0$*

$$G_3(m) = m + 3.$$

Proof. By definition

$$\begin{aligned} G_3(0) &= 1 + G_1(G_1(0)) \\ &= 3, \end{aligned}$$

and

$$\begin{aligned} G_3(m+1) &= 1 + G_1(G_3(m) - 1) \\ &= 1 + (m+3) \\ &= (m+1) + 3 \end{aligned}$$

by induction. □

Lemma 6. *For all $m \geq 0$*

$$G_5(m) = 3m + 7.$$

Proof. By definition

$$\begin{aligned} G_5(0) &= 1 + G_3(G_3(0)) \\ &= 7, \end{aligned}$$

and

$$\begin{aligned} G_5(m+1) &= 1 + G_3(G_5(m) - 1) \\ &= 1 + (3m + 7 - 1) + 3 \\ &= 3(m+1) + 7 \end{aligned}$$

by Lemma 5 and induction. □

Lemma 7. *For all* $m \geq 0$

$$G_7(m) = \frac{1}{2}(7 \cdot 3^{m+2} - 5).$$

Proof. By definition

$$\begin{aligned} G_7(0) &= 1 + G_5(G_5(0)) \\ &= 29, \end{aligned}$$

and

$$\begin{aligned} G_7(m+1) &= 1 + G_5(G_7(m) - 1) \\ &= 1 + 3 \cdot (\frac{1}{2}(7 \cdot 3^{m+2} - 5) - 1) + 7 \\ &= \frac{1}{2}(7 \cdot 3^{m+3} - 5) \end{aligned}$$

by Lemma 6 and induction. □

A comparison of Lemmas 1 to 3 and Lemmas 5 to 7 as well as the form of the recursion involved suggests that the second parameter of Ackermann's function plays a similar role as the number of states in Green's construction. We will need some lemmas concerning Ackermann's function for fixed values of the first parameter.

Lemma 8. *For all* $y \geq 0$

$$A(1, y) = 2.$$

Proof. By definition $A(1, 0) = 2$, and

$$\begin{aligned} A(1, y+1) &= A(A(0, y+1), y) \\ &= A(1, y) \\ &= 2 \end{aligned}$$

by induction. □

Lemma 9. *For all* $y \geq 0$

$$A(2, y) = 4.$$

Proof. By definition $A(2, 0) = 4$, and

$$\begin{aligned} A(2, y+1) &= A(A(1, y+1), y) \\ &= A(2, y) \\ &= 4 \end{aligned}$$

by Lemma 8 and induction. □

Lemma 10. *For all* $y \geq 0$

$$A(4, y) = A(3, y+1).$$

Proof. By definition and Lemma 9 we have

$$\begin{aligned} A(3, y+1) &= A(A(2, y+1), y) \\ &= A(4, y). \end{aligned}$$

□

Lemma 11. *For all* $x, y \geq 0$

$$A(x, y) < A(x+1, y).$$

Proof. For $x \leq 1$ we have $A(x, 0) = x + 1 < x + 2 \leq A(x+1, 0)$, for $x \geq 2$ we have $A(x, 0) = x + 2 < x + 3 = A(x+1, 0)$. Further $A(0, y+1) = 1 < 2 = A(1, y+1)$ and

$$\begin{aligned} A(x+1, y+1) &= A(A(x, y+1), y) \\ &< A(A(x+1, y+1), y) \\ &= A(x+2, y+1) \end{aligned}$$

by induction on x and y. □

Lemma 12. *For all* $m, n \geq 0$

$$G_{2n+3}(m) > m + n + 2.$$

Proof. We have $G_3(m) = m + 3 > m + 2$ by Lemma 5, and

$$\begin{aligned} G_{2n+3}(0) &= 1 + G_{2n+1}(G_{2n+1}(0)) \\ &\geq 1 + (n-1) + 2 + (n-1) + 2 \\ &= 2n + 3 \\ &> n + 2 \end{aligned}$$

by induction on n. Further for $n > 0$ we have

$$\begin{aligned} G_{2n+3}(m+1) &= 1 + G_{2n+1}(G_{2n+3}(m) - 1) \\ &> 1 + (m + n + 2) - 1 + (n + 1) \\ &= 2n + m + 3 \\ &\geq (m + 1) + n + 2 \end{aligned}$$

by induction on m and n. □

Lemma 13. *For all* $m, n \geq 0$

$$G_{2n+1}(m) < G_{2n+1}(m + 1).$$

Proof. We have $G_1(m) = m + 1 < m + 2 = G_1(m + 1)$, and

$$\begin{aligned} G_{2n+3}(m+1) &= 1 + G_{2n+1}(G_{2n+3}(m) - 1) \\ &> 1 + G_{2n+3}(m) - 1 \\ &= G_{2n+3}(m) \end{aligned}$$

by Lemma 12 and induction. □

Theorem 1. *For all* $m, n \geq 0$

$$G_{2n+3}(m) > A(m, n).$$

Proof. We have $G_3(m) = m+3 > m+2 \geq A(m, 0)$ by Lemma 5 and Lemma 1. For $n > 0$ we have $G_{2n+3}(0) > n + 2 > 1 = A(0, n)$ by Lemma 12, and

$$\begin{aligned} G_{2n+3}(m+1) &= 1 + G_{2n+1}(G_{2n+3}(m) - 1) \\ &\geq 1 + G_{2n+1}(A(m, n)) \\ &> A(A(m, n), n - 1) \\ &= A(m + 1, n) \end{aligned}$$

by Lemma 13 and induction on m and n. □

The preceding theorem does not give a meaningful lower bound for Class G machines on empty tape, since $A(0, n) = 1$ for all n. In order to derive such a bound we prove another lemma.

Lemma 14. *For all* $m, n \geq 0$

$$G_{2n+3}(m) > G_{2n+1}(m + 1).$$

Proof. We have $G_{2n+3}(0) = 1 + G_{2n+1}(G_{2n+1}(0)) > G_{2n+1}(1)$ by Lemma 13 and the definition of $G_{2n+1}(m)$. Further

$$G_{2n+3}(m + 1) = 1 + G_{2n+1}(G_{2n+3}(m) - 1) > G_{2n+1}(m + 2)$$

by Lemma 12 and Lemma 13. □

Theorem 2. *For all $m, n \geq 0$*

$$A(m+4,n)-4 > G_{2n+1}(m).$$

Proof. We have $G_1(m) = m+1 < m+2 = A(m+4,0)-4$ by definition. For $n > 0$ we have

$$\begin{aligned}
G_{2n+1}(0) &= 1 + G_{2n-1}(G_{2n-1}(0)) \\
&< A(A(4,n-1),n-1)-4 \\
&= A(A(3,n),n-1)-4 = A(4,n)-4
\end{aligned}$$

by induction on n and Lemma 11. Further

$$\begin{aligned}
G_{2n+1}(m+1) &= 1 + G_{2n-1}(G_{2n+1}(m)-1) \\
&< 1 + G_{2n-1}(A(m+4,n)-5) \\
&< 1 + A(A(m+4,n)-1,n-1)-4 \\
&\leq A(A(m+4,n),n-1)-4 \\
&= A(m+5,n)-4
\end{aligned}$$

by induction on m and n, Lemma 13, and Lemma 11. □

We finally combine Theorem 2, Theorem 1, and Lemma 14.

Corollary 1. *For all $n \geq 0$*

$$A(4,2n+1) > G_{4n+3}(0) > A(n,n).$$

Since $A(n,n)$ eventually exceeds all primitive recursive functions of the single parameter n, we conclude:

Corollary 2. *The function $G_{2n+1}(0)$ (of the single parameter n) is not primitive recursive.*

12.4 Improved Constructions

The performance of the Class G machines on blank input tapes is poor for small numbers of states, e.g. $G_1(0) = 1$, $G_3(0) = 3$, $G_5(0) = 7$, $G_7(0) = 29$. Using a fixed portion of a Turing machine for optimizing the initialization and iteration of a Class G sub-machine, Green could improve his bounds to a surprising strength.

Let us, starting from a Class G machine G_{2n+1} with $n \geq 1$, describe the construction of the machines BB_{2n+4} and BB_{2n+5}.

For BB_{2n+4}:

- Take the transition table of G_{2n+1}.
- Renumber all states by adding 3.
- Add as the first 3 states:

	0	1
1	$1L2$	$1R(2n+5)$
2	$0L3$	$1L3$
3	$0L4$	$0L3$

- Replace the entry $1R(2n+5)$ for state 4 on input 1 by $1R1$.
- Replace the entry $1R(2n+5)$ for state 6 on input 1 by $1R4$.

For BB_{2n+5}:

- Take the transition table of G_{2n+1}.
- Renumber all states by adding 4.
- Add as the first 4 states:

	0	1
1	$1L4$	$1L2$
2	$1L(2n+6)$	$1L1$
3	$0L2$	$1L4$
4	$0L5$	$0L4$

- Replace the entry $1R(2n+6)$ for state 5 on input 1 by $1R3$.
- Replace the entry $1R(2n+6)$ for state 7 on input 1 by $1R5$.

The analysis Green gives introduces new auxiliary functions $B_n(m)$. In contrast, I will show that the score of the Class *BB* machines can be described using functions of the G-variety alone.

Lemma 15. *Let $G^{(k)}_{2n+1}(m)$ denote the k-fold application of G_{2n+1} to m, then*

$$G_{2n+3}(m) = 1 + G^{(m+2)}_{2n+1}(0).$$

Proof. For $m = 0$ the equation is the definition. Otherwise by induction

$$\begin{aligned} G_{2n+3}(m+1) &= 1 + G_{2n+1}(G_{2n+3}(m) - 1) \\ &= 1 + G_{2n+1}(G^{(m+2)}_{2n+1}(0)) \\ &= 1 + G^{(m+3)}_{2n+1}(0). \end{aligned}$$

□

Theorem 3. *If started on an initially empty tape, the machine BB_{2n+4} generates $G_{2n+3}(G_{2n+3}(1)-1)+1$ ones.*

Proof. We trace the computation of BB_{2n+4}, making use of the fact that the submachine consisting of all but the first three states computes G_{2n+1}.

$$\begin{aligned}
{}^{\omega}0[10]00 &\vdash {}^{\omega}0[20]100\\
&\vdash {}^{\omega}0[30]0100\\
&\vdash {}^{\omega}0[40]00100\\
&\vdash^{*} {}^{\omega}01^{G_{2n+1}(0)}[40]0100\\
&\vdash^{*} {}^{\omega}01^{G_{2n+1}(G_{2n+1}(0))}[40]100\\
&\vdash^{*} {}^{\omega}01^{G_{2n+1}(G_{2n+1}(G_{2n+1}(0)))}[41]00\\
&= {}^{\omega}01^{G_{2n+3}(1)-1}[41]00\\
&\vdash {}^{\omega}01^{G_{2n+3}(1)}[10]0\\
&\vdash {}^{\omega}01^{G_{2n+3}(1)-1}[21]10\\
&\vdash {}^{\omega}01^{G_{2n+3}(1)-2}[31]110\\
&\vdash^{*} {}^{\omega}0[40]0^{G_{2n+3}(1)}110\\
&\vdash^{*} {}^{\omega}01^{G_{2n+1}^{(G_{2n+3}(1)+1)}(0)}[41]10\\
\text{(Lemma 15)} &= {}^{\omega}01^{G_{2n+3}(G_{2n+3}(1)-1)-1}[41]10\\
&\vdash {}^{\omega}01^{G_{2n+3}(G_{2n+3}(1)-1)}[11]0\\
&\vdash {}^{\omega}01^{G_{2n+3}(G_{2n+3}(1)-1)+1}[(2n+5)0]
\end{aligned}$$

□

The behavior of *BB*-type machines with odd numbers of states is more complex than that just discussed. In their analysis I will use the following property of Green's bound.

Lemma 16. *For each $n \geq 0$, the function $G_{2n+1}(m)$ reverses the parity of its argument, i.e.,*

$$m + G_{2n+1}(m) \equiv 1 \pmod 2.$$

Proof. The claim holds for $n = 0$, since $m + G_1(m) = 2m+1$. We proceed by induction and assume that the lemma holds for n. For $m = 0$ we have

$$\begin{aligned}
G_{2n+3}(0) &= 1 + G_{2n+1}(G_{2n+1}(0))\\
&\equiv G_{2n+1}(0) + G_{2n+1}(G_{2n+1}(0)) + G_{2n+1}(G_{2n+1}(0))\\
&\equiv 0 + G_{2n+1}(0)\\
&\equiv 1 \pmod 2
\end{aligned}$$

and further

$$\begin{aligned} G_{2n+3}(m+1) &= 1 + G_{2n+1}(G_{2n+3}(m) - 1) \\ &\equiv G_{2n+3}(m) - 1 + 2G_{2n+1}(G_{2n+3}(m) - 1) \\ &\equiv m + G_{2n+3}(m) + G_{2n+3}(m) \\ &\equiv m \pmod 2. \end{aligned}$$

Therefore $1 = (m+1) + G_{2n+3}(m+1) \pmod 2$. □

Theorem 4. *If started on an initially empty tape, the machine* BB_{2n+5} *generates* $G_{2n+5}(G_{2n+3}(0) - 1)$ *ones.*

Proof. We trace the computation of BB_{2n+5}:

$$\begin{aligned}
{}^{\omega}0[10]0 &\vdash {}^{\omega}0[40]10 \\
&\vdash {}^{\omega}0[50]010 \\
&\vdash^* {}^{\omega}01^{G_{2n+1}(0)}[50]10 \\
&\vdash^* {}^{\omega}01^{G_{2n+1}(G_{2n+1}(0))}[51]0 \\
&= {}^{\omega}01^{G_{2n+3}(0)-1}[51]0 \\
&\vdash {}^{\omega}01^{G_{2n+3}(0)}[30] \\
&\vdash {}^{\omega}01^{G_{2n+3}(0)-1}[21]0 \\
\text{(Lemma 16)} &\vdash^* {}^{\omega}0[10]1^{G_{2n+3}(0)}0 \\
&\vdash {}^{\omega}0[40]1^{G_{2n+3}(0)+1}0 \\
&\vdash {}^{\omega}0[50]01^{G_{2n+3}(0)+1}0 \\
&\vdash^* {}^{\omega}01^{G_{2n+3}(0)-1}[51]1^{G_{2n+3}(0)}0 \\
&\vdash {}^{\omega}01^{G_{2n+3}(0)}[31]1^{G_{2n+3}(0)-1}0 \\
&\vdash {}^{\omega}01^{G_{2n+3}(0)-1}[41]1^{G_{2n+3}(0)}0 \\
&\vdash^* {}^{\omega}0[50]0^{G_{2n+3}(0)+1}1^{G_{2n+3}(0)}0 \\
&\vdash^* {}^{\omega}01^{G_{2n+1}^{(G_{2n+3}(0)+2)}(0)}[51]1^{G_{2n+3}(0)-1}0 \\
\text{(Lemma 15)} &= {}^{\omega}01^{G_{2n+3}(G_{2n+3}(0))-1}[51]1^{G_{2n+3}(0)-1}0 \\
&\vdash^* {}^{\omega}01^{G_{2n+3}^{(G_{2n+3}(0)+1)}(0)-1}[51]0 \\
\text{(Lemma 15)} &= {}^{\omega}01^{G_{2n+5}(G_{2n+3}(0)-1)-2}[51]0 \\
&\vdash {}^{\omega}01^{G_{2n+5}(G_{2n+3}(0)-1)-1}[30] \\
&\vdash {}^{\omega}01^{G_{2n+5}(G_{2n+3}(0)-1)-2}[21]0 \\
\text{(Lemma 16)} &\vdash^* {}^{\omega}0[20]1^{G_{2n+5}(G_{2n+3}(0)-1)-1}0 \\
&\vdash {}^{\omega}0[(2n+6)0]1^{G_{2n+5}(G_{2n+3}(0)-1)}0
\end{aligned}$$

□

Acknowledgements

I wish to thank Amir M. Ben-Amram and Zeyi Wang-Petersen for suggesting several improvements. Thanks are also due to Karen Thomsen, Department of Information Technology of the Technical University of Denmark, Lyngby, for her help in obtaining a copy of Green's paper. The research on which this paper is based was supported by "Deutsche Akademie der Naturforscher Leopoldina", grant BMBF-LPD 9901/8-1 of "Bundesministerium für Bildung und Forschung".

References

1. A.H. Brady. The conjectured highest scoring machines for Rado's $\Sigma(k)$ for the value $k = 4$, *IEEE Transactions on Electronic Computers*, EC-15 (1966), 802–803.
2. A.H. Brady. Solution of the non-computable "Busy Beaver" game for $k = 4$, *ACM Computer Science Conference (Washington DC, 1975)*, 1975, 27.
3. A.H. Brady. The determination of the value of Rado's noncomputable function $\Sigma(k)$ for four-state Turing machines, *Mathematics of Computation*, 40 (1983), 647–665.
4. A.H. Brady. The Busy Beaver Game and the meaning of Life, *The Universal Turing Machine: A Half-Century Survey* (R. Herken, ed.), Springer-Verlag, Berlin, 1994, 237–254.
5. M. Buro. Ein Beitrag zur Bestimmung von Rados $\Sigma(5)$ oder Wie fängt man fleißige Biber? Technical Report 146, Rheinisch-Westfälische Technische Hochschule Aachen, 1990.
6. Gregory J. Chaitin. Computing the Busy Beaver function, *Information, Randomness & Incompleteness: Papers on Algorithmic Information Theory*, World Scientific, Series in Computer Science–Vol. 8, 1987, 74–76.
7. M.W. Green. A lower bound on Rado's Sigma Function for binary Turing machines, *Switching Circuit Theory and Logical Design, Proceedings of the Fifth Annual Symposium (Princeton, N.J., 1964)*, The Institute of Electrical and Electronics Engineers, Inc., 1964, 91–94.
8. J.E. Hopcroft and J.D. Ullman. *Introduction to Automata Theory, Languages, and Computation*, Addison-Wesley, Reading, Mass., 1979.
9. S. Lin and T. Rado. Computer studies of Turing machine problems, *Journal of the Association for Computing Machinery*, 12 (1965), 196–212.
10. H. Marxen and J. Buntrock. List of record machines, Available at: `http://www.drb.insel.de/~heiner/BB/bb-list`.
11. H. Marxen and J. Buntrock. Attacking the Busy Beaver 5, *Bulletin of the European Association for Theoretical Computer Science (EATCS)*, 40 (1990), 274–251.
12. P. Michel. Busy beaver competition and Collatz-like problems, *Arch. Math. Logic*, 32 (1993), 351–367.
13. T. Rado. On non-computable functions, *The Bell System Technical Journal*, 41 (1962), 877–884.

14. T. Rado. On a simple source for non-computable functions, *Proceedings of the Symposium on Mathematical Theory of Automata (New York, 1962)*, volume XII of *Microwave Research Institute Symposia Series*, Polytechnic Press of the Polytechnic Institute of Brooklyn, 1963, 75–81.
15. B. Weimann, K. Casper and W. Fenzl. Untersuchungen über haltende Programme für Turingmaschinen mit 2 Zeichen und bis zu 5 Befehlen, *GI Gesellschaft für Informatik e.V.: 2. Jahrestagung (Karlsruhe, Germany, 1972)*, Lecture Notes in Economics and Mathematical Systems, Springer, Berlin, 1973, 72–81.

13

Introduction to Unification Grammars

Shuly Wintner

Department of Computer Science
University of Haifa
Mount Carmel, 31905 Haifa, Israel
E-mail: shuly@cs.haifa.ac.il

13.1 Motivation

Human languages are among Nature's most extraordinary phenomena. Their structural complexity, the ease with which they are acquired by young children, the amazing similarities among unrelated languages – all these are the focus of current research in *linguistics*. In the second half of the 20th century, the field of linguistics has undergone a revolution: the themes that are subject for study, the vocabulary with which they are expressed, the methods and techniques for investigating them – have changed dramatically. While the traditional aims of linguistic research used to be the description of particular languages, sometimes with respect to other, related languages, modern linguistics has broadened the domain it is investigating. Motivated by the assumption that the language faculty is innate, linguists are seeking the *universal* principles that underlie all the languages of the world; they are looking for structural generalizations that hold across languages, as well as across various utterance types in a single language; they investigate constraints on language usage and the reasons for such constraints; and they attempt to delimit the class of natural languages by formal means.

The revolution in linguistics, attributed mainly to Noam Chomsky, has influenced and inspired research in formal languages, especially in programming languages. One of the earliest fields of study in computer science was human cognitive processes, in particular natural language processing (NLP). The main objective of NLP is to computationally simulate processes related to the human linguistic faculty. The ordinary instruments in such endeavors are heuristic techniques that aid in constructing practical applications. A slightly different scientific field has obtained the name *computational linguistics*. It studies the structure of natural languages from a formal, mathematical and computational, point of view. It is concerned with all subfields of the traditional linguistic research, but it approaches these fields from a unique point of departure: it describes linguistic phenomena in a way that, at least in principle, is computationally implementable.

S. Wintner: *Introduction to Unification Grammars*, Studies in Computational Intelligence (SCI) **25**, 321–342 (2006)
www.springerlink.com

Early attempts to formally describe the structure of natural languages [7] resulted in the development of formal grammars, of which *context-free grammars* (CFGs) are a notable example. Context-free grammars, however, have been found inadequate for this purpose; not only because some natural languages are, apparently, beyond the expressive power of CFGs [30, 19], but also because CFGs do not provide sufficiently succinct descriptions of the phenomena linguists are usually interested in [27]. The main drawback of CFGs (and other phrase-structure formalisms, including early incarnations of tree-adjoining grammars [14]) in this respect is that the symbols they manipulate are *atomic*, with no inherent structure, whereas the linguistic entities that have to be modeled by them are inherently structured. For example, words have properties such as *number, gender, tense* or *root*; phrases are categorized and sub-categorized; phonemes are described as sets of features; etc.

A natural way of extending phrase-structure grammars to rectify this problem is by incorporating into the grammar *properties* of symbols, further specifying the crude categorial information. This is achieved by replacing the atomic symbols of CFGs by *feature structures*, which provide a natural representation for the kind of linguistic information grammars specify. In this view, the lexicon associates with each word a set of feature structures, rather than a set of atomic terminal symbols, and grammar rules are defined over feature structures, rather than over atomic non-terminal symbols. With this extension, rule application has to be amended, too, and comparing two symbols cannot be simply reduced to identity checking, as is the case in CFGs. Rather, an information comparison and combination operation, *unification*, is used to decide whether a rule can be applied to some form.

The move from CFGs to unification grammars provides an additional dimension for linguistic generalizations. It facilitates the expression of linguistic constraints by means of a single rule, instead of a collection of 'similar' rules. More significantly, it also results in a dramatic increase in expressive power: the set of languages which can be defined by unification grammars is much larger than the set of context-free languages; in terms of the [7] hierarchy of languages, this is the set of recursively enumerable ('type-0') languages. Consequently, computational processing of unification grammars is significantly less efficient than processing with CFGs.

This chapter defines feature structures (section 13.2), unification (section 13.3) and unification grammars (section 13.4) from a mathematical and computational (rather than linguistic) point of view. Clearly, this is a very superficial description, and references for further reading are provided in section 13.5.

13.2 Feature Structures

Feature structures are graphically represented using a notation known as *attribute-value matrices* (AVMs); both terms are used interchangeably here.

An AVM is a *syntactic* object, which can be either *atomic* or a *complex*. Each AVM is associated with a variable, the idea being that two occurrences of the same variable denote one and the *same* value. An atomic feature structure is a variable, associated with an *atom*, drawn from a fixed set ATOMS; these will usually be non-structured linguistic notions such as *singular, plural* or *accusative*. A complex feature structure is a variable, associated with a finite, possibly empty set of pairs, where each pair consists of a *feature* and a *value*. Features are drawn from a fixed (per grammar), pre-defined set FEATS; values are, recursively, AVMs themselves. Thus, FEATS and ATOMS, as well as the (infinite) set of variables, are parameters for the collection of AVMs, and are referred to as the *signature* over which feature structures are defined.

Definition 1. *Given a signature consisting of a finite set* ATOMS *of atoms, a finite set* FEATS *of features and an enumerable set* VARS *of variables, the set* AVMS *of* ***AVMs*** *is the least set such that*

1. $\boxed{i}a \in$ AVMS *for every* $\boxed{i} \in$ VARS *and* $a \in$ ATOMS; $\boxed{i}$ *is said to be* ***associated with*** a
2. $\boxed{i}[\,] \in$ AVMS *for every* $\boxed{i} \in$ VARS; $\boxed{i}$ *is said to be* ***associated with*** $[\,]$
3. *for every* $\boxed{i} \in$ VARS, $f_1, \ldots, f_n \in$ FEATS *and* $A_1, \ldots, A_n \in$ AVMS, $n \geq 1$,

$$A = \boxed{i}\begin{bmatrix} f_1 : A_1 \\ \vdots \\ f_n : A_n \end{bmatrix} \in \text{AVMS}; \ \boxed{i} \text{ is said to be } \textbf{associated with} \begin{bmatrix} f_1 : A_1 \\ \vdots \\ f_n : A_n \end{bmatrix}$$

and $A_1, \ldots, A_n$ *are sub-AVMs of* A.

In each clause of the definition, the occurrence of the variable $\boxed{i}$ is said to be *associated with* whatever appears to its right (note that in the third clause there may be other occurrences of the same variable in $A_1, \ldots, A_n$). Observe that any sub-AVM of an AVM is associated with a variable, including atoms and empty AVMs. The set of variables occurring in some AVM A is *Tags*(A).

Example 1. In the following AVM, the set of variables includes $\boxed{3}$, $\boxed{5}$, $\boxed{6}$ and $\boxed{8}$. FEATS = {AGR,NUM,PERS} and ATOMS = $\{pl, third\}$.

$$\boxed{3}\begin{bmatrix} \text{AGR} : \boxed{5}\begin{bmatrix} \text{NUM} : \boxed{6}pl \\ \text{PERS} : \boxed{8}third \end{bmatrix} \end{bmatrix}$$

Meta-variables F, G, H are used to range over FEATS and A, B, C to range over feature structures. If

$$A = \boxed{k}\begin{bmatrix} \text{F}_1 : A_1 \\ \vdots \quad \vdots \\ \text{F}_n : A_n \end{bmatrix}$$

is a complex feature structure then the *domain* of A is $dom(A) = \{\text{F}_i \mid 1 \leq i \leq n\}$. When A is atomic, $dom(A)$ is undefined. A special case of a feature structure is the *empty* one, denoted $[\,]$, whose domain is the empty set of features. Note that $|dom(A)| = n$. That is, for every $1 \leq i, j \leq n$ such that $i \neq j$, $\textsc{f}_i \neq \textsc{f}_j$. This uniqueness of value assignment by A to each $\textsc{f}_i$ (known also as the *functional* character of feature structures) is very essential. In particular, the order of the "rows" in an AVM is not important, and any permutation of this order will result in an identical AVM. The *value* of a feature $\textsc{f}_i$ in A is $val(A, \textsc{f}_i) = A_i$. If $\textsc{f} \notin dom(A)$ then $val(A, \textsc{f})$ is undefined.

Exercise 1. In the AVM A of example 1, what is $dom(val(A, \textsc{agr}))$?

Since variables are used to denote value sharing, there is not much sense in associating the same variable with two different values. Therefore, an AVM is *well-formed* if for every variable occurring in it, all its occurrences are associated with the same value.

Example 2. The following AVM, A, is well-formed since both occurrences of $\boxed{1}$ are associated with the same value, a. B is not well-formed due to the two incompatible assignments of $\boxed{1}$; C is not well-formed because the values associated with $\boxed{1}$ are different:

$$A = \boxed{3}\begin{bmatrix} \textsc{f}: \boxed{1} a \\ \textsc{g}: \boxed{2}\begin{bmatrix}\textsc{h}: \boxed{1} a\end{bmatrix} \end{bmatrix}, B = \boxed{4}\begin{bmatrix} \textsc{f}: \boxed{1}\begin{bmatrix}\textsc{h}: \boxed{2} a\end{bmatrix} \\ \textsc{g}: \boxed{1} b \end{bmatrix}, C = \boxed{3}\begin{bmatrix} \textsc{f}: \boxed{1}\begin{bmatrix}\textsc{h}: \boxed{4} a\end{bmatrix} \\ \textsc{g}: \boxed{1}\begin{bmatrix}\textsc{k}: \boxed{5} b\end{bmatrix} \end{bmatrix}$$

Exercise 2. Let

$$A = \boxed{1}\begin{bmatrix} \textsc{g}: \boxed{2}[\,] \\ \textsc{f}: \boxed{2}[\,] \end{bmatrix}$$

Is A well-formed?

In the sequel, all AVMs are assumed to be well-formed. Furthermore, since multiple occurrences of the same variable always are associated with the same values, only one instance of this value is explicated, leaving the other occurrences of the same variable unspecified. As another convention, whenever a variable is associated with the empty AVM the AVM itself is omitted and only the variable is specified. Finally, if a variable occurs only once in an AVM, it is usually not depicted (as it carries no additional information).

Example 3. Consider the following feature structure, A:

$$\begin{bmatrix} \textsc{f}: \boxed{3} a \\ \textsc{g}: \boxed{4}\begin{bmatrix}\textsc{h}: \boxed{3} a\end{bmatrix} \\ \textsc{h}: \boxed{2}[\,] \end{bmatrix}$$

Notice that it is well-formed, since the only variable occurring more than once ($\boxed{3}$) is associated with the same value (a) in both its occurrences. Therefore, one occurrence of the value can be left implicit, yielding:

$$\begin{bmatrix} \textsc{f}: \boxed{3} \\ \textsc{g}: \boxed{4}\begin{bmatrix}\textsc{h}: \boxed{3} a\end{bmatrix} \\ \textsc{h}: \boxed{2}[\,] \end{bmatrix}$$

Next, the tag $\boxed{2}$ is associated with the empty feature structure, which can be omitted:

$$\begin{bmatrix} \text{F}: \boxed{3} \\ \text{G}: \boxed{4}\begin{bmatrix}\text{H}: \boxed{3} a\end{bmatrix} \\ \text{H}: \boxed{2} \end{bmatrix}$$

Finally, the tag $\boxed{4}$ occurs only once, so it can be omitted:

$$\begin{bmatrix} \text{F}: \boxed{3} \\ \text{G}: \begin{bmatrix}\text{H}: \boxed{3} a\end{bmatrix} \\ \text{H}: \boxed{2} \end{bmatrix}$$

A *path* is a (possibly empty) sequence of features that can be used to pick a value in a feature structure. Angular brackets '$\langle \ldots \rangle$' are used to depict paths explicitly. For example, in the AVM A of example 3, the single feature $\langle \text{F} \rangle$ constitutes a path; and so does the sequence $\langle \text{G,H} \rangle$, since $\langle \text{G} \rangle$ can be used to pick the value $\begin{bmatrix}\text{H}: \boxed{3} a\end{bmatrix}$ (because $val(A, \text{G}) = \begin{bmatrix}\text{H}: \boxed{3} a\end{bmatrix}$). In more complex AVMs, with deeper levels of embedding, longer paths can be used recursively to reach more deeply nested features. The set of all the paths of A is denoted $\Pi(A)$.

The notion of values is extended from features to paths: $val(A, \pi)$ is the value obtained by following the path π in A; this value (if defined) is again a feature structure. If A_i is the value of some path π in A then A_i is said to be a *sub-AVM* of A. The *empty path* is denoted ϵ, and $val(A, \epsilon) = A$ for every feature structure A.

Example 4. Let

$$A = \begin{bmatrix} \text{AGR}: \begin{bmatrix} \text{NUM}: pl \\ \text{PERS}: third \end{bmatrix} \end{bmatrix}$$

Then $dom(A) = \{\text{AGR}\}$, and

$$val(A, \text{AGR}) = \begin{bmatrix} \text{NUM}: pl \\ \text{PERS}: third \end{bmatrix}$$

The paths of A are $\Pi(A) = \{\epsilon, \langle \text{AGR} \rangle, \langle \text{AGR,NUM} \rangle, \langle \text{AGR,PERS} \rangle\}$. The values of these paths are: $val(A, \epsilon) = A$, $val(A, \langle \text{AGR,NUM} \rangle) = pl$, $val(A, \langle \text{AGR,PERS} \rangle) = third$. Since there is no path $\langle \text{NUM,AGR} \rangle$ in A, $val(A, \langle \text{NUM,AGR} \rangle)$ is undefined.

Exercise 3. Show an AVM A in which the paths $\pi_1 = \langle \text{F,G} \rangle$ and $\pi_2 = \langle \text{G,F} \rangle$ are defined and $val(A, \pi_1) = val(A, \pi_2)$.

Exercise 4. Can you show an AVM A such that $\Pi(A)$ is empty?

Exercise 5. For an arbitrary AVM A, what is $val(A, \epsilon)$?

Exercise 6. Prove that if $A' = val(A, \pi)$ then all the paths of A' are suffixes of paths in A.

Exercise 7. Show an AVM A for which $val(A, \pi) = A$ for every path π. Show that this AVM is unique, up to the names of its variables.

When two paths in an AVM lead to sub-AVMs that are associated with the same variable, the paths are said to be *reentrant*; reentrancy in some feature structure A is denoted by the symbol '$\overset{A}{\leftrightsquigarrow}$'. For example, in the feature structure

$$A_3 = \begin{bmatrix} \text{SUBJ}: & \begin{bmatrix} \text{AGR}: \boxed{1}\begin{bmatrix}\text{NUM}: \textit{plural}\end{bmatrix}\end{bmatrix} \\ \text{OBJ}: & \begin{bmatrix} \text{AGR}: \boxed{1}\end{bmatrix} \end{bmatrix}$$

the paths $\langle\text{SUBJ}, \text{AGR}\rangle$ and $\langle\text{OBJ}, \text{AGR}\rangle$ are reentrant, that is, $\langle\text{SUBJ}, \text{AGR}\rangle \overset{A_3}{\leftrightsquigarrow} \langle\text{OBJ}, \text{AGR}\rangle$. The distinction between "being the same entity", or being identical, and being "similar but distinct" entities, is referred to in the literature as the distinction between *token identity* and *type identity*, respectively.

Example 5. Consider the following feature structure:

$$A = \begin{bmatrix} \text{SUBJ}: & \begin{bmatrix} \text{AGR}: \begin{bmatrix} \text{NUM}: & \textit{sg} \\ \text{PERS}: & \textit{third} \end{bmatrix} \end{bmatrix} \\ \text{OBJ}: & \begin{bmatrix} \text{AGR}: \begin{bmatrix} \text{NUM}: & \textit{sg} \\ \text{PERS}: & \textit{third} \end{bmatrix} \end{bmatrix} \end{bmatrix}$$

In A, the features SUBJ and OBJ have equal, but not token-identical, values. In contrast, consider the following AVM:

$$B = \begin{bmatrix} \text{SUBJ}: \boxed{4} \begin{bmatrix} \text{AGR}: \begin{bmatrix} \text{NUM}: & \textit{sg} \\ \text{PERS}: & \textit{third} \end{bmatrix} \end{bmatrix} \\ \text{OBJ}: \boxed{4} \end{bmatrix}$$

In B, the values of SUBJ and OBJ are not only type-identical; they are also *token-identical*; in effect, there is only *one and the same* value *shared* by both features in B. The two features simply 'point' to this one value, in much the same way as *pointers* in programming languages might be set to refer to the same location in memory.

An important consequence of token identity is the following: if the value of one of two features having a shared value is modified, this modification is immediately "sensed" by the other feature and its value changes too, to the same new value. This is demonstrated in example 6.

Example 6. Consider again the following feature structures:

$$A = \begin{bmatrix} \text{SUBJ}: & \begin{bmatrix} \text{AGR}: \begin{bmatrix} \text{NUM}: & \textit{sg} \\ \text{PERS}: & \textit{third} \end{bmatrix} \end{bmatrix} \\ \text{OBJ}: & \begin{bmatrix} \text{AGR}: \begin{bmatrix} \text{NUM}: & \textit{sg} \\ \text{PERS}: & \textit{third} \end{bmatrix} \end{bmatrix} \end{bmatrix}$$

$$B = \left[\begin{array}{l} \text{SUBJ} : \boxed{4} \left[\text{AGR} : \left[\begin{array}{l} \text{NUM} : sg \\ \text{PERS} : third \end{array}\right] \right] \\ \text{OBJ} : \boxed{4} \end{array}\right]$$

Observe that $val(A, \text{SUBJ}) = val(A, \text{OBJ})$ and $val(B, \text{SUBJ}) = val(B, \text{OBJ})$. However, while $val(B, \text{SUBJ})$ and $val(B, \text{OBJ})$ are token-identical, this is not the case for A: $val(A, \text{SUBJ})$ and $val(A, \text{OBJ})$ are only type-identical. Suppose that by some action (as shown below), a feature CASE with the value *nom* is added to the value of SUBJ in both A and B. The outcome of such an operation has different effects on the two feature structures. In A, it will result in:

$$A' = \left[\begin{array}{ll} \text{SUBJ} : & \left[\begin{array}{l} \text{AGR} : \left[\begin{array}{l} \text{NUM} : sg \\ \text{PERS} : third \end{array}\right] \\ \text{CASE} : nom \end{array}\right] \\ \text{OBJ} : & \left[\text{AGR} : \left[\begin{array}{l} \text{NUM} : pl \\ \text{PERS} : third \end{array}\right] \right] \end{array}\right]$$

In particular, note that $val(A', \text{SUBJ}) \neq val(A', \text{OBJ})$. However, in B the effect would be:

$$B' = \left[\begin{array}{l} \text{SUBJ} : \boxed{4} \left[\begin{array}{l} \text{AGR} : \left[\begin{array}{l} \text{NUM} : sg \\ \text{PERS} : third \end{array}\right] \\ \text{CASE} : nom \end{array}\right] \\ \text{OBJ} : \boxed{4} \end{array}\right]$$

Here, $val(B', \text{SUBJ})$ and $val(B', \text{OBJ})$ are token-identical, implying $val(B', \text{SUBJ}) = val(B', \text{OBJ})$. Since the values are token identical, the operation has affect over the two reentrant features.

Unification-based theories of language make extensive use of reentrancy to encode grammatical relations. In fact, it is instructive to think of reentrancies as the unification equivalent of *movement* in generative linguistics. Feature structures can be arbitrarily complex, and by sharing a single value between two distinct paths, a theory can express the notion of this value being 'moved' from the end of one path to the end of another. However, there is no notion of direction for such 'movement': the shared value is the common value of both paths, symmetrically.

A special case of reentrancy is *cyclicity*: an AVM can contain a path whose value is the AVM itself. In other words, an AVM can be reentrant with a substructure of itself:

$$A = \left[\text{F} : \boxed{2} \left[\begin{array}{l} \text{G} : a \\ \text{H} : \boxed{2} \end{array}\right] \right]$$

Notice that in the face of cycles there is no way to fully depict (in some finite way) an AVM, unless the convention of omitting multiple occurrences of shared values is adopted (as above).

Exercise 8. What is the set of paths of the feature structure A above? What is its cardinality?

Exercise 9. Prove: A is acyclic iff $\Pi(A)$ is finite.

An additional dimension through which linguistic generalizations can be expressed is obtained by introducing *typing* to feature structures. The motivation for typing stems from the observation that certain features are more appropriate to some feature structures than to others. For example, consider the two AVMs below:

$$A = \begin{bmatrix} \text{CAT}: & v \\ \text{SUBCAT}: & elist \\ \text{AGR}: & \begin{bmatrix} \text{NUM}: & pl \\ \text{PERS}: & 3rd \end{bmatrix} \end{bmatrix}, \qquad B = \begin{bmatrix} \text{NUM}: & pl \\ \text{PERS}: & 3rd \end{bmatrix}$$

In neither of these structures is the feature GENDER defined, but this feature is certainly more appropriate for B than for the top level of A (although adding it to the AVM that is the value of AGR in A *is* natural). Informally, one would like to claim that the feature GENDER is appropriate for feature structures that model agreement properties (such as B or $val(A, \text{AGR})$ above), but not for feature structures that model signs (such as A). This is not a mathematical claim; it is a claim concerning the intended interpretation of feature structures. To formalize this notion, feature structures are classified, and the names of classes are called *types* (sometimes referred to as *sorts*). When each feature structure is associated with a certain type, it becomes possible to define what features are appropriate for a given type, t; these features will be appropriate for every feature structure of type t. For example, the feature structure A above can be assigned the type *sign*, as it represents a linguistic sign; whereas the feature structure B can be assigned the type *agr*, to indicate the fact that it is a collection of agreement features. Now the feature GENDER could be defined as appropriate for the type *agr* but not for the type *sign*.

The move to typed feature structures necessitates an extension of the *signature* over which feature structures are defined: in addition to the set FEATS of features, a (finite) set TYPES of types is required. Incidentally, the introduction of types renders the set of atoms, ATOMS, redundant, as atoms can be represented using types.

Once types and feature appropriateness are introduced, it is possible to extend their use: it should be clear that for different features, different types of values are appropriate. For example, the feature AGR must be restricted to bear only values of the type *agr*; whereas for the feature GENDER, for example, such values must not be appropriate. Furthermore, a typed system can include a mechanism for *forcing* the occurrence of a certain feature in a TFS, not just for *allowing* it.

As mentioned above, types can be viewed as naming classes of feature structures; a natural outcome of this view is the imposing of *order* on types, to reflect the natural set inclusion order on sets of feature structures (of those types). It is possible to extend appropriateness to respect this order, such that if a feature is appropriate for a type, it is also appropriate for all its subtypes.

A typed system allows the grammar designer to impose restrictions on feature structures that reflect real-world or linguistic constraints. It enables better generalizations to be stipulated as it gives the grammar writer an additional dimension in which to state such generalizations: linguistic constraints can be defined in terms of types rather than in terms of features. For example, most of the "rules" of the linguistic theory HPSG are stipulated as constraints on types: if a feature structure is of a certain type, than certain constraints must hold.

While the use of types in linguistic theories is primarily motivated by linguistic considerations, there are several additional, mathematical and computational, advantages to their use. The incorporation of types into programming languages has become the standard. Types provide means for several compile-time and run-time errors to be detected. They allow – in particular when combined with appropriateness – for several optimizations in implementations of feature structure based formalisms, resulting in more efficient processing. Finally, as in programming languages, the use of types aids in grammar organization and modularization. For simplicity, however, the rest of this chapter discusses untyped feature structures only.

13.3 Subsumption and Unification

Feature structures are used to represent linguistic *information*, and the amount of information stored within different feature structures can be compared, thus inducing a natural (partial) order on the structures. This relation is called *subsumption* and is denoted '$\sqsubseteq$'.

Definition 2. *If A and B are feature structures over the same signature, then A* ***subsumes*** *B ($A \sqsubseteq B$; also, A is* ***more general than*** *B, and B is* ***subsumed by****, or is* ***more specific than****, A) if the following conditions hold:*

1. *if A is an atomic AVM then B is an atomic AVM with the same atom;*
2. *for every $\text{F} \in \text{FEATS}$, if $\text{F} \in dom(A)$ then $\text{F} \in dom(B)$, and $val(A, \text{F})$ subsumes $val(B, \text{F})$; and*
3. *if two paths are reentrant in A, they are also reentrant in B: if $\pi_1 \overset{A}{\leftrightsquigarrow} \pi_2$ then $\pi_1 \overset{B}{\leftrightsquigarrow} \pi_2$.*

The second condition of the definition is recursive; the recursion ends when an empty value or an *atomic* value is encountered. Note the direction of the ordering: greater structures are more specific, i.e., contain more information. This extra information can be manifested as additional features having a defined value, or more specific values for an existing feature, or more reentrancies.

Example 7. The empty feature structure, [ ], is the most general feature structure, subsuming all others, including atoms (as it encodes no information at all). Since it contains no features (and, hence, no paths), it satisfies the conditions of the subsumption definition vacuously. In particular, it subsumes a feature structure that determines the value of the NUM feature to be *sg*:

$$[\,] \sqsubseteq [\text{NUM} : sg]$$

In the same way,

$$[\text{NUM} : [\,]] \sqsubseteq [\text{NUM} : sg]$$

since the empty feature structure subsumes the atomic feature structure *sg*.

Adding more information, say, about the PERSON feature, results in a more specific feature structure:

$$[\text{NUM} : sg] \sqsubseteq \begin{bmatrix} \text{NUM} : sg \\ \text{PERS} : third \end{bmatrix}$$

Another way to add information is through reentrancies:

$$\begin{bmatrix} \text{NUM1} : sg \\ \text{NUM2} : sg \end{bmatrix} \sqsubseteq \begin{bmatrix} \text{NUM1} : \boxed{1} sg \\ \text{NUM2} : \boxed{1} \end{bmatrix}$$

Subsumption is a *partial* relation: not every pair of feature structures is comparable. For example, it is impossible to say which of the following feature structures encodes more information:

$$[\text{NUM} : sg] \overset{\not\sqsubseteq}{\not\sqsupseteq} [\text{NUM} : pl]$$

A different case of incomparability is caused by the existence of different features in the two structures. Since each of the AVMs in the following example has a feature that is not defined in the other, none is more specific than the other:

$$[\text{NUM} : sg] \overset{\not\sqsubseteq}{\not\sqsupseteq} [\text{PERS} : third]$$

While subsumption informally encodes an order of information content among AVMs, sometimes the informal notion can be misleading. For example, the following two AVMs informally encode the same information, pertaining to the agreement features of some entity; formally, however, neither subsumes the other.

$$\begin{bmatrix} \text{NUM} : sg \\ \text{PERS} : third \end{bmatrix} \overset{\not\sqsubseteq}{\not\sqsupseteq} \begin{bmatrix} \text{AGR} : \begin{bmatrix} \text{NUM} : sg \\ \text{PERS} : third \end{bmatrix} \end{bmatrix}$$

Subsumption has the following properties:

Least element: the empty feature structure subsumes every feature structure: for every feature structure A, $[\,] \sqsubseteq A$

Reflexivity: for every feature structure A, $A \sqsubseteq A$

Transitivity: If $A \sqsubseteq B$ and $B \sqsubseteq C$ than $A \sqsubseteq C$.

Antisymmetry: If $A \sqsubseteq B$ and $B \sqsubseteq A$ then $A = B$.

Since subsumption is a partial, reflexive, transitive and antisymmetric relation, it is a *partial order.*

Exercise 10. Show that

$$\begin{bmatrix} \text{F} : \boxed{1} a \\ \text{G} : \boxed{1} \end{bmatrix} \sqsubseteq \begin{bmatrix} \text{F} : \boxed{2} a \\ \text{G} : \boxed{2} \end{bmatrix}$$

Exercise 11. Which of the following feature structures subsumes the other?

$$A = \begin{bmatrix} \text{F} : \boxed{3} \begin{bmatrix} \text{F} : \boxed{3} \end{bmatrix} \end{bmatrix}, \qquad B = \boxed{2} \begin{bmatrix} \text{F} : \boxed{2} \end{bmatrix}$$

Exercise 12. Prove: If $A \sqsubseteq B$ then $\Pi(A) \subseteq \Pi(B)$.

Exercise 13. Prove: If $A \sqsubseteq B$ and π_1, π_2 are reentrant in A then π_1, π_2 are reentrant in B.

Motivated by the view of feature structures as representing information, it is useful to have an operation that combines, whenever possible, the information that is encoded in two feature structures. The *unification* operation, denoted '$\sqcup$', is defined over pairs of feature structures, and yields the most general feature structure that is more specific than both operands, if one exists: $A = B \sqcup C$ if and only if A is the most general feature structure such that $B \sqsubseteq A$ and $C \sqsubseteq A$. If such a structure exists, the unification *succeeds*, and the two arguments are said to be *unifiable* (or *consistent*). If none exists, the unification *fails*, and the operands are said to be *inconsistent* (denoted $B \sqcup C = \top$). In terms of the subsumption ordering, non-failing unification returns the least upper bound (*lub*) of its arguments. Example 8 sheds light on different aspects of the unification operation.

Example 8. Unification combines consistent information:

$$\begin{bmatrix} \text{NUM} : sg \end{bmatrix} \sqcup \begin{bmatrix} \text{PERS} : third \end{bmatrix} = \begin{bmatrix} \text{NUM} : sg \\ \text{PERS} : third \end{bmatrix}$$

Different atoms are inconsistent: $\begin{bmatrix} \text{NUM} : sg \end{bmatrix} \sqcup \begin{bmatrix} \text{NUM} : pl \end{bmatrix} = \top$
Atoms and non-atoms are inconsistent: $\begin{bmatrix} \text{NUM} : sg \end{bmatrix} \sqcup sg = \top$
Unification is absorbing:

$$\begin{bmatrix} \text{NUM} : sg \end{bmatrix} \sqcup \begin{bmatrix} \text{NUM} : sg \\ \text{PERS} : third \end{bmatrix} = \begin{bmatrix} \text{NUM} : sg \\ \text{PERS} : third \end{bmatrix}$$

Empty feature structures are identity elements:

$$[\,] \sqcup \begin{bmatrix} \text{AGR} : \begin{bmatrix} \text{NUM} : sg \end{bmatrix} \end{bmatrix} = \begin{bmatrix} \text{AGR} : \begin{bmatrix} \text{NUM} : sg \end{bmatrix} \end{bmatrix}$$

Reentrancy causes two consistent values to coincide:

$$\begin{bmatrix} \text{F} : \begin{bmatrix} \text{NUM} : sg \end{bmatrix} \\ \text{G} : \begin{bmatrix} \text{PERS} : third \end{bmatrix} \end{bmatrix} \sqcup \begin{bmatrix} \text{F} : \boxed{1} \\ \text{G} : \boxed{1} \end{bmatrix} = \begin{bmatrix} \text{F} : \boxed{1} \begin{bmatrix} \text{NUM} : sg \\ \text{PERS} : third \end{bmatrix} \\ \text{G} : \boxed{1} \end{bmatrix}$$

Unification acts differently depending on whether the values are type-identical:

$$\begin{bmatrix} \text{F} : \begin{bmatrix} \text{NUM} : sg \end{bmatrix} \\ \text{G} : \begin{bmatrix} \text{NUM} : sg \end{bmatrix} \end{bmatrix} \sqcup \begin{bmatrix} \text{F} : \begin{bmatrix} \text{PERS} : third \end{bmatrix} \end{bmatrix} = \begin{bmatrix} \text{F} : \begin{bmatrix} \text{NUM} : sg \\ \text{PERS} : third \end{bmatrix} \\ \text{G} : \begin{bmatrix} \text{NUM} : sg \end{bmatrix} \end{bmatrix}$$

...or token-identical:

$$\begin{bmatrix} \text{F} : \boxed{1} \begin{bmatrix} \text{NUM} : sg \end{bmatrix} \\ \text{G} : \boxed{1} \end{bmatrix} \sqcup \begin{bmatrix} \text{F} : \begin{bmatrix} \text{PERS} : third \end{bmatrix} \end{bmatrix} = \begin{bmatrix} \text{F} : \boxed{1} \begin{bmatrix} \text{NUM} : sg \\ \text{PERS} : third \end{bmatrix} \\ \text{G} : \boxed{1} \end{bmatrix}$$

One of the properties of unification is that it *binds* variables together. When two feature structures are unified, and each of them is associated with some variable, then the two variables are said to be bound to each other, which means that they both share one and the same value (the result of the unification). For all practical purposes, this is equivalent to there being only a single variable. However, the *scope* of the variables must be taken into account: all other occurrences of the same variables in this scope are effected. As an example, consider the following two feature structures A and B:

$$A = \begin{bmatrix} \text{F} : \boxed{1} \begin{bmatrix} \text{NUM} : sg \end{bmatrix} \end{bmatrix} \qquad B = \begin{bmatrix} \text{F} : \boxed{2} \begin{bmatrix} \text{PERS} : third \end{bmatrix} \end{bmatrix}$$

When unifying A and B, the obtained result is:

$$A \sqcup B = \begin{bmatrix} \text{F} : \boxed{1}\boxed{2} \begin{bmatrix} \text{NUM} : sg \\ \text{PERS} : third \end{bmatrix} \end{bmatrix}$$

Of course, since the variables $\boxed{1}$ and $\boxed{2}$ occur nowhere else, they can be simply omitted and the result is equal to:

$$A \sqcup B = \begin{bmatrix} \text{F} : \begin{bmatrix} \text{NUM} : sg \\ \text{PERS} : third \end{bmatrix} \end{bmatrix}$$

However, had either $\boxed{1}$ or $\boxed{2}$ occurred elsewhere (for example, as the value of some feature G in A), their values would have been modified as a result of the unification. Consider the following example:

$$\begin{bmatrix} \text{F} : \boxed{1} \begin{bmatrix} \text{NUM} : sg \end{bmatrix} \\ \text{G} : \boxed{1} \end{bmatrix} \sqcup \begin{bmatrix} \text{F} : \boxed{2} \begin{bmatrix} \text{PERS} : third \end{bmatrix} \end{bmatrix} = \begin{bmatrix} \text{F} : \boxed{3} \begin{bmatrix} \text{NUM} : sg \\ \text{PERS} : third \end{bmatrix} \\ \text{G} : \boxed{3} \end{bmatrix}$$

The (recursive) unification of the values of the feature F in both operands causes the variables $\boxed{1}$ and $\boxed{2}$ to be bound to each other. The AVMs these variables are associated with are unified, yielding the following AVM:

$$\begin{bmatrix} \text{NUM} : sg \\ \text{PERS} : third \end{bmatrix}$$

Then, a fresh variable can be associated with the above AVM in the result.

Properties of unification include:

Idempotency: $A \sqcup A = A$
Commutativity: $A \sqcup B = B \sqcup A$
Associativity: $A \sqcup (B \sqcup C) = (A \sqcup B) \sqcup C$
Absorption: If $A \sqsubseteq B$ then $A \sqcup B = B$
Monotonicity: If $A \sqsubseteq B$ then for every C, $A \sqcup C \sqsubseteq B \sqcup C$ (if both exist).

Exercise 14. Let:

$$A = \begin{bmatrix} \text{F} : a \end{bmatrix}$$

$$B = \begin{bmatrix} \text{G} : \begin{bmatrix} \text{F} : a \end{bmatrix} \end{bmatrix}$$

$$C = \begin{bmatrix} \text{F} : a \\ \text{G} : \begin{bmatrix} \text{F} : a \end{bmatrix} \end{bmatrix}$$

$$D = \begin{bmatrix} \text{F} : a \\ \text{G} : \boxed{1} \begin{bmatrix} \text{F} : a \end{bmatrix} \end{bmatrix}$$

$$E = \begin{bmatrix} \text{F} : \boxed{1} \begin{bmatrix} \text{F} : a \end{bmatrix} \\ \text{G} : \boxed{1} \end{bmatrix}$$

$$F = \begin{bmatrix} \text{F} : \boxed{1} \\ \text{G} : \boxed{1} \end{bmatrix}$$

Which of the following holds?

$$\begin{array}{lll} A \sqcup B = C & A \sqcup C = D & A \sqcup F = C \\ A \sqcup F = E & B \sqcup F = E & C \sqcup D = D \\ C \sqcup D = E & D \sqcup D = E & E \sqcup F = E \end{array}$$

Exercise 15. Prove or refute: If A and B contain no reentrancies, then so does $A \sqcup B$ (if it is defined).

Exercise 16. Prove or refute: If A and B are acyclic, then so is $A \sqcup B$ (if it is defined).

The unification operation was originally defined for first-order terms (FOTs). There is an important difference between FOT unification and feature structure unification: an FOT is said to be *ground* if no further unifications can change its value. There is no real equivalent to the concept of groundness for non-atomic feature structures, since unifications can *always* add new features to a structure (unless it is atomic). Another important difference between FOTs and feature structures has to do with the result of the unification operation. For FOTs, unification results in a *substitution*, that when applied to both unificands yields the same FOT. In feature structures, however, unification returns a new feature structure – as if the substitution has already been applied to one of the unificands. In fact, there is no clear analog to the notion of substitution in the domain of feature structures.

13.4 Unification Grammars

AVMs were introduced above as a means of extending atomic symbols (such as terminal and non-terminal symbols of a context-free grammar) with structural (linguistic) information. In the same vein, it is possible to extend AVMs to sequences thereof, called *multi-AVMs* below, thereby defining an equivalent of *phrase-structure rules* over feature structures.

Multi-AVMs can be viewed as sequences of AVMs, with the important observation that some sub-structures can be shared among two or more AVMs. In other words, the scope of variables is extended from a single AVM to a multi-AVM: the same variable can be associated with two sub-structures of different AVMs. An immediate consequence is that if some sub-structure of an element on a multi-AVM (that is, a sub-structure of some AVM in it) is modified, this modification might effect other AVMs in the same "sequence". The *length* of a multi-AVM is the number of its elements. Meta-variables σ, ρ range over multi-AVMs.

Definition 3. *Given a signature consisting of a finite set* ATOMS *of atoms, a finite set* FEATS *of features and an enumerable set* VARS *of variables, a* ***multi-AVM*** *of* ***length*** n *is a sequence* $\sigma = \langle A_1, \ldots, A_n \rangle$ *such that for each* i, $1 \leq i \leq n$, A_i *is an AVM over the signature and the variable associated with* A_i *differs from the variable associated with* A_j *if* $i \neq j$.

Example 9. Following is a multi-AVM of length 3:

$$\begin{bmatrix} \text{CAT} : np \\ \text{AGR} : \boxed{1} \end{bmatrix} \begin{bmatrix} \text{CAT} : d \\ \text{AGR} : \boxed{1} \end{bmatrix} \begin{bmatrix} \text{CAT} : n \\ \text{AGR} : \boxed{1} \end{bmatrix}$$

In this example the feature CAT is used for encoding the category of a constituent. Its values, drawn from the set ATOMS, correspond to the familiar non-terminal symbols of CFGs. The feature AGR encodes a bundle of agreement feature, whichever features the determiner and the noun must agree on. The (single) value of the three AGR features is shared among the determiner, the noun and the noun phrase.

In a unification grammar, rules are expressed as multi-AVMs. The multi-AVM of example 9 can be viewed as a rule which forms noun phrases by combining a determiner with a noun, imposing agreement between the two daughters. Furthermore, the concept of *derivation* and, subsequently, *language* of unification grammars is defined in terms of multi-AVMs. As shown below, derivations are manipulations of multi-AVMs, *forms*, obtained by successively applying grammar rules.

Note the interpretation of agreement emerging from example 9. Agreement is viewed as a bundle of features, that are constrained in two (or more) *independent* ways, which have to be mutually compatible. Thus, the object which is the value of $\boxed{1}$ in the above example is constrained independently by the "D-subtree" and the "N-subtree", and the rule imposes identical resulting values. This view should be contrasted with another, more 'directional', view of agreement, according to which there is one phrase that *determines* the value of the bundle of features in question, while a second phrase *accommodates* these values. The undirectional view of agreement is a typical view of unification-based grammar formalisms, which have a constraint satisfaction view in general, in which undirectionality is inherent. In contrast, transformation-based linguistic theories employ an essential concept of *movement*, that is inherently directional.

The notion of subsumption can be naturally extended from AVMs to multi-AVMs:

Definition 4. *If σ and ρ are two multi-AVMs of the same length, n, then $\sigma \sqsubseteq \rho$ if the following conditions hold:*

1. *every element of σ subsumes the corresponding element of ρ; and*
2. *if two paths are reentrant in σ, they are also reentrant in ρ.*

Note that the second condition refers to reentrancies that possibly relate paths originating in different elements of each multi-AVM.

Example 10. Let σ be the multi-AVM:

$$\begin{bmatrix} \text{F} : \begin{bmatrix} \text{G} : a \\ \text{H} : \boxed{1} \end{bmatrix} \end{bmatrix} \quad \begin{bmatrix} \text{G} : c \end{bmatrix} \quad \begin{bmatrix} \text{F} : \begin{bmatrix} \text{H} : b \\ \text{G} : \boxed{1} d \end{bmatrix} \\ \text{H} : a \end{bmatrix}$$

and ρ be:

$$\begin{bmatrix} \text{F} : \begin{bmatrix} \text{G} : a \\ \text{H} : d \end{bmatrix} \end{bmatrix} \quad \begin{bmatrix} \text{G} : c \end{bmatrix} \quad \begin{bmatrix} \text{F} : \begin{bmatrix} \text{H} : b \\ \text{G} : d \end{bmatrix} \\ \text{H} : a \end{bmatrix}$$

Then σ does not subsume ρ, but $\rho \sqsubseteq \sigma$. This is because of the reentrancy between the values $val(\sigma, 1, \langle F, H \rangle)$ and $val(\sigma, 3, \langle F, G \rangle)$; this constraint does not occur in ρ, contributing to ρ's being more general (less specific).

In the same way, the notion of unification can be extended to multi-AVMs (of the same length): ρ is the unification of σ_1 and σ_2 (denoted $\rho = \sigma_1 \sqcup \sigma_2$) if σ_1, σ_2 and ρ are of the same length, and ρ is the most general multi-AVM that is more specific than both σ_1 and σ_2.

A different unification operation on multi-AVMs, *unification in context*, unifies an element of one multi-AVM with an element of the other, resulting in possible modifications to both operands. Informally, the two AVMs (associated with the head of some rule and with the selected element in some form) are unified in their respective *contexts*: the body of the rule and the entire form. When some variable $\boxed{1}$ in the form is unified with some variable $\boxed{2}$ in the rule, all occurrences of $\boxed{1}$ in the form and of $\boxed{2}$ in the rule are modified: they are all set to the unified value. Note, however, that while the rule can be affected by the unification operation, it is only an *instance* of the rule which is used for each derivation step; derivation cannot alter the grammar itself.

A *unification grammar* consists of a set of rules, each of which is a multi-AVM; a lexicon, which associates a set of AVMs with each word; and a start symbol, which is an AVM.

Definition 5. *A* ***unification grammar*** $G = (\mathcal{R}, A_s, \mathcal{L})$ *over a signature* ATOMS *of atoms and* FEATS *of features is a finite set of* ***rules*** *(multi-AVMs)* $\mathcal{R}$*, a* ***start symbol*** A_s *which is an AVM and a* ***lexicon*** $\mathcal{L}$ *which associates with each word* w *a set of AVMs* $\mathcal{L}(w)$*.*

Rules are applied successively on *forms*, which are themselves multi-AVMs, starting with the form which consists of the start symbol only. *Derivation* is a binary relation over forms. Let σ be a form and ρ a grammar rule; application of ρ to σ consists of the following steps:

- Matching the rule's head with some element of the form, the *selected element*;
- Replacing the selected element in the form with the body of the rule, producing a new form.

This process is more involved than CFG derivation due to the presence of feature structures. *Matching* consists of unifying the head of the rule with the selected element; the rule is only applicable to the form if the unification does not fail. Due to possible reentrancies among elements of the form or the rule, the *replacement* operation requires *unification in context* of the head of the rule and the selected element. The replacement operation inserts the modified rule body into the modified form, replacing the selected element of the form. The full derivation relation is, as usual, the reflexive-transitive closure of rule application. A form is *sentential* if it is derivable from the start symbol.

Example 11. Consider the form

$$\sigma = \begin{bmatrix} \text{CAT} : np \\ \text{AGR} : \boxed{1} \end{bmatrix} \begin{bmatrix} \text{CAT} : vp \\ \text{AGR} : \boxed{1} \end{bmatrix}$$

and the rule

$$\rho = \begin{bmatrix} \text{CAT} : vp \\ \text{AGR} : \boxed{2} \end{bmatrix} \rightarrow \begin{bmatrix} \text{CAT} : v \\ \text{AGR} : \boxed{2} \end{bmatrix} \begin{bmatrix} \text{CAT} : np \\ \text{AGR} : \boxed{3} \end{bmatrix}$$

The unification in the context of the rule's head with the second element of σ succeeds, and identifies the values of $\boxed{2}$ (occurring in the head of the rule) and $\boxed{1}$ (occurring in the selected element of α). The rule's body can then replace the selected element, resulting in a new sentential form:

$$\begin{bmatrix} \text{CAT} : np \\ \text{AGR} : \boxed{1} \end{bmatrix} \begin{bmatrix} \text{CAT} : v \\ \text{AGR} : \boxed{1} \end{bmatrix} \begin{bmatrix} \text{CAT} : np \\ \text{AGR} : \boxed{3} \end{bmatrix}$$

Exercise 17. Show a form and a rule such that a single derivation step of applying the rule to some element of the form causes both the form and the rule to be modified.

Exercise 18. Show a form and a rule such that a single derivation step of applying the rule to some element of the form fails because some atomic value is found incompatible with a non-atomic feature structure.

Similarly, the concept of *derivation trees* can be naturally extended to unification grammars. Example 12 depicts such a tree, where the scope of variables is extended to the entire tree.

Example 12. An example derivation tree:

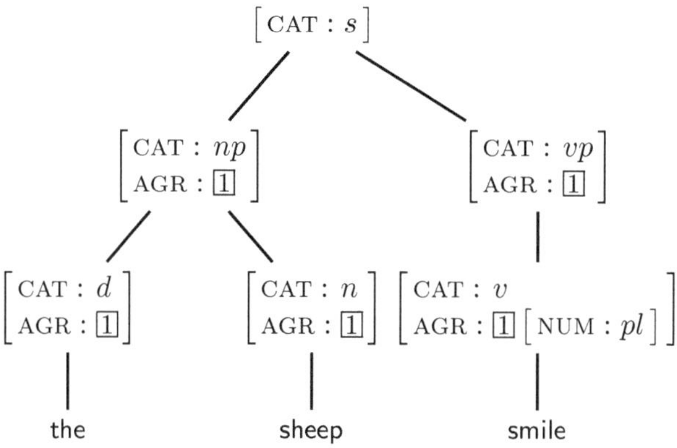

To determine whether a sequence of words, $w = a_1 \cdots a_n$, is in the language of a grammar G, $L(G)$, consider a derivation in G whose first form consists of the start symbol (a feature structure, viewed as a form of length 1), and whose last form is σ'. Let σ be a form obtained by concatenating $A_1, \ldots, A_n$, where each A_i is a lexical entry of the word a_i. Of course, the length of σ is required to be exactly n (the length of w). Notice that as words can be ambiguous, there can be more than one member in the lexical entry of a word; the definition only requires that one such AVM is selected for each word. $w \in L(G)$ if and only if σ' is a multi-AVM that is unifiable with σ: $\sigma \sqcup \sigma'$ does not fail. The reason for taking σ' rather than σ in the definition has to do with the properties of unification. A derivation might end with an underspecified multi-AVM, where additional information has to be gathered from the lexicon; but the reentrancies in the final form of the derivation must be respected: arbitrary instantiations for the variables in this final form cannot be permitted. Rather, if a variable is instantiated to some value (by one lexical entry), all other occurrences of the same variable must be consistent with that value. The current definition takes care of this requirement.

Example 13. The following unification grammar generates the language $\{ww \mid w \in \{a, b\}^+\}$. The signature of the grammar consists in the features CAT, FIRST and REST and the atoms *s, ap, bp, at, bt* and *elist.* The terminal symbols are a and b. The start symbol is the left-hand side of the first rule.

$$\begin{array}{lll}
\begin{bmatrix}\text{CAT}: s\end{bmatrix} & \rightarrow & \begin{bmatrix}\text{FIRST}: \boxed{1}\\ \text{REST}: \boxed{2}\end{bmatrix} \begin{bmatrix}\text{FIRST}: \boxed{1}\\ \text{REST}: \boxed{2}\end{bmatrix}\\[2ex]
\begin{bmatrix}\text{FIRST}: ap\\ \text{REST}: \begin{bmatrix}\text{FIRST}: \boxed{1}\\ \text{REST}: \boxed{2}\end{bmatrix}\end{bmatrix} & \rightarrow & \begin{bmatrix}\text{CAT}: at\end{bmatrix} \quad \begin{bmatrix}\text{FIRST}: \boxed{1}\\ \text{REST}: \boxed{2}\end{bmatrix}\\[2ex]
\begin{bmatrix}\text{FIRST}: bp\\ \text{REST}: \begin{bmatrix}\text{FIRST}: \boxed{1}\\ \text{REST}: \boxed{2}\end{bmatrix}\end{bmatrix} & \rightarrow & \begin{bmatrix}\text{CAT}: bt\end{bmatrix} \quad \begin{bmatrix}\text{FIRST}: \boxed{1}\\ \text{REST}: \boxed{2}\end{bmatrix}\\[2ex]
\begin{bmatrix}\text{FIRST}: ap\\ \text{REST}: elist\end{bmatrix} & \rightarrow & \begin{bmatrix}\text{CAT}: at\end{bmatrix}\\[2ex]
\begin{bmatrix}\text{FIRST}: bp\\ \text{REST}: elist\end{bmatrix} & \rightarrow & \begin{bmatrix}\text{CAT}: bt\end{bmatrix}\\[2ex]
\begin{bmatrix}\text{CAT}: at\end{bmatrix} & \rightarrow & a\\[2ex]
\begin{bmatrix}\text{CAT}: bt\end{bmatrix} & \rightarrow & b
\end{array}$$

In this grammar, the feature structure associated with a phrase encode, in a list notation, the characters that constitute the phrase.

Exercise 19. Show a derivation tree for the string *aabaab* with the above grammar.

Exercise 20. Show that the string *ab* is *not* derivable with the above grammar.

Exercise 21. Design a unification grammar for the language $L = \{a^n b^n c^n \mid n > 0\}$.

Exercise 22. Design a unification grammar for the language $L = \{a^n b^m c^n d^m \mid 1 \leq n \leq m\}$.

Exercise 23. Design a unification grammar for the language $L = \{a^n b^m c^n d^m e^n f^m \mid 1 \leq n \leq m\}$.

Exercise 24. Show that every CFG is weakly equivalent to a unification grammar; that is, for every CFG G there exists a unification grammar G' such that $L(G) = L(G')$.

Exercise 25. A *restricted feature structure* contains a set of *features*, each associated with a *value*, where the features are drawn from a fixed, finite set FEATS and the values are drawn from a fixed, finite set ATOMS (e.g., structures are not nested). An *augmented CFG* is a four-tuple $\langle \Sigma, V, S, P \rangle$ where Σ is an alphabet of terminal symbols, V is a set of *restricted feature structure*, $S \in V$

and P is a set of production rules over V^* (as in unification grammars). Prove that augmented CFGs are equivalent in their weak generative power to CFGs, that is, for every augmented CFG G there exists a CFG G' such that $L(G) = L(G')$ (this is just half of the proof; the reverse direction is trivial).

Unification grammars not only provide better linguistic generalizations than CFGs, they actually define a completely different class of languages. While CFGs are limited to defining context-free languages, unification grammars can generate the entire class of *recursively enumerable languages*, or in other words, can simulate the operation of a *Turing machine*. In this respect, unification grammars are very similar to programming languages.

The extended expressive power is a double-edged sword. It provides the grammar designer with extremely powerful means for expressing linguistic phenomena; but it is also equally easy to design grammars which never terminate: the computational process of *parsing*, whose aim is to produce all the grammatical structures that a grammar induces on a given sentence, is not guaranteed to terminate for all grammars.

In spite of their inherent computational complexity, unification grammars have become widespread in computational linguistics due to their appealing expressivity and ability to provide detailed accounts of a wide variety of linguistic phenomena, from phonology and morphology to syntax and semantics. Several computational frameworks provide the grammar designer with environments for grammar development; among the most common are LKB [8] and ALE [6]. These systems have been successfully used for developing large-scale, linguistically motivated grammars of a variety of natural languages.

13.5 Further Reading

The extension of CFGs by adding features to lexical items and grammar categories dates back to the 1970's [16] and was incorporated into various linguistic theories, starting with functional grammars, later called unification grammars and then functional-unification grammars [17] and lexical-functional grammars (LFG) [15, 9]. Later, it was the underlying formalism for Generalized Phrase-Structure Grammars (GPSG) [10]. The best introductory text with an informal, elementary discussion of feature structures is [31]. In addition to giving a rather informal characterization of feature structures, subsumption and unification, [31] defines the simplest unification-based grammatical formalism, PATR [33], and demonstrates how it can be used to encode grammars in different linguistic theories. A linguistically motivated comparative survey of three theories, namely Government and Binding, GPSG and LFG is given in [28]. Non-recursive feature structures were added to the tree-adjoining grammar (TAG) formalism [14] by [37], and to some variants of categorial grammars by [36] and [41]. A clear introduction to contemporary phrase structure grammars, with an extensive description of both GPSG and HPSG, including their handling of various linguistic phenomena, is [2]. A more recent exposition of syntactic theory in terms of (typed) unification grammars is [26].

Multi-AVMs were used implicitly in many implementations of unification based grammars, but their formal definition [34], as well as a discussion of their mathematical properties [39, 38], was done much later.

The unification operation was originally defined for first-order terms by [25]. A variety of algorithms for term unification were since presented, among which the most popular is that of [20], extended to cyclic terms by [12]. The origin of graph unification is due to [18], while an adaptation of the general unification algorithms for feature structures was done by [1] and [21].

Linguistic motivation for the use of typed AVMs first given by [1], who provides also mathematical and computational motivation for using types. This work influenced the development of the linguistic theory Head-Drive Phrase-Structure grammars (HPSG) [23, 24], which remains the main theory which employs TFSs. TFSs are directly defined also by [21], and a view of TFSs as a generalization of first-order terms is presented in [4]. A logical formulation of TFSs, starting in the basic notions and ending in a complete characterization of TFS based grammars and their languages, is given by [5]. [40] show how typing can be beneficial even for systems which employ untyped feature structures.

Parsing with unification grammars was developed along with the increased popularity of these grammars in the beginning of the 1980's. A large variety of parsing algorithms have been adapted to unification grammars [35, 3, 34]. [22] define parsing as a special case of logical deduction, and show how definite clause grammars in particular and other unification formalisms in general can be parsed using the Earley deduction proof procedure. The paradigm of parsing as deduction was later extended to many parsing strategies and a variety of grammar formalisms by [29].

However, it was immediately observed that general unification grammars are Turing-equivalent; this is first mentioned by [15] and is formally proven by [13] and [32]. The main reason is that with unification grammars, ϵ-rules (rules whose body is empty) and unit-rules (whose body consists of a single element) cannot be removed from the grammar as is the case with context-free grammars. This leads to derivations in which unbounded amount of material is generated without expanding the frontier word. To rule out offensive grammars, [22] define the concept of *off-line parsability* for definite clause grammars, having been introduced (unnamed) by [15] for LFG grammars. Several alternative definitions for this concept are available in the literature, each capturing slightly different properties of the constraint on grammars; a thorough survey is provided by [11].

References

1. H. Aït-Kaci. *A lattice-theoretic approach to computation based on a calculus of partially ordered types.* Ph.D. thesis, University of Pennsylvania, 1984.
2. R. D. Borsley. *Modern phrase structure grammar.* Blackwell, Oxford, 1996.
3. G. Bouma and G. van Noord. Head-driven parsing for lexicalist grammars: Experimental results. *Proceedings of the 6th Meeting of the European Chapter of the Association for Computational Linguistics*, Utrecht, 1993, 71–80.
4. B. Carpenter. Typed feature structures: A generalization of first-order terms. *Logic Programming – Proceedings of the 1991 International Symposium* (V. Saraswat and U. Kazunori, eds.), Cambridge, MA. MIT Press, 1991, 187–201.
5. B. Carpenter. *The Logic of Typed Feature Structures.* Cambridge Tracts in Theoretical Computer Science, Cambridge University Press, 1992.
6. B. Carpenter and G. Penn. ALE: The attribute logic engine – user's guide. Technical report, Lucent Technologies and Universität Tübingen, May, 1999.
7. N. Chomsky. Three models for the description of language. *I. R. E. transactions on information theory, Proceedings of the symposium on information theory*, IT-2 (1956), 113–123.
8. A. Copestake. The (new) LKB system. Technical report, Stanford University, September, 1999.
9. M. Dalrymple, R. M. Kaplan, J. T. Maxwell and A. Zaenen, eds. *Formal Issues in Lexical-Functional Grammar*, volume 47 of *CSLI lecture notes.* CSLI, Stanford, CA, 1995.
10. G. E. Gazdar, E. Klein, J. K. Pullum and I. A. Sag. *Generalized Phrase Structure Grammar.* Harvard University Press, Cambridge, Mass., 1985.
11. E. Jaeger, N. Francez and S. Wintner. Unification grammars and off-line parsability. *Journal of Logic, Language and Information*, 13, 4 (2004).
12. J. Jaffar. Efficient unification over infinite terms. *New Generation Computing*, 2 (1984), 207–219.
13. M. Johnson. *Attribute-Value Logic and the Theory of Grammar*, volume 16 of *CSLI Lecture Notes.* CSLI, Stanford, California, 1988.
14. A. K. Joshi, L. Levy, and M. Takahashi. Tree Adjunct Grammars. *Journal of Computer and System Sciences*, 1975.
15. R. Kaplan and J. Bresnan. Lexical functional grammar: A formal system for grammatical representation. *The Mental Representation of Grammatical Relations* (J. Bresnan ed.), MIT Press, Cambridge, Mass., 1982, 173–281.
16. M. Kay. Functional unification grammar. *5th Annual Meeting of the Berkeley Linguistic Society*, Berkeley, CA, 1979.
17. M. Kay. Unification grammar. Technical report, Xerox Palo Alto Research Center, Palo Alto, CA, 1983.
18. M. Kay. Parsing in functional unification grammar. *Natural Language Parsing: Psychological, Computational and Theoretical Perspectives* (D. Dowty, L. Karttunen, and A. Zwicky, eds.), Cambridge University Press, Cambridge, 1985, 251–278.
19. A. Manaster-Ramer. Dutch as a formal language. *Linguistics and Philosophy*, 10 (1987), 221–246.
20. A. Martelli and U. Montanari. An efficient unification algorithm. *ACM Transactions on Programming Languages and Systems*, 4, 2 (1982), 258–282.
21. D. Moshier. *Extensions to Unification Grammars for the Description of Programming Languages.* Ph.D. thesis, University of Michigan, Ann Arbor, 1988.

22. F. C. N. Pereira and D. H. D. Warren. Parsing as deduction. *Proceedings of the 21st Annual Meeting of the Association for Computational Linguistics*, 1983, 137–144.
23. C. Pollard and I. A. Sag. *Information Based Syntax and Semantics.* Number 13 in CSLI Lecture Notes, CSLI, 1987.
24. C. Pollard and I. A. Sag. *Head-Driven Phrase Structure Grammar.* University of Chicago Press and CSLI Publications, 1994.
25. J. A. Robinson. A machine-oriented logic based on the resolution principle. *Journal of the ACM*, 12 (1965), 23–41.
26. I. A. Sag and T. Wasow. *Syntactic Theory: A Formal Introduction.* CSLI, Stanford, 1999.
27. W. J. Savitch, E. Bach, W. Marsh and G. Safran-Naveh, eds. *The formal complexity of natural language*, volume 33 of *Studies in Linguistics and Philosophy.* D. Reidel, Dordrecht, 1987.
28. P. Sells. *Lectures on contemporary syntactic theories: an introduction to government-binding theory, generalized phrase structure grammar, and lexical-functional grammar*, volume 3 of *CSLI lecture notes.* CSLI, Stanford, CA, 1988.
29. S. M. Shieber, Y. Schabes and F. Pereira. Principles and implementation of deductive parsing. *Journal of Logic Programming*, 24, 1-2 (1995), 3–36.
30. S. M. Shieber. Evidence against the context-freeness of natural language. *Linguistics and Philosophy*, 8 (1985), 333–343.
31. S. M. Shieber. *An Introduction to Unification Based Approaches to Grammar.* Number 4 in CSLI Lecture Notes, CSLI, 1986.
32. S. M. Shieber. *Constraint-Based Grammar Formalisms.* MIT Press, Cambridge, Mass., 1992.
33. S. M. Shieber, H. Uszkoreit, F. C. N. Pereira, J. J. Robinson and M. Tyson. The formalism and implementation of PATR-II. *Research on Interactive Acquisition and Use of Knowledge.* SRI International, Menlo Park, Cal., 1983.
34. K. Sikkel. *Parsing Schemata.* Texts in Theoretical Computer Science – An EATCS Series. Springer Verlag, Berlin, 1997.
35. M. Tomita, ed. *Generalized LR Parsing.* Kluwer Academic Publishing, 1991.
36. H. Uszkoreit. Categorial unification grammars. *Proceedings of the 5th International Conference on Computational Linguistics*, Bonn, 1986, 187–194.
37. K. Vijay-Shanker and A. K. Joshi. Unification Based Tree Adjoining Grammars. *Unification-based Grammars* (J. Wedekind ed.), MIT Press, Cambridge, Massachusetts, 1991.
38. S. Wintner. *An Abstract Machine for Unification Grammars.* Ph.D. thesis, Technion – Israel Institute of Technology, Haifa, Israel, January, 1997.
39. S. Wintner and N. Francez. Parsing with typed feature structures. *Proceedings of the Fourth International Workshop on Parsing Technologies*, Prague, 1995, 273–287.
40. S. Wintner and A. Sarkar. A note on typing feature structures. *Computational Linguistics*, 28, 3 (2002), 389–397.
41. H. Zeevat, E. Klein and J. Calder. An introduction to unification categorial grammar. *Edinburgh Working Papers in Cognitive Science, Vol. 1: Categorial Grammar, Unification Grammar, and Parsing* (N. J. Haddock, E. Klein, and G. Morill, eds.), 1987.

14

Introduction to Petri Net Theory *

Hsu-Chun Yen

Dept. of Electrical Engineering
National Taiwan University
Taipei, Taiwan, R.O.C.
E-mail: yen@cc.ee.ntu.edu.tw

Summary. This paper gives an overview of Petri net theory from an algorithmic viewpoint. We survey a number of analytical techniques as well as decidability/complexity results for various Petri net problems.

14.1 Introduction

Petri nets, introduced by C. A Petri in 1962 [54], provide an elegant and useful mathematical formalism for modelling concurrent systems and their behaviors. In many applications, however, modelling by itself is of limited practical use if one cannot analyze the modelled system. As a means of gaining a better understanding of the Petri net model, the decidability and computational complexity of typical automata theoretic problems concerning Petri nets have been extensively investigated in the literature in the past four decades.

In this paper, we first give an overview of a number of analytical techniques known to be useful for reasoning about either structural or behavioral properties of Petri nets. However, due to the intricate nature of Petri nets, none of the available analytical techniques is a panacea. To understand the limitations and capabilities of analyzing Petri nets from an algorithmic viewpoint, we also summarize a variety of decidability/complexity results reported in the literature for various Petri net problems including *boundedness, reachability, containment*, *equivalence*, and more. Among them, Lipton [40] and Rackoff [56] have shown exponential space lower and upper bounds, respectively, for the boundedness problem. As for the containment and the equivalence problems, Rabin [2] and Hack [19], respectively, have shown these two problems to be undecidable. In spite of the efforts made by many researchers over the years, many analytical questions concerning Petri nets remain unanswered.

The quest for solving the general reachability problem for Petri nets has constituted perhaps the most important and challenging line of research in

* This work was supported in part by NSC Grant 94-2213-E-002-087, Taiwan.

H.-C. Yen: *Introduction to Petri Net Theory*, Studies in Computational Intelligence (SCI) **25**, 343–373 (2006)
www.springerlink.com

the Petri net community in the past. Knowing that the problem requires exponential space [40], the decidability issue of the problem was left unsolved for a long period of time until Mayr [42, 43] finally provided an answer in the affirmative (see also [38]). Before Mayr's proof, a number of attempts were made to investigate the problem for restricted classes of PNs, in hope of gaining more insights and developing new tools in order to conquer the general PN reachability problem (see, e.g., [16, 21, 39, 41, 48, 61]). A common feature of those attempts is that decidability of reachabiltiy for those restricted classes of Petri nets was built upon their reachability sets being *semilinear.* As semilinear sets precisely correspond to the those characterized by Presburger Arithmetic (a decidable theory), decidability of the reachability problem follows immediately. In view of the importance of the role played by the concept of semilinearity in Petri net theory, we devote a section in this paper to surveying analytical techniques and complexity results for subclasses of Petri nets exhibiting semilinear reachability sets. As for the general reachability problem, the only known algorithm is nonprimitive recursive (see [38, 42, 43]). The exact complexity of the reachability problem remains the most challenging open problem in Petri net theory.

It is well-known that the computational power of Petri nets is strictly weaker than that of Turing machines, making them inadequate for modelling certain real-world systems such as prioritized systems [1]. To overcome this shortcoming, a number of extended Petri nets have been introduced to enhance the expressive capabilities of Petri nets. Among them are *colored Petri nets*, *Petri nets with inhibitor arcs*, *timed Petri nets*, *prioritized Petri nets*, and more. With the above extended Petri nets powerful enough to simulate Turing machines, all nontrivial problems for such Petri nets become undecidable. A natural and interesting question to ask is: are there Petri nets whose powers lie between conventional Petri nets and Turing machines? It turns out that the so-called *reset nets* and *transfer nets* are two such witnesses. The quest for such 'weaker' extensions has attracted considerable attentions in recent years.

This paper gives an overview of basic analytical techniques and decidability/complexity results for various Petri net problems. Our survey is by no means comprehensive; the interested reader is refer to [12, 33, 49] for other survey articles concerning the decidability and complexity issues of Petri nets. See also [7, 53, 58] for more about Petri nets and their related problems.

The rest of this paper is organized as follows. Section 14.2 gives the basic notations and terminologies of Petri nets and their equivalent models. Section 14.3 is devoted to the definitions of various Petri net problems that are of interest in the Petri net community. An overview of analytical techniques known to be useful for reasoning about Petri net behaviors is presented in Section 14.4. Decidability and complexity results concerning various Petri net problems for general Petri nets and for subclasses of Petri nets are given in Section 14.5 and Section 14.6, respectively. Finally, in Section 14.7 we briefly discuss the computational power of various extended Petri nets.

14.2 Preliminaries

Let Z (N) denote the set of (nonnegative) integers, and Z^k (N^k) the set of vectors of k (nonnegative) integers. For a k-dimensional vector v, let $v(i), 1 \leq i \leq k$, denote the ith component of v. For a $k \times m$ matrix A, let $A(i,j)$, $1 \leq i \leq k, 1 \leq j \leq m$, denote the element in the ith row and the jth column of A. We let $|S|$ be the number of elements in set S. Given a vector x, we let x^T denote the *transpose* of x. Given an alphabet (i.e., a finite set of symbols) Σ, we write Σ^* to denote the set of all finite-length strings (including the empty string λ) using symbols from Σ.

14.2.1 Petri Nets

A *Petri net* (PN, for short) is a 3-tuple (P, T, φ), where P is a finite set of *places*, T is a finite set of *transitions*, and φ is a *flow function* $\varphi : (P \times T) \cup (T \times P) \rightarrow N$. A *marking* is a mapping $\mu : P \rightarrow N$. (μ assigns tokens to each place of the net.) Pictorially, a PN is a directed, bipartite graph consisting of two kinds of nodes: *places* (represented by circles within which each small black dot denotes a *token*) and *transitions* (represented by bars or boxes), where each arc is either from a place to a transition or vice versa. In addition, each arc is annotated by either $\varphi(p,t)$ or $\varphi(t,p)$, where p and t are the two endpoints of the arc. See Figure 14.1 for an example, in which all the arc labels are one and are therefore omitted.

A transition $t \in T$ is *enabled* at a marking μ iff $\forall p \in P$, $\varphi(p,t) \leq \mu(p)$. If a transition t is enabled, it may *fire* by removing $\varphi(p,t)$ tokens from each input place p and putting $\varphi(t,p')$ tokens in each output place p'. We then write $\mu \stackrel{t}{\longmapsto} \mu'$, where $\mu'(p) = \mu(p) - \varphi(p,t) + \varphi(t,p)$, $\forall p \in P$.

Example 1. Figure 14.1 depicts a PN (P, T, φ) with $P = \{p_1, p_2, p_3, p_4, p_5\}$ and $T = \{t_1, t_2, t_3, t_4\}$, modelling a simple *producer-consumer* system. We can view a marking μ as a 5-dimensional column vector in which the ith component is $\mu(p_i)$. In Figure 14.1, transition t_2 is enabled at marking $\mu = (1,0,0,1,0)$ since the only input place of t_2 (i.e., p_1) satisfies $\varphi(p_1, t_2) \leq \mu(p_1)$. After firing the transition t_2, the PN reaches a new marking $\mu' = (0,1,1,1,0)$, i.e., $(1,0,0,1,0) \stackrel{t_1}{\longmapsto} (0,1,1,1,0)$. □

A sequence of transitions $\sigma = t_1 ... t_n$ is a *firing sequence* from μ_0 iff $\mu_0 \stackrel{t_1}{\longmapsto} \mu_1 \stackrel{t_2}{\longmapsto} \cdots \stackrel{t_n}{\longmapsto} \mu_n$ for some markings $\mu_1, ..., \mu_n$. (We also write '$\mu_0 \stackrel{\sigma}{\longmapsto} \mu_n$'.) We write '$\mu_0 \stackrel{\sigma}{\longmapsto}$' to denote that σ is enabled and can be fired from μ_0, i.e., $\mu_0 \stackrel{\sigma}{\longmapsto}$ iff there exists a marking μ such that $\mu_0 \stackrel{\sigma}{\longmapsto} \mu$. The notation $\mu_0 \stackrel{*}{\longmapsto} \mu$ is used to denote the existence of a σ such that $\mu_0 \stackrel{\sigma}{\longmapsto} \mu$. A *marked* PN is a pair $((P, T, \varphi), \mu_0)$, where (P, T, φ) is a PN, and μ_0 is called the *initial marking*. Throughout the rest of this paper, the word 'marked' will be omitted if it is clear from the context.

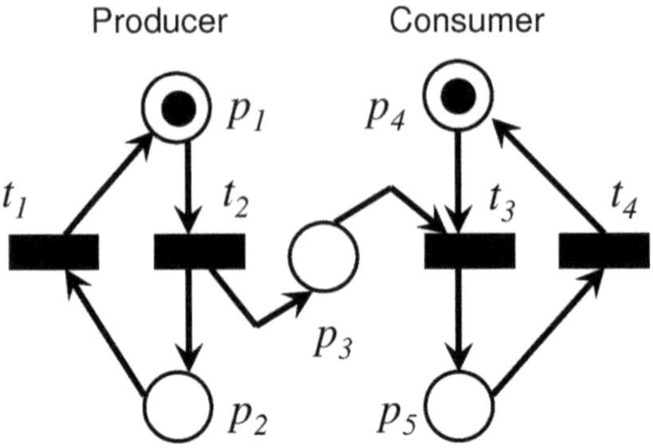

Fig. 14.1. A Petri net.

By establishing an ordering on the elements of P and T (i.e., $P = \{p_1, ..., p_k\}$ and $T = \{t_1, ..., t_m\}$), we define the $k \times m$ *incidence matrix* $[T]$ of (P, T, φ) so that $[T](i,j) = \varphi(t_j, p_i) - \varphi(p_i, t_j)$. Note that $\varphi(t_j, p_i)$, $\varphi(p_i, t_j)$, and $[T](i,j)$, respectively, represent the number of tokens removed, added, and changed in place i when transition j fires once. Thus, if we view a marking μ as a k-dimensional column vector in which the ith component is $\mu(p_i)$, each column of $[T]$ is then a k-dimensional vector such that if $\mu_0 \stackrel{\sigma}{\longmapsto} \mu$, then the following *state equation* holds:

$$\mu_0 + [T] \cdot \#_\sigma = \mu,$$

where $\#_\sigma$ is an m-dimensional vector with its jth entry denoting the number of times transition t_j occurs in σ.

Example 2. Consider the PN in Figure 14.1. As marking $\mu = (0,1,2,1,0)$ is reachable from the initial marking $\mu_0 = (1,0,0,1,0)$ through the firing sequence $\sigma = t_2 t_1 t_2$, it is easy to verify (as the following equation shows) that the state equation $\mu_0 + [T] \cdot \#_\sigma = \mu$ holds.

$$\begin{pmatrix} 1 \\ 0 \\ 0 \\ 1 \\ 0 \end{pmatrix} + \begin{pmatrix} 1 & -1 & 0 & 0 \\ -1 & 1 & 0 & 0 \\ 0 & 1 & -1 & 0 \\ 0 & 0 & -1 & 1 \\ 0 & 0 & 1 & -1 \end{pmatrix} \begin{pmatrix} 1 \\ 2 \\ 0 \\ 0 \end{pmatrix} = \begin{pmatrix} 0 \\ 1 \\ 2 \\ 1 \\ 0 \end{pmatrix}$$

□

As the following example shows, the existence of a solution for the state equation of a PN is necessary but not sufficient to guarantee reachability.

Example 3. Consider a PN $\mathcal{P}=(P, T, \varphi)$ with $P = \{p_1, p_2, p_3\}$, $T = \{t_1, t_2\}$ and $\varphi(p_1, t_1) = -1$, $\varphi(t_1, p_2) = 1$, $\varphi(p_2, t_2) = -1$, $\varphi(t_2, p_1) = 1$, and $\varphi(t_2, p_3) = 1$. Let the initial marking be $(0,0,0)$ and the final marking be $(0,0,1)$. The associated state equation is

$$\begin{pmatrix} 0 \\ 0 \\ 0 \end{pmatrix} + \begin{pmatrix} -1 & 1 \\ 1 & -1 \\ 0 & 1 \end{pmatrix} \begin{pmatrix} x_1 \\ x_2 \end{pmatrix} = \begin{pmatrix} 0 \\ 0 \\ 1 \end{pmatrix}$$

Clearly (1,1) is a solution to the above equation although $(0,0,1)$ is not reachable from $(0,0,0)$.

□

For ease of expression, the following notations will be used extensively throughout the rest of this paper. (Let σ, σ' be transition sequences, p be a place, and t be a transition.)

- $\#_\sigma(t)$ represents the number of occurrences of t in σ. (For convenience, we sometimes treat $\#_\sigma$ as an m-dimensional vector assuming that an ordering on T is established ($|T| = m$).)
- $\Delta(\sigma) = [T] \cdot \#_\sigma$ defines the *displacement* of σ. (Notice that if $\mu_0 \overset{\sigma}{\longmapsto} \mu$, then $\Delta(\sigma) = \mu - \mu_0$.) For a place $p \in P$, we write $\Delta(\sigma)(p)$ to denote the component of $\Delta(\sigma)$ corresponding to place p.
- $Tr(\sigma) = \{t | t \in T, \#_\sigma(t) > 0\}$, denoting the set of transitions used in σ.
- $p^\bullet$=$\{t | \varphi(p,t) \geq 1, t \in T\}$ is the set of output transitions of p;
  $t^\bullet$=$\{p | \varphi(t,p) \geq 1, p \in P\}$ is the set of output places of t.
- $^\bullet p$=$\{t | \varphi(t,p) \geq 1, t \in T\}$ is the set of input transitions of p;
  $^\bullet t$=$\{p | \varphi(p,t) \geq 1, p \in P\}$ is the set of input places of t.

Given $\mu_0 \overset{\sigma}{\longmapsto} \mu$, a sequence σ' is said to be a *rearrangement* of σ if $\#_\sigma = \#_{\sigma'}$ and $\mu_0 \overset{\sigma'}{\longmapsto} \mu$.

Let $\mathcal{P} = ((P,T,\varphi),\mu_0)$ be a marked PN. The *reachability set* of $\mathcal{P}$ is $R(\mathcal{P},\mu_0) = \{\mu \mid \exists \sigma \in T^*, \mu_0 \overset{\sigma}{\longmapsto} \mu\}$.

14.2.2 Vector Addition Systems (With States), Vector Replacement Systems

Vector addition systems (VAS) were introduced by Karp and Miller [36], and were later shown by Hack [18] to be equivalent to PNs. An n-dimensional VAS is a pair $G = (x, W)$, where $x \in N^n$ is called the *start point* (or *start vector*) and W is a finite set of vectors (called *addition vectors*) in Z^n. The *reachability set* of the VAS G is the set $R(G) = \{z \mid$ for some j, $z = x + v_1 + ... + v_j$, where, for all $1 \leq i \leq j$, each $v_i \in W$ and $x + v_1 + ... + v_i \geq 0\}$.

An n-dimensional *vector addition system with states* (VASS) [21] is a VAS (x, W) together with a finite set T of transitions of the form $p \rightarrow (q, v)$, where q and p are states and v is in W. The meaning is that such a transition can be applied at point y in state p and yields the point $y + v$ in state q, provided that $y + v \geq 0$. The VASS is specified by $G = (x, W, T, p_0)$, where p_0 is the starting state.

Example 4. For the PN shown in Figure 14.1, the corresponding VAS $\langle x, W \rangle$ is:

- $x = (1, 0, 0, 1, 0)$,
- $W = \{(1, -1, 0, 0, 0), (-1, 1, 1, 0, 0), (0, 0, -1, -1, 1), (0, 0, 0, 1, -1)\}$.

Such a VAS can also be regarded as a VASS with a single state. □

A $k \times m$ *vector replacement system (VRS)* [37] is a triple (w_0, U, W), where $w_0 \in N^k$ *(start vector)*, $U \in N^{k \times m}$ *(check matrix)*, and $W \in Z^{k \times m}$ *(addition matrix)* such that, for any i, j with $1 \leq i \leq m$ and $1 \leq j \leq k$, we have $U_i(j) + W_i(j) \geq 0$. Here U_i (respectively, W_i) is the i-th column vector of U (respectively, W). A vector $W_i \in W$ is said to be *enabled* in a vector $x \in N^k$ if and only if $x \geq U_i$; as $U_i + W_i \geq 0$, adding W_i to x yields $x + W_i \in N^k$. For a VRS $G = (w_0, U, W)$, $R(G)$ denotes the set of vectors from N^k that can be reached from w_0 by iteratively adding vectors from W enabled in the vector computed so far.

It is known that Petri net, VAS, VASS, and VRS are computationally equivalent. In fact, given an n-dimensional VASS G, we can effectively construct an $(n+3)$-dimensional VAS G' that simulates G [21].

14.3 Petri Net Problems

What follows are problems that are of particular importance and interest in the study of PNs.

- The *reachability* problem: given a PN $\mathcal{P}$ (with initial marking μ_0) and a marking μ, deciding whether $\mu \in R(\mathcal{P}, \mu_0)$.

- The *boundedness* problem: given a PN $\mathcal{P}$ (with initial marking μ_0), deciding whether $|R(\mathcal{P}, \mu_0)|$ is finite or not.

- The *covering* problem: given a PN $\mathcal{P}$ (with initial marking μ_0) and a marking μ, deciding whether there exists a $\mu' \in R(\mathcal{P}, \mu_0)$ such that $\mu' \geq \mu$.

- The *equivalence* problem: given two PNs $\mathcal{P}_1$ (with initial marking μ_1) and $\mathcal{P}_2$ (with initial marking μ_2), deciding whether $R(\mathcal{P}_1, \mu_1) = R(\mathcal{P}_2, \mu_2)$.

- The *containment* problem: given two PNs $\mathcal{P}_1$ (with initial marking μ_1) and $\mathcal{P}_2$ (with initial marking μ_2), deciding whether $R(\mathcal{P}_1, \mu_1) \subseteq R(\mathcal{P}_2, \mu_2)$.

- The *model checking* problem: given a PN $\mathcal{P}$ (with initial marking μ_0) and a temporal formula ϕ (expressed in some temporal logic), deciding whether $\mathcal{P}, \mu_0 \models \phi$ (i.e., $(\mathcal{P}, \mu_0)$ satisfies ϕ).

- The *liveness* problem: given a PN $\mathcal{P}$ (with initial marking μ_0), deciding whether for every $t \in T, \mu \in R(\mathcal{P}, \mu_0)$, there exists a sequence of transitions σ such that $\mu \stackrel{\sigma \cdot t}{\longmapsto}$, i.e., t is enabled after firing σ from μ.

- Others include *home-state, reversibility, self-stabilization, fairness, regularity, synchronic distance, controllability* ... and more.

Example 5. Consider the two PNs shown in Figure 14.2. Clearly, Figure 14.2(a) is bounded with respect to the given initial marking. However, if the upper-leftmost place contains two tokens, then the PN becomes unbounded. That is, being boundedness or not depends on the initial marking for the PN in Figure 14.2(a). The PN in Figure 14.2(b), on the other hand, remains bounded no matter what the initial marking of the PN is. Such a PN is called *structurally bounded.*

□

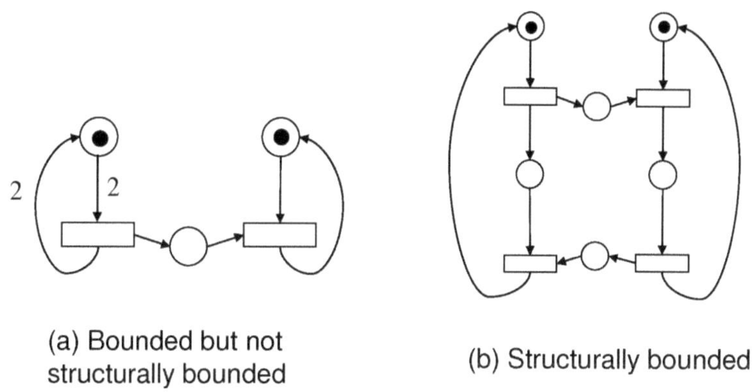

Fig. 14.2. Structural vs. behavioral boundedness.

To capture the essence of a transition being 'live' in various application areas, a hierarchy of liveness notions was defined in the literature (see [49]). Transition t in a PN ($\mathcal{P}$, μ_0) is said to be:

1. *Dead (L0-live)* if t can never be fired in any firing sequence from μ_0;
2. *L1-live (potentially firable)* if t can be fired at least once in some firing sequence from μ_0;
3. *L2-live* if, given any positive integer k, t can be fired at least k times in some firing sequence from μ_0;

4. *L3-live* if t appears infinitely often in some firing sequence from μ_0;
5. *L4-live* or *live* if t is *L1-live* for every marking μ in $R(\mathcal{P}, \mu_0)$;

A PN is said to be *L0, L1, L2, L3*, and *L4*-live if each of its transitions is *L0, L1, L2, L3*, and *L4*-live, respectively.

Example 6. Consider the PN shown in Figure 14.3. For any $k \in N$, $(1,0,0) \overset{(t_1)^k t_2 (t_3)^k}{\longmapsto}$; hence, t_3 is *L2-live*. However, it is reasonably easy to see that t_3 is not *L3-live* as there is no computation along which t_3 is fired infinitely many times. In fact, the following implications hold: L4-liveness (the strongest) $\Longrightarrow$ L3-liveness $\Longrightarrow$ L2-liveness $\Longrightarrow$ L1-liveness. □

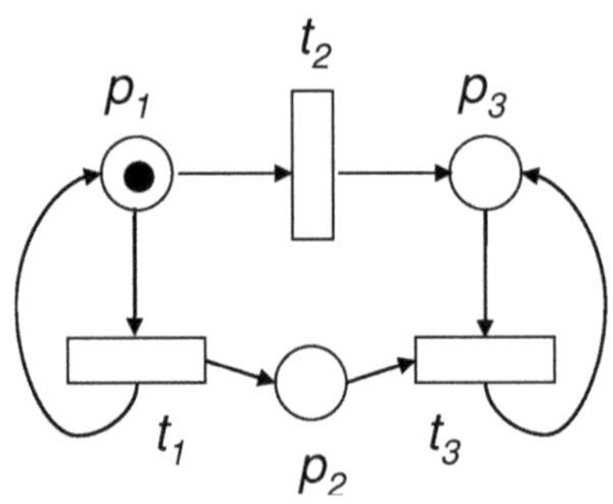

Fig. 14.3. L2 vs. L3 liveness.

14.4 Analytical Techniques

In this section, we summarize various techniques useful for analyzing PN properties. Our focus is on *algebraic techniques*, *structural analysis*, and *state-space analysis*. Other techniques such as *simulation* and *synthesis/reduction* are beyond the scope of our discussion. Structural analysis is mainly designed for reasoning about properties of PNs that are independent of the initial markings. State-space analysis, on the other hand, allows us to infer properties of PNs that are sensitive to the initial markings.

14.4.1 Algebraic Techniques

In the framework of using algebraic techniques for reasoning about PNs, solving a PN problem is reduced to finding a solution for an algebraic (in)equation associated with the PN. Due to the nature of this technique, the method is in general efficient (in most cases, polynomial in the size of the PN). Unfortunately, this technique generally provides only necessary or sufficient information for either inferring desired properties or ruling out dangerous conditions.

State Equation:

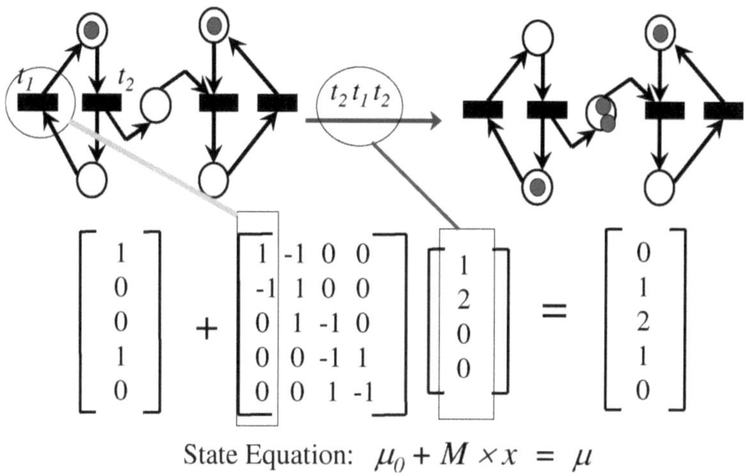

Fig. 14.4. State equation.

Figure 14.4 highlights the idea behind the technique based on *state equations*. If μ is reachable from μ_0 through transition sequence σ, then $\#_\sigma$ corresponds to an integer solution to the state equation $\mu_0 + [T] \cdot \#_\sigma = \mu$, where $[T]$ is the incidence matrix of the PN. The technique relies on relating the PN reachability analysis to *integer linear programming*, which is a well-established formalism. Unfortunately, a direct application of the state equation technique to general PN problems is normally not feasible, as the existence of a solution to a state equation is necessary but not sufficient for witnessing reachability. There are, however, various subclasses of PNs for which an extended state equation is sufficient and necessary to capture reachability of the underlying PN. More will be said about this in our subsequent discussion.

Place Invariant:

A *place invariant* of PN $\mathcal{P} = (P, T, \varphi)$ is a mapping Inv_P: $P \rightarrow Z$ (i.e., assigning *weights* to places) such that $\forall \mu, \mu'$ and $t \in T$, if $\mu \overset{t}{\longmapsto} \mu'$, then $\sum_{p \in P} Inv_P(p)\mu(p) = \sum_{p \in P} Inv_P(p)\mu'(p)$. In words, the firing of any transition does not change the weighted sum of tokens in the PN. Consider the PN shown in Figure 14.5. It is reasonably easy to observe that $(1, 2, 1)$ is a P-invariant. Other P-invariants include $(1, 1, 0), (2, 5, 3), (-2, 1, 3)$. Note that any linear combination of P-invariants is a P-invariant. Any solution of the equation $X \cdot [T] = 0$ is a P-invariant, where X is a row vector. For instance,

$$(1,2,1)\begin{pmatrix} -1 & 1 \\ 1 & -1 \\ -1 & 1 \end{pmatrix} = \begin{pmatrix} 0 \\ 0 \end{pmatrix},$$

as (1,2,1) is a P-invariant. Using the so-called Farkas Algorithm, the minimal P-invariants (i.e., bases) of a PN can be calculated. Nevertheless, in the worst case the number of minimal P-invariants is exponential in the size of the PN, indicating that Farkas Algorithm may require exponential worst-case time.

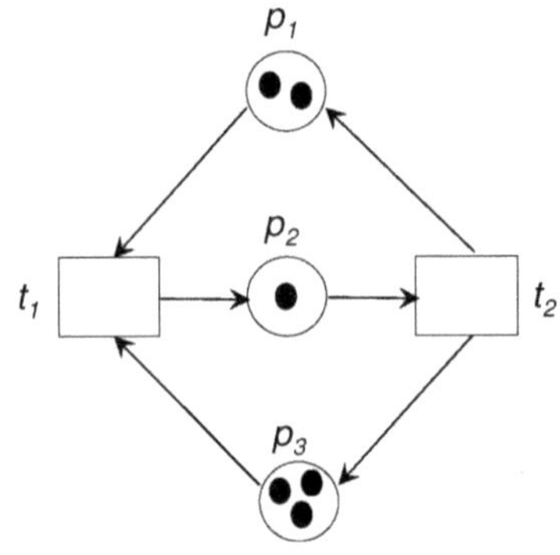

Fig. 14.5. A PN with a P-invariant (1,2,1).

Suppose μ is a reachable marking (from the initial marking μ_0) through a firing sequence σ. Clearly, $\mu_0 + [T]\#_\sigma = \mu$. (Here μ_0 and μ are column vectors.) Let X be a P-invariant. Then
$X \cdot \mu = X \cdot (\mu_0 + [T] \cdot \#_\sigma) = X \cdot \mu_0 + X \cdot [T] \cdot \#_\sigma = X \cdot \mu_0$.
Recall that in the example in Figure 14.5, $(1,1,0)$ is a P-invariant. For every reachable marking μ, we have $\mu(p_1) + \mu(p_2) = \mu_0(p_1) + \mu_0(p_2)$, meaning that the total number of tokens in p_1 and p_2 together remain unchanged during the course of the PN computation. Hence, if the PN starts from the initial marking $(1,0,1)$, then the property of *mutual exclusion* for places p_1 and p_2 can be asserted as $\mu(p_1) + \mu(p_2) = \mu_0(p_1) + \mu_0(p_2) = 1$ for all reachable marking μ. It is easy to see that if there exists a P-invariant X with $X(p) > 0$, for all $p \in P$, then the PN is guaranteed to be structurally bounded. Hence, *place invariants* can be used for reasoning about structural boundedness.

Transition Invariant:

A *transition invariant* of PN $\mathcal{P} = (P, T, \varphi)$ is a mapping $Inv_T : T \to N$ (i.e., assigning nonnegative *weights* to transitions) such that $\Sigma_{t \in T} Inv_T(t)(\Delta(t)) = 0$. In words, firing each transition the number of times specified in the T-invariant brings the PN back to its starting marking. Again consider the PN shown in Figure 14.5 in which $(2,2)$ (i.e., $Inv_T(t_1) = Inv_T(t_2) = 2$) is clearly a T-invariant, so is (n,n), for arbitrary $n \geq 0$. Like P-invariants, any linear combination of T-invariants is a T-invariant. It is easy to see that T-invariants correspond to the solutions of the following equation: $[T] \cdot X^T = 0$ (where X is a row vector representing a T-invariant). The existence of a T-invariant is a necessary condition for a bounded PN to be live. To see this, suppose $\mathcal{P} = ((P, T, \varphi), \mu_0)$ is a live and bounded PN. Clearly due to $\mathcal{P}$ being live, there exists an infinite path $\mu_0 \stackrel{\sigma_1}{\longmapsto} \mu_1 \stackrel{\sigma_2}{\longmapsto} \cdots \mu_i \stackrel{\sigma_{i+1}}{\longmapsto} \mu_{i+1} \cdots$ such that $Tr(\sigma_i) = T$ for all $i \geq 1$ (i.e., σ_i uses all the transitions in T). Since $\mathcal{P}$ is also bounded, there exist $h > j \geq 0$ such that $\mu_j = \mu_h$; hence, $\#_{(\sigma_{j+1} \cdots \sigma_h)}$ constitutes a T-invariant.

14.4.2 Structural Analysis

Given a PN $\mathcal{P} = (P, T, \varphi)$, a subset of places $S \subseteq P$ is called a *trap* (resp., *siphon*) if $S^\bullet \subseteq {}^\bullet S$ (resp., ${}^\bullet S \subseteq S^\bullet$). (Here ${}^\bullet S$ and $S^\bullet$ denote the sets of input and output transitions of S, respectively.) Intuitively, a trap S represents a set of places in which every transition consuming a token from S must also deposit a token back into S. In contrast, if a transition is going to deposit a token to a place in a siphon S, the transition must also remove a token from S.

Suppose S is a siphon in a live PN without isolated places (i.e., places without input/output transitions), then S must be marked in the initial marking μ_0 (i.e., $\mu_0(p) > 0$, for some place $p \in S$). Otherwise, none of the transitions in ${}^\bullet S$ or $S^\bullet$ is fireable – violating the assumption that the PN being live.

The concept of a *siphon* plays an important role in the liveness analysis for the class of *free-choice* PNs. A PN $\mathcal{P} = (P, T, \varphi)$ is called a *free-choice* PN if the following conditions hold: (1) $\forall t \in T, p \in P, \varphi(p, t) \leq 1$ and $\varphi(t, p) \leq 1$, (2) $\forall t_1, t_2 \in T, ({}^\bullet t_1 \cap {}^\bullet t_2 \neq \emptyset \implies {}^\bullet t_1 = {}^\bullet t_2)$. In words, if two transitions share some input places, then they share all their input places. Known as Commoner's Theorem (see, e.g., [7]), a free-choice PN is live iff every nonempty siphon contains an initially marked trap. The reader is referred to [7] for more about free-choice PNs.

Example 7. Consider the free-choice PN shown in Figure 14.6. It is easy to see that $\{p_1, p_2, p_5, p_6, p_7\}$ is a siphon which does not contain any initially marked trap. Hence, the PN is not live due to Commoner's Theorem. In fact, $(1,0,0,0,0,0,0) \stackrel{t_1}{\longmapsto} (0,0,0,0,0,1,1) \stackrel{t_2}{\longmapsto} (0,1,0,0,0,0,1) \stackrel{t_5}{\longmapsto} (0,1,0,0,1,0,0)$ reaches a dead marking. □

14.4.3 State Space Analysis

Reachability Graph Analysis:

The so-called *reachability graph analysis* is perhaps the simplest and the most straightforward approach for analyzing the behavior of a PN. As its name suggests, such a technique relies on exhaustively generating all the reachable

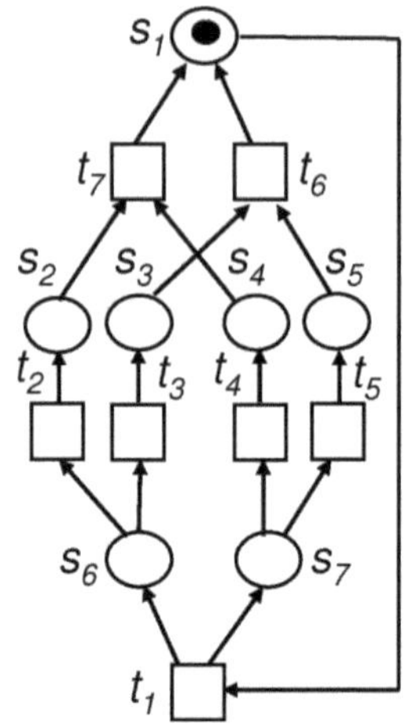

Fig. 14.6. A free-choice PN which is bounded but not live.

markings from a given initial marking, in hope of deducing PN properties by examining the structure of the reachability graph. Figure 14.7 displays a portion of the reachability graph associated with the PN in Figure 14.1.

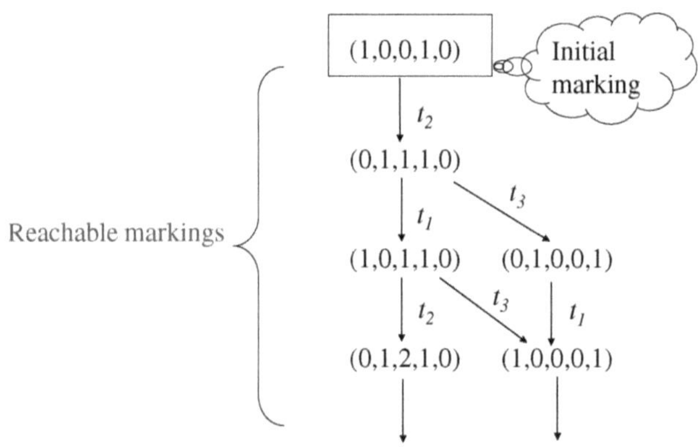

Fig. 14.7. Portion of the reachability graph of the Petri net in Figure 14.1.

In spite of its simplicity, the applicability of the technique of reachability graph analysis is rather limited; it can only be applied to bounded (i.e., finite) PNs with small reachability sets. Even for bounded PNs which exhibit finite reachbility graphs, the technique is 'expensive' in the sense that it suffers from the *state explosion phenomenon* as the sizes of the reachability sets grow beyond any primitive recursive function even for bounded PNs in the worst case. More will be said about this later.

Coverability Graph Analysis:

Coverability graph analysis offers an alternative to the technique of reachability graph analysis by abstracting out certain details to make the graph finite. To understand the intuition behind coverability graphs, consider Figure 14.7 which shows (part of) the reachability graph of the PN in Figure 14.1. Consider the path $(1,0,0,1,0) \stackrel{t_2}{\longmapsto} (0,1,1,1,0) \stackrel{t_1}{\longmapsto} (1,0,1,1,0)$ along which the third coordinate gains an extra token in the end (i.e., $(1,0,1,1,0) > (1,0,0,1,0)$). Clearly the third coordinate can be made arbitrarily large by repeating t_2t_1 for a sufficient number of times, as $(1,0,0,1,0) \stackrel{t_2t_1}{\longmapsto} (1,0,1,1,0) \stackrel{t_2t_1}{\longmapsto} (1,0,2,1,0) \stackrel{t_2t_1}{\longmapsto} \cdots \stackrel{t_2t_1}{\longmapsto} (1,0,n,1,0) \stackrel{t_2t_1}{\longmapsto} \cdots$, for arbitrary n. In order to capture the notion of a place being *unbounded*, we short-circuit the above infinite sequence of computation as $(1,0,0,1,0) \stackrel{t_2}{\longmapsto} (0,1,1,1,0) \stackrel{t_1}{\longmapsto} (1,0,\omega,1,0)$, where ω is a symbol denoting something being arbitrarily large. One can regard ω as "infinity" having the property that $\omega > n$ for any integer n, $\omega + n = n + \omega = \omega$, $\omega - n = \omega$ and $\omega \geq \omega$. A coverability graph relates each node to a *general marking* ($\in (N \cup \{\omega\})^{|P|}$) of the original PN. The corresponding coverability graph of the PN in Figure 14.1 is depicted in Figure 14.8.

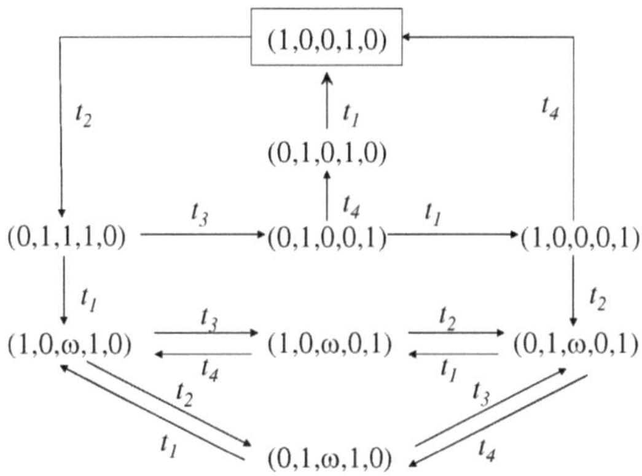

Fig. 14.8. Coverability graph of the Petri net in Figure 14.1.

The algorithm for generating the coverability graph of a PN is shown below (see [36]):

Coverability graph aglorithm
Input: A Petri net $\mathcal{P} = (P, T, \varphi)$ with the initial marking μ_0
Output: The coverability graph $G_{\mathcal{P}}(\mu_0)$ of PN $(\mathcal{P}, \mu_0)$
(1) Create a node μ_{init} such that $\mu_{init} = \mu_0$, and mark it as "new"
(2) **while** there contains some new node μ **do**
(3) **for** each transition t enabled at μ **do**

    case (i) there is a node $\mu' = \mu + \Delta t$ in $G_{\mathcal{P}}(\mu_0)$
        add an edge $\mu \xrightarrow{t} \mu'$ to $G_{\mathcal{P}}(\mu_0)$
    case (ii) there is a node μ'' from μ_{init} to μ such that $\mu'' < \mu + \Delta t$
        add an "new" node x with
            $x(p) = \omega$ if $\mu''(p) < (\mu + \Delta t)(p)$
            $x(p) = \mu''(p)$, otherwise
        add an edge $\mu \xrightarrow{t} x$
    case (iii) otherwise
        add an "new" node x with $x = \mu + \Delta t$ and an edge $\mu \xrightarrow{t} x$
  end for
  mark μ with "old"
end while
end algorithm

As it turns out, the coverability graph of a PN is always finite ([36]). In addition, a PN is unbounded iff an ω occurs in the corresponding coverability graph, which, in turn, yields a decision procedure for deciding the boundedness property. It should be noted that such a technique does not answer the reachability problem as ω abstracts out the exact number of tokens that a place can accumulate, should the place be potentially unbounded.

A reachability graph (possibly of infinite size) captures the exact information about the set of reachable markings of a PN, whereas a coverability graph (always of finite size) provides an over-approximation of the reachability set, should it be infinite. Again the coverability graph analysis suffers from *state explosion.*

14.5 Complexity Analysis of Various Petri Net Problems

14.5.1 Boundedness and Covering

The boundedness problem was first considered by Karp and Miller in [36], where it was shown to be decidable using the technique of *coverability graph analysis.* The algorithm presented there was basically an unbounded search and consequently no complexity analysis was shown. Subsequently, a lower bound of $O(2^{c \times m})$ space was shown by Lipton in [40], where m represents the dimension of the problem instance and c is some constant. Finally, an upper bound of $O(2^{c \times n \times log n})$ space was given by Rackoff in [56]. Here, however, n represents the size or number of bits in the problem instance and c is a constant.

Lipton's exponential space lower bound was established by constructing a VAS to maintain and store a number, whose value ranges between 0 and 2^{2^k}. This number could then be incremented by 1 (as long as the current value was below the upper limit), decremented by 1 (as long as the current value exceeded 0), and tested for zero. The zero-test was the hard part in

Lipton's construction. See [59] for a refinement of Lipton's lower bound in the framework of multiparameter analysis.

Rackoff's exponential space upper bound was established using induction (on the dimension of the VAS instance). Such a technique has found additional applications in [14, 17, 59]. In particular, Rosier and Yen [59] refined Rackoff's strategy to derive a *multiparameter analysis* of the boundedness problem for VASSs, yielding an upper bound of $O(2^{c \times k \times logk}(l + logn))$ space, where k is the dimension, n is the number of states, and l is the length of the binary representation of the largest number mentioned in the VASS. A direct consequence of the above is that when the dimension of a VASS (VAS, or PN) is a fixed constant, then the boundedness problem is solvable in polynomial space. As the strategy behind Rackoff's proof is interesting and important in its own right (as was witnessed by its additional applications in [14, 17, 59]), in what follows we briefly describe the key steps of the proof.

Let (v, A) be a VAS of size n. It is well known that (v, A) is unbounded iff there is a computation $v \overset{*}{\longmapsto} v' \overset{*}{\longmapsto} v''$ such that $v'' > v'$. Rackoff's strategy relies on showing that if such a path exhibiting unboundedness exists, then there is 'short' witness.

A $w \in Z^k$ is called *i-bounded* (resp., *i-r bounded*) if $0 \leq w(j)$, $\forall 1 \leq j \leq i$ (resp. $0 \leq w(j) \leq r$, $\forall 1 \leq j \leq i$). Let $p = w_1 w_2 \cdots w_m$ be a sequence of vectors in Z^k. Sequence p is said to be

- *i-bounded* (resp., *i-r bounded*) if every member of p is *i-bounded* (resp., *i-r bounded*).
- *self-covering* if there is a $1 \leq j \leq m$ such that $w_m > w_j$.
- an *i-loop* if $w_m(j) = w_1(j), \forall 1 \leq j \leq i$.

Let $m(i, v)$ be the length of the shortest i-bounded self-covering path in (v, A); =0 if no such path exists. Also let $g(i) = max\{m(i, v) : v \in Z^k\}$. Note that $g(i)$ represents the length of the longest i-bounded self-covering path from any starting vector in Z^k. The key in Rackoff's proof relies on finding a bound for $g(i)$ inductively, where $i = 1, 2, ..., k$.

To derive $g(i)$, it was shown in [56] that if there is an $i - r$ bounded self-covering path in (v, A), then there is a 'short witness' of length bounded by r^{n^c}, where c is a constant independent of (v, A). The proof of this result relies on rearranging as well as chopping off unnecessary i-loops along an $i - r$ bounded self-covering path, using a result in [3] concerning bounds of integer solutions of linear (in)equations. Using the above result, it could be shown that $g(0) \leq 2^{n^c}$ and $g(i + 1) \leq (2^n g(i))^{n^c}$, $0 \leq i \leq k - 1$. To see this, let $p : v_1 \cdots v_m$ be any $(i + 1)$-bounded self-covering path. Consider two cases:

- Case 1: Path p is $(i + 1) - (2^n g(i))$-bounded. Then the length of p is $\leq (2^n g(i))^{n^c}$, which follows immediately from the earlier result concerning short witnessing paths.
- Case 2: Otherwise, let v_h be the first vector along p that is not $(2^n g(i))$ bounded. By removing $(i+1)$-loops, the prefix $v_1 ... v_h$ can be shortened (if

necessary) to make the length $\leq (2^n g(i))^{i+1}$. With no loss of generality, we assume the $(i+1)$st position to be the coordinate whose value exceeds $2^n g(i)$ at v_h. Recalling the definition of $g(i)$, there is a self-covering path, say l, of length $\leq g(i)$ from v_h. By appending l to $v_1...v_h$ (i.e., replacing the original suffix path $v_h...v_m$ by l), the new path is an $(i+1)$-bounded self-covering path, because the value of the (i+1)st coordinate exceeds $2^n g(i)$ and the path l (of length bounded by $\leq g(i)$) can at most subtract $(2^n g(i))$ from coordinate $i+1$. (Note that the application of an addition vector can subtract at most 2^n from a given coordinate.)

By solving the recurrence relation $g(0) \leq 2^{n^c}$ and $g(i+1) \leq (2^n g(i))^{n^c}$, $0 \leq i \leq k-1$, the length of the shortest path witnessing unboudedness is $\leq 2^{2^{c \times n \times logn}}$. A nondeterministic search immediately yields a $O(2^{c \times n \times logn})$ space complexity for the boundedness problem.

The complexity (both upper and lower bounds) of the covering problem can be derived along a similar line of that of the boundedness problem. See [56] for details.

14.5.2 Reachability

Of various problems of interest in the study of PNs, the reachability problem is perhaps the one that has attracted the most attention in the PN community in the past four decades. One reason is that the problem has many real-world applications; furthermore, it is the key to the solutions of several other PN problems (such as liveness).

Before the decidability question of the reachability problem for general PNs was answered in the affirmative by Mayr [42, 43] in the early 1980's (see also [38]), a number of attempts were made to investigate the problem for restricted classes of PNs, in hope of gaining more insights and developing new tools in order to conquer the general PN reachability problem. Before Mayr's proof, Sacerdote and Tenney [60] claimed the reachability problem to be decidable; yet they failed to provide a convincing proof. What follows are notable milestones along this line of research.

In 1974, van Leeuwen [61] first showed the reachability problem to be decidable for 3-dimensional PNs. Hopcroft and Pansiot [21] later extended van Leeuwen's finding to 5-dimensional PNs in 1979. About the same time Landweber and Robertson [39] as well as Grabowski [16], Mayr [41] and Muller [48] considered PNs on which either structural or behavioral constraints are imposed, and showed the reachability problem to be decidable for the classes of *conflict-free* and *persistent* PNs. A important common feature of the above attempts is that the decidability result was built upon showing the reachability set to be *semilinear*. As it turns out, semilinearity is also preserved for the so-called *normal* [62], *sinkless* [62], and *communication-free* PNs (also known as *BPP nets*) [28, 10].

Although the reachability problem for general PNs is known to be decidable, no complexity analysis was given in [42, 43], (nor in [38]). The best known lower bound for the problem is exponential space hard, which is identical to that of the boundedness problem. (In fact, Lipton's lower bound proof works for both the reachability and the boundedness problems.) In view of the importance of the reachability problem, finding the exact complexity of the problem remains one of the most important open problems in Petri net theory.

14.5.3 Containment and Equivalence

In the late 1960's, Rabin first showed the containment problem for PNs to be undecidable. Even though the original work of Rabin was never published, a new proof was presented at a talk at MIT in 1972 [2]. In 1975, Hack [19] extended Rabin's result by showing the equivalence problem of PNs to be undecidable as well. Both undecidability proofs were based on *Hilbert's Tenth Problem* [6], a famous undecidable problem.

Hilbert's Tenth Problem is the problem of, given a polynomial $P(x_1, ..., x_n)$ over n variables with integer coefficients, deciding whether $P(x_1, ..., x_n) = 0$ has integer solutions. Reducing from Hilbert's Tenth Problem, it is not hard to see that the so-called *polynomial graph inclusion problem*, i.e., given two polynomials P and Q, deciding whether $\{(x_1, ..., x_n, y) \mid y \leq P(x_1, ..., x_n),\ \text{with}\ x_1, ..., x_n, y \in N\} \subseteq \{(x_1, ..., x_n, y) \mid y \leq Q(x_1, ..., x_n),\ \text{with}\ x_1, ..., x_n, y \in N\}$, is undecidable. The key behind Rabin's and Hack's proofs relies on showing PNs to be capable of *weakly computing* polynomials. By weakly computing a polynomial $P(x_1, ..., x_n)$, we mean a PN $\mathcal{P}$ with n designated places $p_1, ..., p_n$ (holding the n input values of P) and a designated 'output' place q can be constructed in such a way that for arbitrary input values $v_1, ..., v_n \in N$, starting from $v_1, ..., v_n$ tokens in places $p_1, ..., p_n$, respectively, $\mathcal{P}$ has the ability to deposit y tokens in place q for some $y \leq P(v_1, ..., v_n)$ when halting. With such capabilities of weakly computing polynomials, two PNs $\mathcal{P}$ and $\mathcal{Q}$ can be constructed from two given polynomials P and Q such that $R(\mathcal{P}) \subseteq R(\mathcal{Q})$ (or $R(\mathcal{P}) = R(\mathcal{Q})$) iff the answer to the polynomial graph inclusion problem regarding polynomials P and Q is positive.

It turns out that the equivalence problem remains undecidable with respect to several other notions of *equivalence*, including *trace equivalence*, *language equivalence* as well as *bisimulation equivalence* [32] for labelled PNs. In fact, it follows from [27] that all the equivalences under the interleaving semantics are undecidable.

14.5.4 Liveness

In [18], several variants of the reachability problem (including the general one) were shown to be recursively equivalent. Among them is the *single-place zero reachability problem*, i.e., the problem of determining whether a marking with no tokens in a designated place can be reached. Hack [18] also showed the *single-place zero reachability problem* to be recursively equivalent to the liveness problem. As a result, deciding whether a PN is live or not is decidable. Like the general reachability problem, the exact computational complexity of the liveness problem remains open.

14.5.5 Model Checking

For some time, *temporal logic* has been considered a useful formalism for reasoning about systems of concurrent programs. A typical problem involving temporal logic is the *model checking* problem, i.e., the problem of determining whether a given structure defines a model of a correctness specification expressed in the temporal logic.

Before getting into the study of model checking for PNs, we require the following basic definitions of *linear-time temporal logic* (LTL) and a *branching-time temporal logic* called *computation tree logic* (CTL).

An *LTL well-formed formula* is defined as

$$F \mid \neg f \mid f \wedge g \mid \bigcirc f \mid f\, \mathcal{U}\, g$$

where F is a predicate, and f and g are well-formed formulas. For convenience, we also write $f \vee g \equiv \neg(\neg f \wedge \neg g)$, $\Diamond f \equiv true\, \mathcal{U}\, f$ and $\Box f \equiv \neg \Diamond \neg f$. Intuitively, $\bigcirc$, $\mathcal{U}$, $\Diamond$, and $\Box$ are temporal operators denoting *next-time*, *until*, *eventually*, and *always*, respectively. LTL formulas are interpreted on *computation paths*. With respect to a computation path $s_0 \to s_1 \to \cdots$, the intuitive meanings of the the above formulas are:

1. F: the atomic predicate F holds at s_0,
2. $\neg f$: formula f does not hold at s_0,
3. $f \wedge g$: both f and g hold at s_0,
4. $\bigcirc f$: f holds at s_1 (i.e., the immediate successor of s_0),
5. $f\, \mathcal{U}\, g$: there exists an $i \geq 0$ such that f holds at $s_0, ..., s_{i-1}$ and g holds at s_i.

A *CTL formula* is defined as

$$F \mid \neg f \mid f \wedge g \mid \exists \bigcirc f \mid \forall \bigcirc f \mid \exists(f\, \mathcal{U}\, g) \mid \forall(f\, \mathcal{U}\, g)$$

where F is a predicate, f and g are CTL formulas, and $\exists$ and $\forall$ are path quantifiers denoting 'there exists a path' and 'for all paths', respectively. Among others, useful abbreviations include: $\exists\Diamond f \equiv \exists(true\, \mathcal{U} f)$; $\forall\Diamond f \equiv \forall(true\, \mathcal{U} f)$; $\forall\Box f \equiv \neg\exists\Diamond(\neg f)$; $\exists\Box f \equiv \neg\forall\Diamond(\neg f)$.

CTL formulas are interpreted on *computation trees*. With respect to a tree rooted at s_0, the intuitive meanings of the formulas mentioned above are:

1. F , $\neg f$ and $f \wedge g$ are the same as those in the LTL case,
2. $\exists \bigcirc f$: there exists a *child* s of s_0 such that f holds at s,
3. $\forall \bigcirc f$: for every *child* s of s_0, f holds at s,
4. $\exists(f\, \mathcal{U}\, g)$: there exists a computation path l from s_0 such that $f\, \mathcal{U}\, g$ holds with respect to l,

5. $\forall(f \,\mathcal{U}\, g)$: for every computation path l emanating from s_0, $f \,\mathcal{U}\, g$ holds with respect to l.

The *model checking problem* for PNs is the problem of, given a PN $\mathcal{P}$ (with initial marking μ_0) and a temporal formula ϕ (expressed in some temporal logic), deciding whether the computation of $\mathcal{P}$ from μ_0 satisfies ϕ.

Model checking Petri nets was first investigated by Howell, Rosier, and Yen in [24, 25]. As was shown in [24], the model checking problem for a fairly simple temporal logic is undecidable, even for the class of conflict-free PNs. Following a number of subsequent work on the study of model checking for PNs (see, e.g., [31, 44, 11, 17]) from a decidability/complexity viewpoint, we now have a reasonably clear picture. Consider two types of atomic predicates: *state-based* and *action-based* predicates. As their names suggest, a state-based predicate applies to the 'markings' of a PN computation, whereas the value of an action-based predicate depends only the 'actions' taken along the computation.

With respect to *state-based* predicates, the model checking problem is undecidable for both linear-time and branching-time temporal logics. The undecidability result holds for branching-time, action-based temporal logics as well. Interestingly, however, model checking for linear-time, action-based temporal logics turns out to be decidable [17].

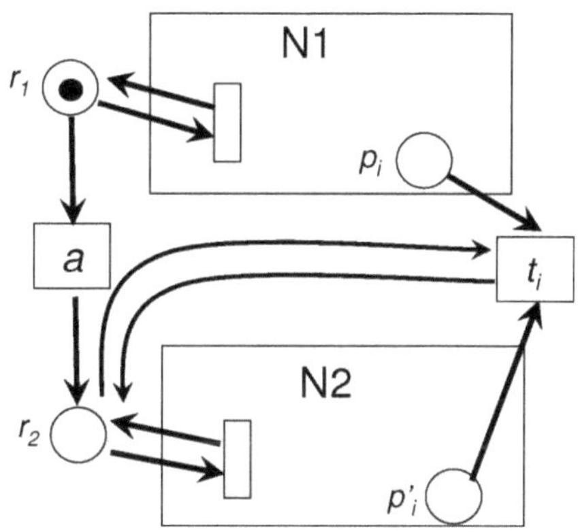

Fig. 14.9. Undecidability proof.

To show the model checking problem for linear-time, state-based temporal logic to be undecidable, we reduce from the containment problem of PNs. Given two PNs N_1 and N_2, we construct a PN N shown in Figure 14.9, in which r_1 and r_2 serve as 'control' places for controlling the firings of transitions in N_1 and N_2, respectively. For each place p_i in N_1, let p'_i be the corresponding place in N_2. A transition t_i is introduced to remove an equal amount of tokens from p_i and p'_i. It is not hard to see that

$$R(N_1) \subseteq R(N_2)$$

$$iff$$

$$\Box(\ ((r_1 = 1 \wedge r_2 = 0) \wedge \bigcirc(r_1 = 0 \wedge r_2 = 1)) \Rightarrow \Diamond(\bigwedge_i (p_i = 0 \wedge p'_i = 0))\ )$$

Note that $((r_1 = 1 \wedge r_2 = 0) \wedge \bigcirc(r_1 = 0 \wedge r_2 = 1))$ holds only at a marking at which transition a fires. Using a similar construction, the undecidability result for either state-based or action-based branching-time temporal logic can be shown.

The idea behind model checking an action-based linear-time temporal formula ϕ for PN $\mathcal{P}$ is the following. (See [17] for details.) Construct a *Buchi automaton* $M_{\neg\phi}$ to capture all the computations satisfying the negation of ϕ. Then it can be shown that $\mathcal{P}$ satisfies ϕ iff the intersection between the sets of computations of $\mathcal{P}$ and $M_{\neg\phi}$ is empty. By using a VASS to capture the 'Cartesian product' of $\mathcal{P}$ (equivalently, a VAS) and $M_{\neg\phi}$ (a finite automaton), the model checking problem is then reduced to finding certain infinite computations in a VASS, which turns out to be decidable.

14.5.6 Self-stabilization

Before the end of this section, let us elaborate a bit about the *self-stabilization* issue of PNs. The notion of *self-stabilization* was introduced by Dijkstra [8] to describe a system having the behavior that regardless of its starting configuration, the system would return to a 'legitimate' configuration eventually. (By a legitimate configuration, we mean a configuration which is reachable from the initial configuration of the system.) The motivation behind self-stabilization is that a self-stabilizing system has the ability to 'correct' itself even in the presence of certain unpredictable errors that force the system to reach an 'illegitimate' configuration during the course of its operations. In this sense, self-stabilizing systems exhibit fault-tolerant behaviors to a certain degree.

Let $\mathcal{S}$ be a (finite or infinite) *system* with c_0 as its *initial configuration.* Also let $R(\mathcal{S}, c_0)$ $(=\{c \mid c_0 \xrightarrow{*} c\})$ denote the set of *reachable configurations* from c_0. A computation σ from configuration c_1 is said to be *non-self-stabilizing* iff one of the following holds:

1. σ is finite ($\sigma : c_1 \xrightarrow{t_1} c_2 \xrightarrow{t_2} \cdots c_{m-1} \xrightarrow{t_{m-1}} c_m$, for some m) such that c_m is a dead configuration and $c_m \notin R(\mathcal{S}, c_0)$, <u>or</u>
2. σ is infinite ($\sigma : c_1 \xrightarrow{t_1} c_2 \xrightarrow{t_2} \cdots c_i \xrightarrow{t_i} c_{i+1} \cdots$) such that $\forall i \geq 1, c_i \notin R(\mathcal{S}, c_0)$.

See Figure 14.10. A system is said to be *self-stabilizing* if for each configuration c, none of the computations emanating from c is non-self-stabilizing. The *self-stabilization problem* is to determine, for a given (finite or infinite) system, whether the system is self-stabilizing.

The self-stabilization problem has only been scarcely studied in the Petri net community. It is known [4] that for bounded ordinary PNs (i.e. PNs without multiple arcs), the problem is PTIME-complete, whereas for bounded general PNs (i.e. PNs with multiple arcs), the problem becomes PSPACE-complete. For general unbounded PNs, the analysis of the self-stabilization problem remains open.

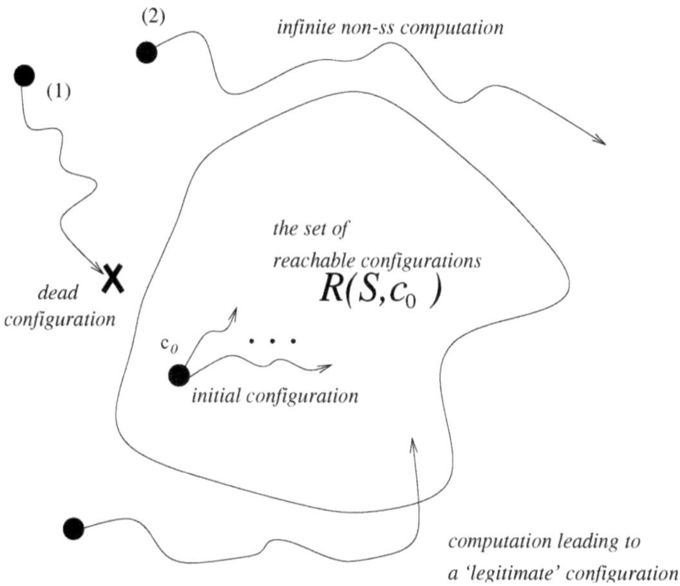

Fig. 14.10. Non-self-stabilizing computations.

14.6 Petri Nets With Semilinear Reachability Sets

The concept of *semilinearity* plays a key role not only in conventional automata theory and formal languages (see, e.g., [29]), but also in the analysis of PNs. A subset L of N^k is a *linear set* if there exist vectors $v_0, v_1, \ldots, v_t$ in N^k such that $L = \{v \mid v = v_0 + m_1 v_1 + \cdots + m_t v_t,\ m_i \in N\}$. The vectors v_0 (referred to as the *constant vector*) and $v_1, v_2, \ldots, v_t$ (referred to as the *periods*) are called the *generators* of the linear set L. For convenience, such a linear set is written as $\mathcal{L}(v_0; v_1, \ldots, v_t)$. A set $SL \subseteq N^k$ is *semilinear* [15] if it is a finite union of linear sets, i.e., $SL = \bigcup_{1 \leq i \leq m} \mathcal{L}_i$, where $\mathcal{L}_i$ ($\subseteq N^k$) is a linear set. The empty set is a trivial semilinear set. Every finite subset of N^k is semilinear – it is a finite union of linear sets whose generators are constant vectors. Figure 14.11 shows an example of a semilinear set, which consists of three linear sets L_1, L_2 and L_3. Clearly semilinear sets are closed under (finite) union. It is also known that they are closed under complementation and intersection. It is worthy of noting that semilinear sets are exactly those that can be expressed by *Presburger Arithmetic* (i.e., first order theory over natural numbers with addition) [55], which is a decidable theory.

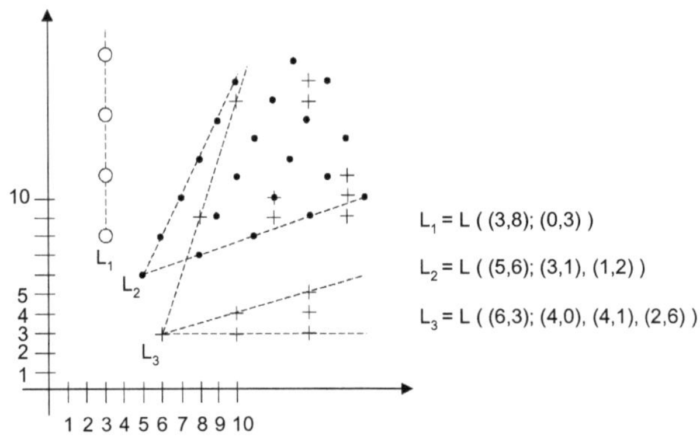

Fig. 14.11. A semilinear set.

It is known [21] that for PNs of dimension 6 (equivalently, 6-dimensional vector addition systems, or 3-dimensional vector addition systems with states) or beyond, their reachability sets may not be semilinear in general.

For subclasses of PNs with semilinear reachability sets, a natural question to ask is: *What is the size of its semilinear representation?* An answer to the above question is key to the complexity analysis of various PN problems. In what follows, we survey several notable examples along the line of research of analyzing the sizes of semilinear representations for PNs with semilinear reachability sets.

Finite PNs:

The reachability sets of finite (i.e., bounded) PNs are trivially semilinear. Mayr and Meyer [45] showed that the containment and equivalence problems for finite VASs are not primitive recursive. Subsequently, McAloon [46] showed that the problems are primitive recursive in the Ackermann function, and Clote [5], using Ramsey theory, showed the finite containment problem to be DTIME (Ackermann) complete. Using a different approach, Howell, Huynh, Rosier and Yen [22] showed an improvement of two levels in the primitive recursive hierarchy over results previously obtained by McAloon, thus answering a question posed by Clote.

2-dimensional VASS or 5-dimensional VAS (PN):

It was first known by Hopcroft and Pansiot [21] that PNs of dimension 5 always have semilinear reachability sets. However, Hopcroft-Pnasiot algorithm does not reveal any upper bound on the size of the semilinear set representation, nor does it tell how quickly the set can be generated. In a subsequent study, Howell, Huynh, Rosier and Yen [22] gave a detailed analysis of the semilinear reachability sets of 2-dimensional VASSs, yielding the following result:

Given a 2-dimensional, n-state VASS V in which the largest integer mentioned can be expressed in l bits, we can construct in $DTIME(2^{2^{c\times l\times n}})$ (for some constant c) a semilinear reachability set representation $SL = \bigcup_{1\leq i\leq k} \mathcal{L}_i(x_i; P_i)$ such that, for some constants d_1, d_2, d_3,

1. $k = O(2^{2^{d_1\times l\times n}})$,
2. $\forall 1 \leq i \leq k$, $||x_i|| = O(2^{2^{d_2\times l\times n}})$
3. $\forall 1 \leq i \leq k$, $|P_i| = O(2^n)$
4. $\forall v \in P_i$, $\forall 1 \leq i \leq k$, $||v|| = O(2^{d_3\times l\times n})$

(Note that $||x||$ denotes the *1-norm* of vector x.)

Using the above result, the reachability, containment and equivalence problems for such VASSs were shown to be solvable in $DTIME(2^{2^{d\times l\times n}})$, for some constant d [22]. A matching lower bound was also established in [22].

Conflict-free, Normal, Sinkless, Communication-free PNs:

In this section, we employ a *decompositional approach* to serve as a unified framework for analyzing a wide variety of subclasses of PNs. Under this framework, answering the reachability question is equated with solving an instance of *integer linear programming*, which is relatively well-studied. To a certain extent, the decompositional approach can be thought of as a generalization of the *state equation* approach mentioned in our earlier discussion.

Before going into the details, the definitions of those subclasses of PNs for which the decompositional approach works are given first.

A *circuit* of a PN is simply a closed path (i.e., a cycle) in the PN graph. The presence of complex circuits is troublesome in PN analysis. In fact, strong evidence has suggested that circuits constitute the major stumbling block in the analysis of PNs. To get a feel for why this is the case, recall that in a PN $\mathcal{P}$ with initial marking μ_0, a marking μ is reachable (from μ_0) in $\mathcal{P}$ *only if* there exists a column vector $x \in \mathrm{N}^k$ satisfying the state equation $\mu_0+[T]\cdot x = \mu$. The converse, however, does not necessarily hold. In fact, lacking a necessary and sufficient condition for reachability in general has been blamed for the high degree of complexity in the analysis of PNs. (Otherwise, one could have tied the reachability analysis of PNs to the *integer linear programming* problem, which is relatively well understood.) There are restricted classes of PNs for which necessary and sufficient conditions for reachability are available. Most notable, of course, is the class of *circuit-free* PNs (i.e., PNs without circuits) for which the equation $\mu_0 + [A] \cdot x = \mu$ is sufficient and necessary to capture reachability. A slight relaxation of the circuit-freedom constraint yields the same necessary and sufficient condition for the class of PNs without *token-free* circuits in every reachable marking [62].

Formally, a *circuit* c of a PN is a sequence $p_1 t_1 p_2 t_2 \cdots p_n t_n p_1$ ($p_i \in P$, $t_i \in T$, $p_i \in {}^\bullet t_i$, $t_i \in {}^\bullet p_{i+1}$), such that $p_i \neq p_j, \forall i \neq j$ (i.e., all nodes except the first and the last are distinct along the closed path). We write $P_c = \{p_1, p_2, \cdots, p_n\}$ (resp., $T_c = \{t_1, t_2, \cdots, t_n\}$) to denote the set of places (resp., transitions) in c, and $tr(c)$ to represent the sequence $t_1 t_2 \cdots t_n$. We define the token count of circuit c in marking μ to be $\mu(c) = \sum_{p \in P_c} \mu(p)$. A circuit c is said to be *token-free* in μ iff $\mu(c) = 0$. Given two circuits c and c', c is said to be *included* (resp., *properly included*) in c' iff $P_c \subseteq P_{c'}$ (resp., $P_c \subset P_{c'}$). We say c is *minimal* iff it does not properly include any other circuit. Circuit c is said to be a

- $\oplus$-circuit iff $\forall i, 1 \leq i \leq n, {}^\bullet t_i = \{p_i\}$ (i.e., t_i has p_i as its unique input place), or
- $\odot$-circuit iff $\forall t \in T, \sum_{p \in P_c} (\varphi(t,p) - \varphi(p,t)) \geq 0$ (i.e., no transition can decrease the token count of c).

In this section, we elaborate on a useful technique, called *decomposition*, using which the reachability relation for various subclasses of PNs is linked with integer linear programming [26, 50, 63, 64]. Among those for which our decompositional approach is applicable include:

- *Conflict-free Petri nets*: A PN $\mathcal{P} = (P, T, \varphi)$ is *conflict-free* [39] iff for every place p, either
  1. $|p^\bullet| \leq 1$, or
  2. $\forall t \in p^\bullet$, t and p are on a self-loop.

  In words, a PN is conflict-free if every place which is an input of more than one transition is on a self-loop with each such transition.
- *Normal Petri nets*: A PN $\mathcal{P} = (P, T, \varphi)$ is *normal* [62] iff for every minimal circuit c and transition $t_j \in T$, $\sum_{p_i \in P_c} a_{i,j} \geq 0$. Intuitively, a PN is normal iff no transition can decrease the token count of a minimal circuit by firing at any marking, Alternatively, every minimal circuit of a normal PN is a $\odot$-circuit.
- *Sinkless Petri nets*: A PN $\mathcal{P} = (P, T, \varphi)$ is *sinkless* [62] iff each minimal circuit of $\mathcal{P}$ is sinkless.
- *BPP nets*: A PN $\mathcal{P} = (P, T, \varphi)$ is a *BPP net* [10, 28] iff $\forall\, t \in$T, $|{}^\bullet t| = 1$, i.e., every transition has exactly one input place, implying that every circuit of a BPP net is a $\oplus$-circuit. BPP nets are also known as *communication-free PNs*.
- *Trap-circuit Petri nets*: A PN $\mathcal{P} = (P, T, \varphi)$ is a *trap-circuit* PN [30] iff for every circuit c in $\mathcal{P}$, P_c is a trap.
- *Extended trap-circuit Petri nets*: A PN $\mathcal{P} = (P, T, \varphi)$ is an *extended trap-circuit* PN [63] iff for every circuit c in $\mathcal{P}$, either P_c is a trap or c is a $\oplus$-circuit.

The containment relationship among the above subclasses of PNs is depicted in Figure 14.12.

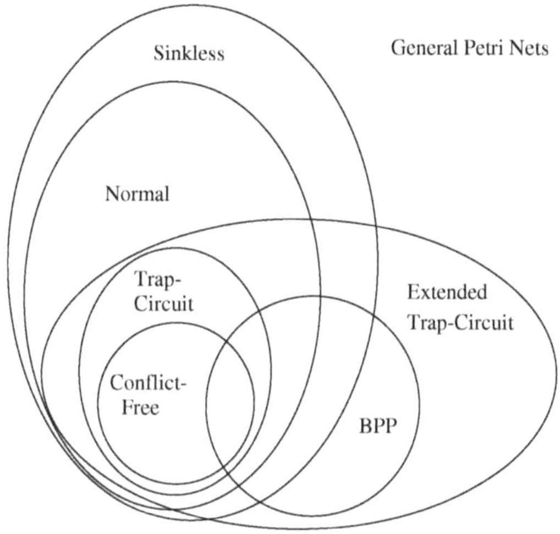

Fig. 14.12. Containment relationships among various Petri net classes.

The idea behind the decompositional technique relies on the ability to decompose a PN $\mathcal{P}=(P,T,\varphi)$ (possibly in a nondeterministic fashion) into sub-PNs $\mathcal{P}_i = (P, T_i, \varphi_i)$ ($1 \leq i \leq n$, $T_i \subseteq T$, and φ_i is the restriction of φ to $(P \times T_i) \cup (T_i \times P)$) such that for an arbitrary computation $\mu_0 \stackrel{\sigma}{\longmapsto} \mu$ of PN $\mathcal{P}$, σ can be rearranged into a canonical form $\sigma_1\sigma_2 \cdots \sigma_n$ with $\mu_0 \stackrel{\sigma_1}{\longmapsto} \mu_1 \stackrel{\sigma_2}{\longmapsto} \mu_2 \cdots \mu_{n-1} \stackrel{\sigma_n}{\longmapsto} \mu_n = \mu$, and for each i, a system of linear inequalities $ILP_i(x,y,z)$ can be set up (based upon sub-PN $\mathcal{P}_i$, where x, y, z are vector variables) in such a way that $ILP_i(\mu_{i-1}, \mu_i, z)$ has a solution for z iff there exists a σ_i in T_i^* such that $\mu_{i-1} \stackrel{\sigma_i}{\longmapsto} \mu_i$ and $z = \#_{\sigma_i}$. See Figure 14.13.

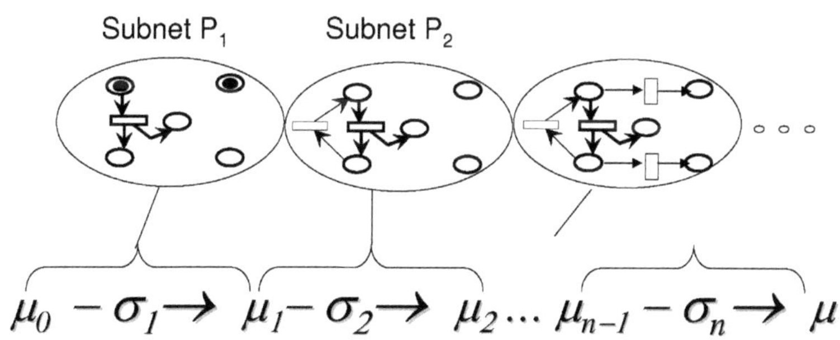

Fig. 14.13. Decompositional approach.

Each of the subclasses of PNs depicted in Figure 14.12 enjoys the merit of having a 'nice' decomposition. In what follows, we elaborate on two notable examples, namely, the classes of *normal* and *sinkless* PNs. For other subclsses of PNs, the reader is referred to [50, 63, 64].

Let $\mu_0 \overset{\sigma_1}{\longmapsto} \mu_1 \overset{\sigma_2}{\longmapsto} \mu_2 \cdots \mu_{n-1} \overset{\sigma_n}{\longmapsto} \mu_n = \mu$ be a computation in a normal (or sinkless) PN $\mathcal{P}$ reaching μ. The rearrangement $\sigma_1\sigma_2 \cdots \sigma_n$ of σ is such that if $\sigma = t_1\sigma'_1 t_2\sigma'_2 \cdots t_n\sigma'_n$ where $t_1, t_2, ..., t_n$ mark the first occurrences of the respective transitions in σ (i.e., $t_i, 1 < i \leq n$, is not in the prefix $t_1\sigma'_1 \cdots t_{i-1}\sigma'_{i-1}$) then σ_i is a permutation of $t_i\sigma'_i$. (In words, the appearance of a new transition triggers the beginning of a new segment.) Furthermore, by letting

- $T_0 = \emptyset$,
- $\forall 1 \leq i \leq n$, $T_i = T_{i-1} \cup \{t_i\}$, for some $t_i \notin T_{i-1}$ enabled at μ_{i-1}, and
- φ_i is the restriction of φ to $(P \times T_i) \cup (T_i \times P)$,

it can be shown [26] that reachability in sub-PN $\mathcal{P}_i$ can be captured by an instance $ILP_i(\mu_{i-1}, \mu_i, z_i)$. That is, marking μ_i is reachable from μ_{i-1} in sub-PN $\mathcal{P}_i$ iff there is a solution with respect to variable z_i in $ILP_i(\mu_{i-1}, \mu_i, z_i)$. As a result, if μ is reachable, then (1) there must exist a canonical computation reaching μ such that the computation can be decomposed into a sequence of sub-computations, say $\sigma_1, \sigma_2, \cdots, \sigma_n$, each coincides with the respective member in the PN $\mathcal{P}$ decomposition; namely $\mathcal{P}_1, \mathcal{P}_2, \cdots, \mathcal{P}_n$, and (2) checking the reachability of PN $\mathcal{P}$ is equivalent to solving the a collection of systems of linear equalities made up of $ILP_i(\mu_{i-1}, \mu_i, z_i)$, for $1 \leq i \leq n$. See [26] for more details.

Based on the decompositional approach, the reachability problem for each of the subclasses of PNs depicted in Figure 14.12 is solvable in NP. (In fact, the problem is NP-complete as the NP-hardness lower bound is easy to show.)

In addition to the aforementioned subclasses of PNs, *single-path PNs* are another class whose semilinear reachability sets have been characterized in detail. See [23] for more.

14.7 Extended Petri Nets

As mentioned in our earlier discussion, the power of conventional PNs is strictly weaker than that of Turing machines. Those using PNs to model real-world systems have often found the expressive power of PNs to be too simple and limited. In many real-time applications, it is often desirable to give certain jobs higher priorities over others, so that critical actions can be finished within their time constraints. One way to do so, for example, is to assign each transition of a process a priority which indicates the degree of importance or urgency. As [1] indicates, the conventional PN model is unable to model prioritized systems. From a theoretical viewpoint, the limitation of PNs is precisely due to the lack of abilities to test potentially unbounded places for zero and then act accordingly. With zero-testing capabilities, it is fairly easy to show the 'extended' PNs to be equivalent to *two-counter machines* (which are Turing equivalent).

To remedy the weakness in expressive power, a number of extended PN models have been proposed in the literature. Among them include *colored PNs* [34], *timed PNs* [47, 57], *PNs with inhibitor arcs* [51], *Prioritized PNs* [20], *PNs under the maximal firing strategy* ... and more. Each of the above extensions allows the PN to faithfully simulate *test-for-zero*, rendering the above extended PNs Turing-equivalent. In what follows, we consider the so-called *PNs under the maximal firing strategy*, which are of interest due to their close connection with the model of *P systems* [52], a model abstracting from the way living cells process their chemical compounds in their compartmental structures.

Under the maximal firing strategy, all the fireable rules are applied in a nondeterministic and maximally parallel fashion at any point in time. Now we show how PNs operating under this new semantics are capable of simulating two-counter machines, which are finite automata augmented with two counters on which the following operations can be applied to each counter: (1) add one to a counter, (2) subtract one from a counter, provided that the counter is not zero, and (3) test a counter for zero and then move accordingly. It is well-known that two-counter machines and Turing machines are computationally equivalent.

The simulations of types (1) and (2) operations of a two-counter machine are straightforward. Now we see how (3) can be faithfully simulated by a PN operating under the maximal firing mode. Consider the PN structure shown in Figure 14.14, in which place C simulates one of the two counters, and the current state of the two-counter machine is recorded by placing a token in the corresponding place, for example, p_1. Consider the following two cases, depending on whether C is empty or not:

1. ($C = 0$): the computation involves $(p_1 = 1) \stackrel{t_1}{\longmapsto} (p_2 = p_3 = 1) \stackrel{t_3}{\longmapsto} (p_3 = p_4 = 1) \stackrel{t_4}{\longmapsto} (p = 1)$
2. ($C = k > 0$): the computation involves $(p_1 = 1; C = k) \stackrel{t_1}{\longmapsto} (p_2 = p_3 = 1; C = k) \stackrel{\{t_3,t_2\}}{\longmapsto} (p_4 = p_5 = 1; C = k - 1) \stackrel{t_5}{\longmapsto} (p' = 1; C = k - 1)$

It is then reasonably easy to see that a token is deposited into p provided that the counter is zero to begin with; otherwise, a token is moved into p' while C is being decremented by one.

With the above extended PNs powerful enough to simulate Turing machines, all nontrivial problems for such PNs become undecidable. A natural and interesting question to ask is: are there PNs whose powers lie between conventional PNs and Turing machines? As it turns out, the quest for such 'weaker' extensions has attracted considerable attentions in recent years. Such PN extensions include *reset nets* [9], and *transfer nets* [13]. A reset net is a PN

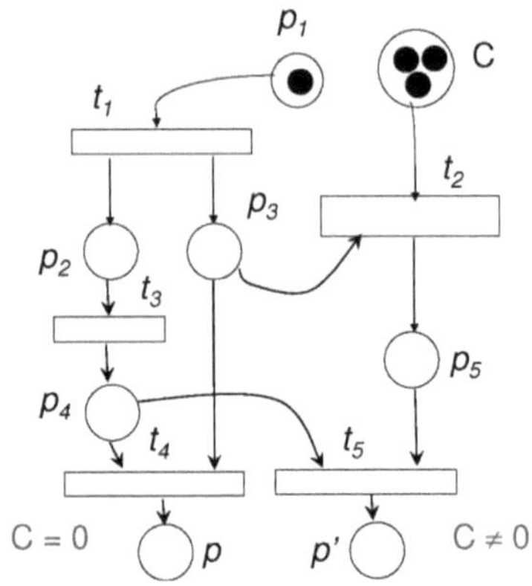

Fig. 14.14. Simulating test-for-zero of a two-counter machine.

(P, T, φ) equipped with a set F_R ($\subseteq T \times P$) of *reset arcs*. When a transition t with $(t, p) \in F_R$ is fired, place p is reset to zero. In a transfer net, a set F_T ($\subseteq P \times T \times P$) of *transfer arcs* is associated with a PN (P, T, φ) such that when t is fired with $(p, t, q) \in F_T$, then the following actions are taken in the given order: (1) removing the enabling tokens, (2) transferring all tokens from p to q, and (3) adding the usual output tokens.

Interestingly and somewhat surprisingly, the boundedness problem is decidable for transfer nets but undecidable for reset nets. The *termination problem* is decidable for both classes, whereas *structural termination* turns out to be undecidable for both. The interested reader is referred to [9, 13] for more about reset and transfer nets. It is of interest to seek additional decidability/complexity results for problems related to such PN extensions as well as characterizing additional PN extensions which are weaker than Turing machines but more powerful than conventional PNs.

References

1. T. Agerwala and M. Flynn. Comments on capabilities, limitations and 'correctness' of Petri nets, *Proc. of 1st Annual Symposium on Computer Architecture*, 1973, 81–86.
2. H. Baker. Rabin's proof of the undecidability of the reachability set inclusion problem of vector addition systems, Computation Structures Group Memo 79, Project MAC, MIT, July 1973.
3. I. Borosh, and L. Treybig. Bounds on positive integral solutions of linear Diophantine equations, *Proc. Amer. Math. Soc.* 55 (1976), 299–304.
4. L. Cherkasova, R. Howell and L. Rosier. Bounded self-stabilizing Petri nets, *Acta Informatica*, 32 (1995), 189–207.
5. P. Clote. On the finite containment problem for Petri nets, *Theoretical Computer Science*, 43 (1986), 99–105.

6. M. Davis. Hilbert's tenth problem is unsolvable, *American Mathematical Monthly*, 80 (1973), 233–269.
7. J. Desel and J. Esparza. *Free Choice Petri Nets*, volume 40 of Cambridge Tracts in Theoretical Computer Science. Cambridge University Press, 1995.
8. E. Dijkstra. Self-stabilizing systems in spite of distributed control, *C. ACM*, 17 (1974), 643–644.
9. C. Dufourd, P. Jancar and P. Schnoebelen. Boundedness of reset P/T nets, *Int'l Colloquium on Automata, Languages and Programming*, LNCS 1644, Springer-Verlag, Berlin, 1999, 301–310.
10. J. Esparza. Petri nets, commutative context-free grammars and basic parallel processes, *Fundamenta Informaticae*, 30 (1997), 24–41.
11. J. Esparza. Decidability of model checking for infinite-state concurrent systems, *Acta Informatica*, 34 (1997), 85–107.
12. J. Esparza. Decidability and complexity of Petri net problems : an introduction, *Lectures on Petri Nets I: Basic Models. Advances in Petri Nets* (G. Rozenberg, W. Reisig, eds.), LNCS 1491, Springer-Verlag, Berlin, 1998, 374–V428.
13. A. Finkel and P. Schnoebelen. Fundamental structures in well-structured infinite transition systems, *Proc. of Latin American Theoretical INformatics (LATIN)*, LNCS 1380, Springer-Verlag, Berlin, 1998, 102–118.
14. S. German and A. Sistla. Reasoning about systems with many processes, *J. ACM* 39, 3 (1992), 675–735.
15. S. Ginsburg. *The Mathematical Theory of Context-Free Languages*, New York: McGraw-Hill, 1966.
16. J. Grabowski. The decidability of persistence for vector addition systems, *Inf. Process. Lett.* 11, 1 (1980), 20–23.
17. P. Habermehl. On the complexity of the linear time mu-calculus for Petri nets, *18th International Conference on Application and Theory of Petri Nets*, LNCS 1248, Springer-Verlag, Berlin, 1997, 102–116.
18. M. Hack. The recursive equivalence of the reachability problem and the liveness problem for Petri nets and vector addition systems, *FOCS*, 1974, 156–164.
19. M. Hack. The equality problem for vector addition systems is undecidable, *C.S.C. Memo 121, Project MAC, MIT*, 1975.
20. M. Hack. Decidability questions for Petri nets, PhD dissertation, Dept. of Electrical Engineering, MIT, 1975.
21. J. Hopcroft and J. Pansiot. On the reachability problem for 5-dimensional vector addition systems, *Theoretical Computer Science*, 8, 2 (1979), 135–159.
22. R. Howell, D. Huynh, L. Rosier and H. Yen. Some complexity bounds for problems concerning finite and 2-dimensional vector addition systems with states, *Theoretical Computer Science*, 46, 2-3 (1986), 107–140.
23. R. Howell, P. Jancar and L. Rosier. Completeness results for single path Petri nets, *Information and Computation* 106 (1993), 253-265.
24. R. Howell and L. Rosier. Problems concerning fairness and temporal logic for conflict-free Petri nets, *Theoretical Computer Science* 64, 3 (1989), 305–329.
25. R. Howell, L. Rosier and H. Yen. A taxonomy of fairness and temporal logic problems for Petri nets, *Theoretical Computer Science*, 82 (1991), 341-372.
26. RHowell, L. Rosier and H. Yen. Normal and sinkless Petri nets, *Journal of Computer and System Sciences* 46 (1993), 1–26.
27. H. Huttel. Undecidable equivalences for basic parallel processes, *TACS 94*, LNCS 789, Springer-Verlag, Berlin, 1994, 454–464.

28. D. Huynh. Commutative grammars: the complexity of uniform word problems, *Information and Control* 57 (1983), 21–39.
29. O. Ibarra. Reversal-bounded multicounter machines and their decision problems, *JACM* 25, 1 (1978), 116–133.
30. A. Ichikawa and K. Hiraishi. Analysis and control of discrete event systems represented by Petri nets, LNCS 103, Springer-Verlag, 1987, 115–134.
31. P. Jancar. Decidability of a temporal logic problem for Petri nets, *Theoretical Computer Science* 74 (1990), 71–93.
32. P. Jancar. Non-primitive recursive complexity and undecidability for Petri net equivalences, *Theoretical Computer Science* 256 (2001), 23-30.
33. M. Jantzen. Complexity of place/transition nets, *Advances in Petri nets* 86, LNCS 254, Springer-Verlag, Berlin, 1986, 413–435.
34. K. Jensen. Coloured Petri Nets and the Invariant-Method, *Theor. Comp. Science* 14 (1981), 317–336.
35. N. Jones, L. Landweber and Y. Lien. Complexity of some problems in Petri nets, *Theoretical Computer Science* **4**:277-299, 1977.
36. R. Karp and R. Miller. Parallel program schemata, *Journal of Computer and System Sciences* 3 (1969), 167–195.
37. R. Keller. Vector replacement systems: a formalism for modelling asynchronous systems, Tech. Rept. 117, Computer Science Lab., Princeton Univ. 1972.
38. R. Kosaraju. Decidability of reachability in vector addition systems, *Proc. the 14th Annual ACM Symposium on Theory of Computing*, 1982, 267–280.
39. L. Landweber and E. Robertson. Properties of conflict-free and persistent Petri nets, *JACM* 25, 3 (1978), 352–364.
40. R. Lipton. *The reachability problem requires exponential space*, Technical Report 62, Yale University, Dept. of CS., Jan. 1976.
41. E. Mayr. Persistence of vector replacement systems is decidable, *Acta Informattica* 15 (1981), 309–318.
42. E. Mayr. An algorithm for the general Petri net reachability problem, *STOC*, 1981, 238–246.
43. E. Mayr. An algorithm for the general Petri net reachability problem, *SIAM J. Comput.* 13 (1984), 441–460.
44. R. Mayr. Decidability and complexity of model checking problems for infinite-state systems, PhD thesis, Computer Science Dept., TU-Munich, Germany, April 1998.
45. E. Mayr and A. Meyer. The complexity of the finite containment problem for Petri nets, *J. ACM*, 28, 3 (1981), 561–576.
46. K. McAloon. Petri nets and large finite sets, *Theoretical Computer Science* 32 (1984), 173–183.
47. P. Merlin. A study of the recoverability of computing systems, PhD thesis, Dept. of Information and COmputer Science, Univ. of California at Irvine, 1974.
48. H. Muller. Decidability of reachability in persistent vector replacement systems, *9th Symp. on Math. Found. of Computer Science*, LNCS 88, Springer-Verlag, Berlin, 1980, 426–438.
49. T. Murata. Petri nets: properties, analysis and applications, *Proc. of the IEEE*, 77, 4 (1989), 541–580.
50. H. Olsén. Automatic verification of Petri nets in a CLP framework, Ph.D. Thesis, Dept. of Computer and Information Science, IDA, Linköping Univ., 1997.
51. S. Patil. Coordination of asynchronous events, PhD thesis, Dept. of Elec. Eng., MIT, Cambridge, Mass., May. 1970.

52. Gh. Paun. Computing with membranes, *Journal of Computer and System Sciences*, 61, 1 (2000), 108–143.
53. J. Peterson. *Petri Net Theory and the Modeling of Systems*, Prentice Hall, Englewood Cliffs, NJ, 1981.
54. C. Petri. Kommunikation mit Automaten, Dissertation, Rheinisch-Westfalisches Institut fur. Intrumentelle Mathematik an der Universitat Bonn, Bonn. 1962.
55. M. Presburger. Uber die Vollstandigkeit eines gewissen Systems der Arithmetik ganzer Zahlen, in welchem die Addition als einzige Operation hervortritt, *Comptes Rendus du I congres de Mathematiciens des Pays Slaves*, Warszawa, 1929, 92–101.
56. C. Rackoff. The covering and boundedness problems for vector addition systems, *Theoretical Computer Science* 6 (1978), 223–231.
57. C. Ramchandani. Analysis of asynchronous concurrent systems by timed Petri nets, PhD thesis, MIT, Boston, 1974
58. W. Reisig. *Petri Nets: An Introduction*, Springer-Verlag New York, Inc., New York, NY, 1985.
59. L. Rosier and H. Yen. A multiparameter analysis of the boundedness problem for vector addition systems, *Journal of Computer and System Sciences* 32, 1 (1986), 105–135.
60. G. Sacerdote and R. Tenney. The decidability of the reachability problem for vector addition systems, *STOC*, 1977, 61–76.
61. J. van Leeuwen. A partial solution to the reachability problem for vector addition systems, *STOC*, 1974, 303–309.
62. H. Yamasaki. Normal Petri nets, *Theoretical Computer Science* 31 (1984), 307–315.
63. H. Yen. On the regularity of Petri net languages, *Information and Computation* 124, 2 (1996), 168–181.
64. H. Yen. On reachability equivalence for BPP-nets, *Theoretical Computer Science* 179 (1997), 301–317.

MIX
Papier aus verantwortungsvollen Quellen
Paper from responsible sources
FSC® C105338

If you have any concerns about our products,
you can contact us on
ProductSafety@springernature.com

In case Publisher is established outside the EU,
the EU authorized representative is:
Springer Nature Customer Service Center GmbH
Europaplatz 3, 69115 Heidelberg, Germany

Printed by Libri Plureos GmbH
in Hamburg, Germany